Praise for *Aquagenesis* by Richard Ellis

"[A] fascinating and important new book . . . what Ellis has done in *Aquagenesis* is something more ambitious than providing a bestiary of marine life (although that alone would be worth the book's price). Ellis adroitly limns scientific understanding of evolution at work in the sea—always a moving target. He places these magnificent creatures in their context of time and space, and examines the sometimes bitter scientific controversies they've bred. Remarkable in so many ways, this is what really sets Ellis' book apart from similar works on animal evolution. As Ellis approvingly quotes writer Philip Ball: 'The story of life is . . . a story of life at sea.' And this is the story Ellis gives us, in great depth and full of richly entertaining details. Ellis is also a talented illustrator, and his fine drawings of the creatures under discussion help the reader appreciate the astonishing diversity of life over the long haul of marine evolution." —*Los Angeles Times*

"Author and artist Richard Ellis has done a tremendous amount over the years to put things in proportion, to draw our eyes from the land to the sea. In his new book Ellis leaves today's oceans and sails off into the past. The facts come as fast and furious as before. His inventory includes the latest fossil discoveries, and it is graced with a number of illustrations from his own pen. When he is paying homage to giant penguins, glowing squid, chambered nautiluses or monstrous marine lizards, he drives home one of the most important features about the evolution of life in the ocean: its ebullience, which, after four billion years, shows no sign of letting up."
—*The New York Times Book Review*

"A great read." —*The Economist*

"In lucid and engaging prose, Ellis traces the history of life on Earth as animals moved from sea to land and back to sea again. Ellis provides an overview of the prevailing theories of paleontology and evolutionary biology. In the process, he introduces a bevy of interesting creatures—living and extinct . . . Ellis imparts a thorough evolutionary history of the sea." —*Science News*

"With a fact-soaked dive into deep ocean, marine biologist Richard Ellis immerses the reader in an absorbing survey of the soggy evolution of life, from plankton and trilobites to dolphins and whales." —*Boston Herald*

"More lucid science writing from Ellis, who this time cuts a broad swath through the history of marine animals. Ellis covers an incredible land- and waterscape. There's a rogue's gallery of toothy, spiny creatures and an equally long list of sideshow marvels. The theories tendered for the evolution of these creatures are often as fabulous as the creatures themselves. Then there are all the questions that remain unanswered . . . Ellis samples from all these topics with the enthusiasm of a child let loose in a candy shop. As entertaining as a three-ring circus, and as scholarly as any intellectually curious lay reader would wish for." —*Kirkus Reviews*

"Ellis' readers, especially new ones, will delight that breadth doesn't eschew depth in this survey of the evolution of marine life. Ellis covers it all. Ellis' mediation of professional literature is a genuine boon to generalists interested in today's sea creatures and their links to the past. This will capture the attention of anyone interested in the evolution of sea life." —*Booklist*

"This fascinating and scientifically rigorous work by a noted expert on marine biology addresses the beginning of plant and animal life in the sea and the return to the sea of many life forms. Strongly recommended for all."
—*Library Journal*

"Richard Ellis, a leading authority on marine biology and a marine life artist, now takes on the deep mysteries of evolution in the sea, tracing the path from the first microbes to jawless, finless creatures that became the myriad species alive today, including sharks, whales, penguins—and us." —*Pennsylvania Gazette*

PENGUIN BOOKS

AQUAGENESIS

Richard Ellis is a celebrated authority on marine biology and America's leading marine life artist. He is the author of twelve books including *The Book of Sharks, Men and Whales, Deep Atlantic, Imagining Atlantis,* and *The Search for the Giant Squid,* and has written for *Natural History, Audubon, National Geographic,* and many other magazines. His paintings and murals are featured in major museums around the country. Richard Ellis is also a research associate at the American Museum of Natural History. He lives in New York City.

ALSO BY RICHARD ELLIS

The Book of Whales

Dolphins and Porpoises

The Book of Sharks

Great White Shark (with John McCosker)

Men and Whales

Monsters of the Sea

Deep Atlantic

Imagining Atlantis

The Search for the Giant Squid

AQUAGENESIS

THE ORIGIN AND EVOLUTION OF LIFE IN THE SEA

Richard Ellis

With illustrations by the author

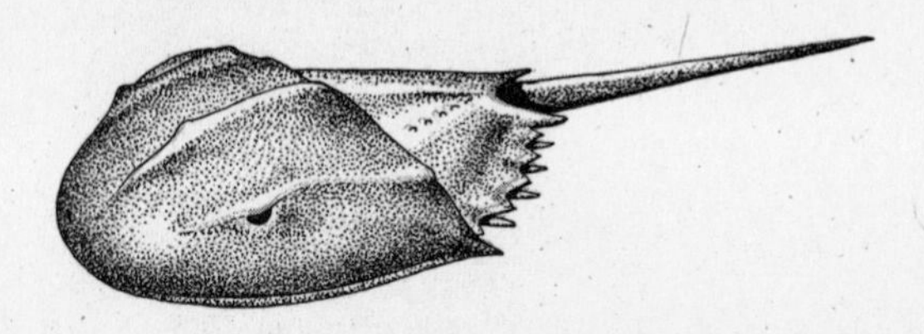

PENGUIN BOOKS

PENGUIN BOOKS

Published by the Penguin Group

Penguin Putnam Inc., 375 Hudson Street, New York, New York 10014, U.S.A.

Penguin Books Ltd, 80 Strand, London WC2R 0RL, England

Penguin Books Australia Ltd, 250 Camberwell Road, Camberwell, Victoria 3124, Australia

Penguin Books Canada Ltd, 10 Alcorn Avenue, Toronto, Ontario, Canada M4V 3B2

Penguin Books India (P) Ltd, 11 Community Centre, Panchsheel Park, New Delhi – 110 017, India

Penguin Books (N.Z.) Ltd, Cnr Rosedale and Airborne Roads, Albany, Auckland, New Zealand

Penguin Books (South Africa) (Pty) Ltd, 24 Sturdee Avenue, Rosebank, Johannesburg 2196, South Africa

Penguin Books Ltd, Registered Offices:
Harmondsworth, Middlesex, England

First published in the United States of America by Viking Penguin, a member of Penguin Putnam Inc. 2001
Published in Penguin Books 2003

1 3 5 7 9 10 8 6 4 2

Previous page: The paradigmatic "living fossil," the horseshoe crab, has been around for 200 million years. The most primitive hominid fossils are perhaps 4 million years old, and *Homo sapiens*, the species of which you are a member, has been around for maybe 100,000 years.

ISBN: 0-670-03023-6 (hc)
ISBN 0 14 20.0156 2 (pbk)
CIP data available

Printed in the United States of America
Set in Adobe Garamond
Designed By Jaye Zimet

Full fathom five thy father lies;
Of his bones are coral made;
Those are pearls that were his eyes;
Nothing of him that doth fade
But doth suffer a sea change
Into something rich and strange.
Sea-nymphs hourly ring his knell . . .
Hark! now I hear them—ding-dong bell.

—*The Tempest,* I.ii

Hence without parents, by spontaneous birth,
Rise the first specks of animated earth. . . .
Organic life beneath the shoreless waves
Was born and nurs'd in ocean's pearly caves;
First forms minute, unseen by spheric glass,
Move on the mud, or pierce the watery mass;
These, as successive generations bloom,
New powers acquire and larger limbs assume;
Whence countless groups of vegetation spring,
And breathing realms of fin and feet and wing.

—Erasmus Darwin,
The Temple of Nature (1802)

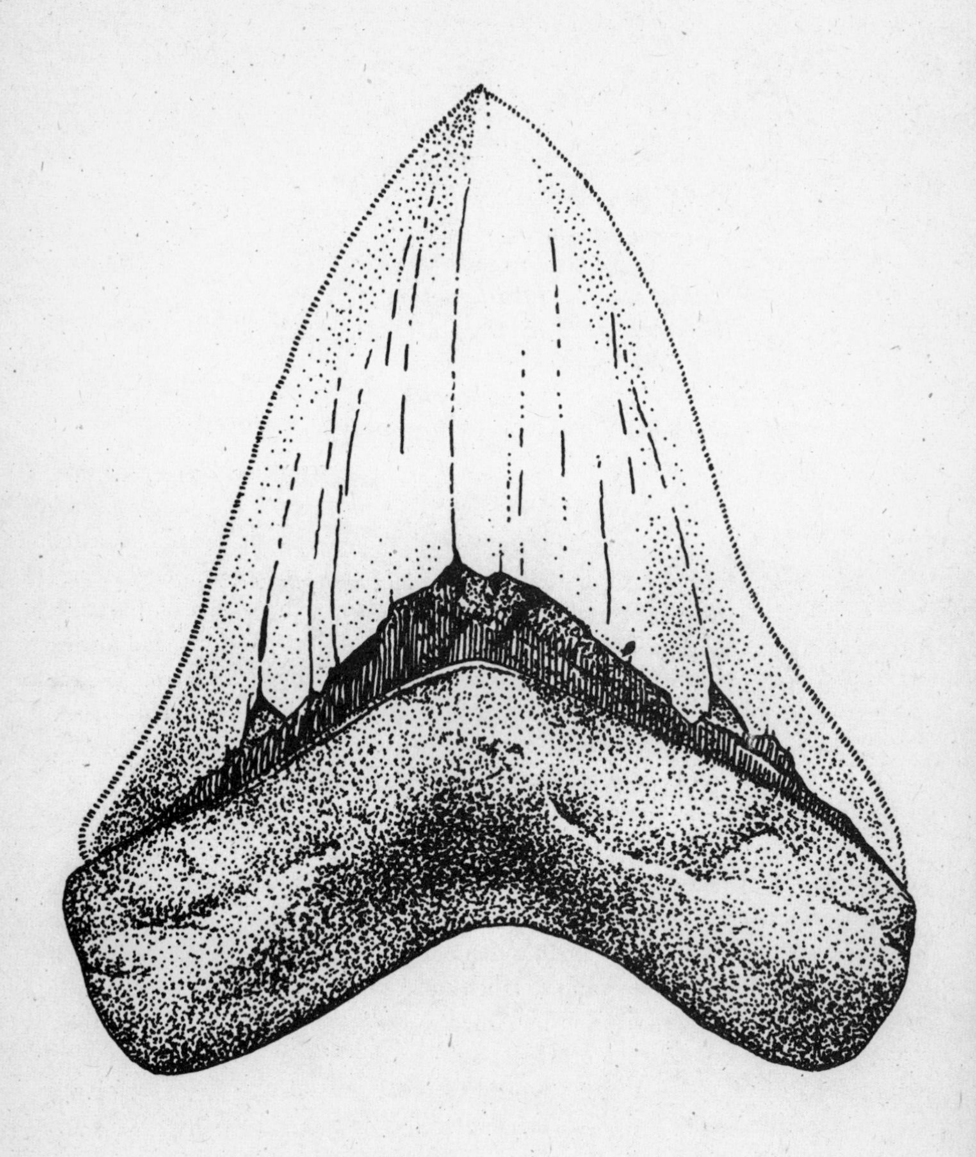

A fossil tooth of Carcharodon megalodon, *shown actual size. This 50-foot-long shark was the oceans' apex predator until a couple of million years ago, when it became extinct. Think of sharing the ocean with a shark large enough to swallow a horse.*

PREFACE AND ACKNOWLEDGMENTS

The ancestors of the living whales, seals, manatees, sea turtles, sea snakes, and penguins were terrestrial, and their living descendants all returned to the sea, to one degree or another. (Some, like the whales, dolphins, and manatees, are fully aquatic, but the seals and penguins spend a portion of their lives on land or ice.) That seemed a most interesting phenomenon, and it was supposed to be the subject of this book. But when I began to trace the phylogenetic origins of aquatic life further and further back, I found myself investigating—but certainly not solving—the eternal conundrum of the origin of life itself. Instead of a book about whales, seals, and manatees, this has become a book about the beginning of life in the sea and the various life forms that developed there, and also those that returned to the sea.

For the first time, I have written a book where opinion plays a considerable role. My earlier books were mostly crammed with facts—my agent once called me a "fact junkie," and from that remark came the *Encyclopedia of the Sea*. In this book facts again abound, but they are tempered, qualified, molded, and influenced by opinion. In many cases, the "facts" are so tenuous that opinion serves as the primary vehicle of discussion. Because interpretation is so much a part of this book, I have relied heavily on the opinions of others, via their published works or through direct communication. Many of the opinions are contradictory. There is argument but no consensus. That I have not been able to say, "X is obviously correct while Y has got it all wrong" goes to the heart of paleontology: there are very few "facts" that everyone accepts, and many of the scientific papers are devoted to the exposition of differences of opinion or interpretation.

A list of the people I dragooned into this project would fill a small volume, but some people did much more than answer my questions; they read portions of the manuscript and made suggestions that undoubtedly improved it. Through the miracle of e-mail—not to mention the Internet searches that provided their addresses—I was able to obtain answers from the people who had written the original books or papers that I was citing. When I was trying to crack the mysteries of bacterial bioluminescence, for example, I wrote to Peter Herring at the Southampton Oceanography Centre in England, Edie Widder at Harbor Branch, John McCosker of the California Academy of Sciences, Woody Hastings at Harvard, Margaret McFall-Ngai at the University of Hawaii, and Lynn Margulis at the University of Massachusetts. I'm sure they all felt that they had already answered the kinds of questions I was asking them, but they took the time to answer, and allowed me to dodge the bullet again. I am grateful to Richard Fortey for explaining chemoautotrophic trilobites, and for writing the wonderful *Trilobite!* Neil Landman, Peter Ward, John Arnold, and Neale Monks all helped increase my understanding of the complexities of cephalopod evolution, and Neale read my essay on early cephalopods and managed to catch some of the more egregious misstatements and misinterpretations. Richard Lund introduced me to the Bear Gulch Formation chondrichthyan fauna, and I am still reeling from the wonder of it all. He also read a preliminary version of my essay on Devonian sharks and chimaerids, and corrected my more flagrant errors. Of course, there is no one, no matter how wise or learned, who could protect me from my own obstinacy or the fascination with the sound of my own words, as *The New Yorker* used to say.

Charles "Rep" Repenning reviewed my discussions of desmostylians, mustelids, and pinnipeds, and made so many helpful suggestions that he ought perhaps to be considered coauthor of those sections. On the complex subject of sea mammal evolution, I communicated with Philip Gingerich, Hans Thewissen, Mark Uhen, Richard Hulbert, John Heyning, Daryl Domning, Ewan Fordyce, and Mike Novacek. Christian de Muizon read my section on the amazing *Odobenocetops,* and corrected it as necessary. Probably my greatest surprise in researching this book was the discovery of the aquatic ape theory, and communication directly with its originator, Elaine Morgan, added a special pleasure to the experience. Despite all the help that was available to me, I take full responsibility for any mistakes or misinterpretations herein.

There is practically a cottage industry dedicated to the solving of the problem of *Helicoprion,* the fossil tooth-spiral that nobody understands. I signed aboard by communicating with Dick Lund, Ray Troll, and Michael Williams, but the critical translation of the 1952 Obruchev paper was provided by Andrei Suntsov of Moscow's Academy of Sciences, Division of Ichthyofauna.

The reader will find many articles in scientific journals in the bibliography of this book. Scientists, whether they are paleontologists, ichthyologists, molecular biologists, or nuclear physicists, publish the results of their research in scientific journals. There are journals for just about everything, but for the dissemination of science, the more important ones are those that are "peer-reviewed" or "refereed." That means that after the paper has been submitted to a particular journal, the editors send it to readers qualified to ascertain the accuracy and potential importance of the paper, and to recommend to the editors whether or not it warrants publication. In refereed journals, therefore, the opinions and conclusions of the author or authors do not enter the literature unchecked. (One paper mentioned in this book had 189 authors, all of whom were listed at the top of the paper; their names took up all of the first page and a good part of the second.) The readers of the journal may be surprised by the author's conclusions, and they may even disagree with them, but they know that before publication, several experts had already looked at the paper in question, but they are no safeguard against author error.

The list of references for this book also includes many books, often written by people who wrote or cowrote the original papers on the subjects. (There is another kind of book, too, which is a collection of articles on related subjects, but these books simply bring together peer-reviewed articles, and are subject to the same qualifications.) The books are usually not peer-reviewed (unless they have been published by a university press, in which case they are), so it is possible for authors, even if they are highly respected scientists, to throw in unsubstantiated ideas, wild speculations, and biased opinions, sometimes referring to other authors, often those who do not agree with them. I am grateful for these "popular" books, for they afford me the opportunity to read an author's summary of his own work, not only the pieces that may eventually—often along with the work of many others—produce some sort of unified theory about symbiosis, the evolution of trilobites, or the origin of whales.

Much of my research was done in a place that serves as a portal to the myriad wonders of science: the library of the American Museum of Natural History in New York City. This book could not possibly have been written without my access to the collections of the library, and indeed, without the help of the staff. I am particularly indebted to Tom Moritz, the library's director, who smoothed the path that would have otherwise have been full of pitfalls and snares, as well as delusions as to my competence to locate things in the vastness of that magical resource.

Popular articles form another category of works that I consulted in writing this book. Some popular articles are written by scientists, and a writer who purports even to touch on the subject of evolution ignores Stephen Jay Gould at his peril. Gould, an invertebrate paleontologist by profession and an essayist by avocation, has written a monthly column for the magazine *Natural History*, the magazine of the American Museum of Natural History, for 25 years and has become the best-known spokesperson in English (and the many other languages into which his books have been translated) for Darwinian evolution—with modifications—not to mention the myriad other subjects to which he has addressed himself in his eloquent and graceful style. These articles have been collected into books, which have spread the Gould gospels further than any magazine (except maybe *Time* or *Newsweek*) could possibly hope to do.

Gould is probably the best-known of the popularizers of science, but many others have written excellent books that have helped me to understand their subject in its broader scope and implications, and have provided insight that I might not have gained from a reading of the fragmentary information presented in the scientific journals.

Some of the recently published popular books that cover a broader terrain than this one are *Life* (by Richard Fortey), *The Variety of Life* (Colin Tudge), *History of Life* (Richard Cowen), *Prehistoric Life* (David Norman), *The Emergence of Life on*

Earth (Iris Fry), and Paul Davies's *The Fifth Miracle*, which is subtitled "The Search for the Origin and Meaning of Life." Niles Eldredge, Stephen Gould's coauthor of the theory of punctuated equilibria, has been almost as prolific as Gould himself, writing a series of books on evolution, a few of which are *Fossils: The Evolution and Extinction of Species*; *Life in the Balance: Humanity and the Biodiversity Crisis*; *The Miner's Canary: Unraveling the Mysteries of Extinction*; *The Pattern of Evolution*; *Time Frames: The Evolution of Punctuated Equilibria*; and, most recently, *The Triumph of Evolution and the Failure of Creationism*.

The contentious and fascinating subject of early life forms can be further investigated in Gould's *Wonderful Life*, and an opposing interpretation can be found in Simon Conway Morris's *Crucible of Creation*. On particular groups, Ricardo Levi-Setti wrote *Trilobites*, and Richard Fortey wrote a book on the same subject with the same title, but with an exclamation point justifiably added. Mark McMenamin wrote *The Garden of the Ediacara*, a title he borrowed from his scientific paper of the same name; and with his wife, Dianna, wrote *Hypersea: Life on Land*. Elaine Morgan's theories on the aquatic ape have appeared almost exclusively in her books: *The Descent of Woman*; *The Aquatic Ape*; *The Scars of Evolution*; and *The Aquatic Ape Hypothesis*. She also published a several articles in *New Scientist*.

In 1956, J. L. B. Smith wrote *Old Fourlegs* about his discovery of the first coelacanth; then Keith Thomson wrote about the state of coelacanth biology up to 1991 (*Living Fossil*); and then, in 1999, largely because a totally unexpected population of coelacanths were found in Indonesia, Samantha Weinberg wrote another book about coelacanths, *A Fish Caught in Time*. Although it is not a "popular" book, Peter Forey's 1998 *History of the Coelacanth Fishes* is the most comprehensive of all coelacanth books, and an absolute necessity for anyone who would attempt to understand these "living fossils." Speaking of which, in 1984, Niles Eldredge and Steven Stanley edited a volume that they called *Living Fossils*, which contains the qualifications of numerous creatures of ancient ancestry to be considered for this elite category. Many authors are so prodigious that it became a full-time job to keep up with their writings. I thought I had a grip on the Ediacarans (not that I know what they were; only that I had read a lot about them), but I hadn't reckoned with the fecundity of Mark McMenamin, Adolf Seilacher, or Ben Waggoner. It is almost impossible to track the literature on the myriad subjects I decided to include in this book, but I made a stab at it, only to find that there was hardly any time to write; there was barely time enough to read.

The illustrations also presented an unprecedented set of problems. For *Book of Whales* and *Dolphins and Porpoises*, I drew and painted various aspects of the lives of cetaceans, from reference material not always easily available, but rarely subject to innovation. We pretty much know what a sperm whale or a bottlenose dolphin looks like, and all I had to do for their portraits was put them in the water. For *Monsters of the Sea,* I drew hardly anything, relying instead on published interpretations of giant cephalopods, mermaids, ship-sinking whales, man-eating sharks, and the always elusive Loch Ness monster. When it came time to illustrate *Deep Atlantic,* I developed a white-on-black style for the various underwater creatures in an attempt to convey the impression of depth, but I couldn't have drawn any of the deepsea denizens if it hadn't been for available photographs, drawings reproduced in various papers, or study specimens in jars. But with many of the animals in this book, there are no photographs of whole animals; there are drawings of largely incomplete skeletons, and no specimens except fossils. So I have had to do what the dinosaur illustrators do: reconstruct the exterior of the animal from the skeleton, and make up the skin, scales, feathers or coloration. In a 2000 article entitled "What did dinosaurs really look like . . . and will we ever know?" William Weed wrote, "Dinosaur skeletons are rarely found intact, and figuring out how the scattered bones fit together is not always clear. Then making the leap of placing tissue and skin on those bones is a process fraught with unknowns."

In this book, you will encounter scientific names that will make your eyes glaze over. More than any other book I've written, the subject matter of this one does not often lend itself to comfortable, easy names. There are seals and sea lions and penguins, of course, and whales, dolphins, and manatees (all of which have scientific names that are used to identify them by related characteristics within their hierarchical group), but then there are all kinds of weird creatures that never had common names. According to the Linnaean system of classification, every living thing is given a binomial, a scientific name of two terms, always in Latin. The first part of the name refers to the organism's genus, and the second part indicates its species. In many cases, I've attempted to translate the Latin (or Greek etymology) to make these names a little more user-friendly, but some are just going to be jawbreakers, translation or no. For example, there is a long-extinct kind of walrus-whale called *Odobenocetops,* which can be roughly translated as "cetacean that walks on its teeth," but no amount of explication is going to make this name roll trippingly off the tongue. Many fossil creatures were never given common names, so you're just going to have to struggle through *Anomalocaris, Vampyroteuthis, Helicoprion,* and *Iniopteryx*—and these are only their first names. The great white shark is *Carcharodon carcharias*—so far so good—but then there is *Harpagofututor volsellorhinus* (a kind of extinct shark), *Ogygiocarella debuchii* (a trilobite), or our old friend the humpback whale, known to science as *Megaptera novaeangliae.* I realize that some of these names are more than a little difficult, and they do not make unknown creatures any more familiar, but you might take some comfort in knowing that scientists rely on these names to nail down the identification of a particular animal; no matter what a biologist's native language, when he wants to make sure you know he's talking about the emperor penguin, he will write its name exactly as you see it here: *Aptenodytes forsteri.*

The group of extinct marine mammals known as desmostylians has been the subject of heated paleontological controversy since the first bones were discovered in 1888. Were they terrestrial or aquatic? Did they move like seals, like hippopotamuses, or like no animal we know today? How and what did they eat? What happened to them? There is no consensus, even though paleontologists have devoted their careers to solving the mystery of the desmostylians. I have been in contact with various desmostylian authorities, and when I sent some preliminary sketches to Charles Repenning (the discoverer of the desmostylian *Paleoparadoxia*—wonderful name!—in California in 1965), he said to me, "Trouble is, everybody is trying to make it look plausible. It isn't. No other animal is built this way." With his counsel, as well as reconstructions by numerous other desmostylogists, I have drawn the somewhat implausible creature on page 199.

Mark Norell is chairman of the department of vertebrate paleontology at the American Museum of Natural History and one of the esteemed dinosaur hunters of the Gobi Desert, so when he tells you that something is important, you pay attention. He said that he thought the various animals' return to the sea was among the more intriguing subjects in evolutionary history, and that was all the encouragement I needed to start me down this long and winding road. I thank him for jump-starting this project, and for providing wisdom and guidance along the way.

Since *The Book of Sharks* in 1975, all my books have passed before the critical eye of my agent, Carl Brandt, whose contributions and observations have served to keep me honest. In this case, none of my prehistoric interpretations would have appeared in print if it were not for the willingness of my editor, Wendy Wolf, to join me in this unexplored world. Even though Stephanie said that *Aquagenesis* sounded like a facial treatment, this book is for her; she is the most important thing that has happened since the early tetrapods took their first tentative steps on land.

CONTENTS

INTRODUCTION

From space, our planet is blue, which belies its solid-sounding name and suggests that a more descriptive denomination would be not Earth, but Water. Assuming, at least for the moment, that life did not arrive here from space, we can pretty safely bet that the oceans are where we come from. As Philip Ball wrote, in his *Biography of Water*, "That we live on land is, in the grander scheme of things, best regarded as an anomaly, or even an eccentricity—albeit with sound evolutionary justification. The story of life is, if we retain a true sense of proportion, a story of life at sea." Even if it took a little longer to create than six days, life in the water has achieved an overwhelming dominance on the planet; the oceans are our prevailing life-support system.

Life as we know it cannot exist without water, and we don't have to look very far for it: the Pacific Ocean, our world's largest feature, covers one third of the Earth. Water is our most abundant liquid, an odorless substance that is colorless in small amounts, but takes on a bluish tinge in large quantities. It is a compound of two parts hydrogen and one part oxygen: H_2O. In its solid form, ice, and its liquid form, it covers more than 70 percent of the planet. The white swirls that usually block the view of the planet from space are clouds, composed, naturally, of droplets of water. The average depth of the oceans is about 12,000 feet, and the deepest spot in the world is located in the Marianas Trench, some 35,840 feet below the surface of the Pacific, 6,838 feet deeper than Mt. Everest is high. (If Everest sat at the bottom of the Marianas Trench, and Edmund Hillary and Tenzing Norgay stood on its peak, they would be under a mile and a half of water.)

Water is the stuff of life on Earth; protoplasm is a suspension of a number of substances in water, and every living thing is made of protoplasm. Water is also the home of nine tenths of the Earth's living things. Most of them breathe dissolved oxygen, not the oxygen in H_2O. The earliest life forms on Earth evolved in water; they were microscopic and soft-bodied, without skeletons or shells, and they left few records of their existence. Five hundred million years ago, by the time the world had reached what we call the Cambrian Period, life began to expand and diversify, assuming some of the myriad forms we recognize today, and many that we know only from their weird fossil remains. Early on, the benefit to living things of some sort of a skeleton became apparent, and some creatures wore theirs on the outside, while others carried theirs within. Animals that are not soft-bodied are divided into two great classes: those with an exoskeleton, including crabs, shellfish, lobsters, and other crustaceans as well as the vast legions of insects on land, and the vertebrates—fishes, sharks, amphibians, reptiles, birds, and mammals—that have gone for the "bones-inside" look. Some aquatic or semiaquatic creatures are gone forever: the marine reptiles—the ichthyosaurs, plesiosaurs, and mosasaurs—not to mention a host of fishes now extinct flourished in ancient seas for hundreds of millions of years, and they are now all gone, leaving not a single descendant.

For the most part, aquatic creatures live their entire lives submerged; they are born, breathe, feed, excrete, move, grow, mate, and reproduce within a single medium. Marine invertebrates such as worms, echinoderms, holothurians, crustaceans, gastropods, and cephalopods as well as fishes and sharks live all their natural lives underwater. A few aquatic creatures—those that are descended from land animals—come all or part of the way out of the water for one reason or another. Sea turtles, pinnipeds, and penguins come ashore to breed, thereby acknowledging their terrestrial heritage. The fully aquatic cetaceans and sirenians poke their noses out of the water only to breathe but are otherwise not designed to venture onto terra firma. Human beings, with their awkward arms and legs, seem poorly engineered for any sort of an aquatic existence, but they have a surprising number of modifications that make them more, rather than less, suited for an aquatic lifestyle. We are the only terrestrial mammals that hold our breath. We have proportionately as much subcutaneous fat as a dolphin. We can swim almost from birth. There are those who believe that we had to have passed through a semiaquatic phase before awkwardly clambering up on our legs to begin the long, uncomfortable march toward enlightenment.

If you do not believe that God made everything on Earth in the first six actual days, you probably think that the theory of evolution, as defined by Charles Darwin in 1859, contains the explanation for how various life forms—microbes, jellyfishes, sea cucumbers, amphibians, reptiles (including dinosaurs), birds, and mammals—evolved to their present state. Along the way, many creatures went extinct, and this probably had to do with climate change, or asteroid impacts, or their being poorly adapted to their environment. In any event, something drove the dinosaurs, the pterodactyls, and the plesiosaurs to extinction; because they disappeared millions of years ago, and because we've been around for only a piddling 100,000 years, we know it wasn't us.

It is not surprising that Darwin's theory has not withstood the test of time unchallenged, but in the end, the most important aspect of the theory is that in 1859, Darwin—and Alfred Russel Wallace, more or less simultaneously—looked carefully at the natural world around them and came up with a way of explaining how it got to be that way. Although economist Thomas Malthus (1776–1834) was concerned with human overcrowding, Darwin applied Malthus's approach to his own study of animals and concluded that the reproductive powers of animals are much greater than is required to maintain their numbers, and, therefore, a large number of offspring must somehow be eliminated if the numbers are to remain constant. There must, he said, be a "struggle for existence" between members of a species, and also between species in competition for available resources. In his book *The Origin of Species by Natural Selection*, Darwin recognized that the best—if not the only—way to account for the vast variety in living things was what he called "descent with modification." He observed that the similarities or differences between animal species were explainable only if animals had common ancestors and could therefore not be independent creations. Darwin then described the gradual process where species that are marginally better adapted to their environment reproduce more successfully than those that do not possess a particular minute advantage, and by this process—which he called "natural selection"—the less-well-adapted species gradually disappear, and are replaced by the more fit. This is the definition of "survival of the fittest." Operating over numerous generations, this process produces sufficient change to give rise to new species.

After returning from his epic voyage in 1836, Darwin remained in Kent, England, where he studied domestic animals, such as fancy pigeons, and reflected upon what he had observed in the wilds of such remote outposts as Tierra del Fuego, Patagonia, the Galápagos Islands, Tasmania, and Mauritius. In any given population of animals (which he recognized as a species), some of them naturally acquire modifications that make them a little better at making a living in their particular environment. If there is no change in that environment, there will be no noticeable change in the species. If, however, the environment changes even a little, as a result of, say, a gradual change in temperature that affects the plant life, the animals must adjust to this change, and those that already have a modification that gives them a little edge will produce offspring that already have the advantageous modification. Those that—by chance—are not so advantaged will be unable to compete with the better-adapted, and will eventually die off. In a given population, this will not be visible, for the species will remain constant over time, but if one element of the population is somehow separated from another and left to develop on its own, a new species might appear.

This concept, long understood but rarely observed, was brought into sharp focus by the evolutionary biologists Rosemary and Peter Grant in their study of various closely related species of finches on the little Galápagos island of Daphne Major. Their observations have been extensively documented in their own published papers, but in 1994, Jonathan Weiner won the Pulitzer Prize for his brilliant discussion of the Grants' findings and their implications. In *The Beak of the Finch,* Weiner tells of the Grants' experiences with some thirteen species of "Darwin's finches," including one that is flightless; one that cohabits with marine iguanas; one, the vampire finch, that lives on blood; and another, the cactus finch, that makes tools. The Grants caught and banded thousands of finches and traced their elaborate lineage, enabling them to document the adjustments that the individual species made, especially in their beaks, in response to environmental changes. During prolonged drought, for instance, those with even slightly longer beaks are better able to reach the tiniest of seeds. Even more fascinating, the Grants have documented changes in DNA among their birds, leading Weiner to declare that "Darwin did not know the strength of his own theory. He vastly underestimated the power of natural selection. Its action is neither rare nor slow. It leads to evolution daily and hourly, all around us, and we can watch."

In *The Intelligent Universe* (1983), Sir Fred Hoyle, an astronomer who, like fellow astronomer Chandra Wickramasinghe, is an unreconstructed opponent of Darwinian logic (and, it must be added, of almost every other mainstream evolutionary precept), wrote:

> Is natural selection really the powerful idea it is popularly supposed to be? The more I thought about it, the more circular the argument seemed to become: "If among a number of varieties of a species one is best fitted to survive in the environment as it happens to be, then it is the variety that is best fitted to survive that will best survive." Surely the rich assembly of plants and animals found on Earth cannot have been produced by a truism of this minor order? The spark plug of evolution must lie elsewhere. It lies in the source of the variations on which natural selection operates. Darwinians believe nowadays that the ultimate source lies in chance miscopyings of genetic information, a view which I believe to be quite erroneous.

Hoyle and Wickramasinghe reject the whole theory of evolution, saying that all genetic changes were induced by periodic storms of gene-mutating viruses from space. Even if we discount such a suggestion, we still have major problems fixing the evolutionary time frame: there simply hasn't been enough time for the tens of millions of life forms now on Earth—and the billions that are extinct—to have developed by the gradual, often accidental process that we call evolution. Darwin recognized this and wrote, "It may be objected that time cannot have sufficed for so great an amount of organic change, all changes having been effected slowly."

Various explanations have been offered to address the time problem. The most controversial is "punctuated equilibria," a theory promulgated in 1972 by the paleontologists Stephen Jay Gould and Niles Eldredge, two of the best-known and most widely read students of current evolutionary theory. In this interpretation, species remain in equilibrium for a long time, but then change rapidly in response to some climatic or other environmental catastrophe—an asteroid impact, a wobble in the Earth's rotation, a change in global sea level, a climate change—that punctuates their equilibrium. Then, those that are better able to cope with the changes come to dominate the population at large. The unfit—that is, less well adapted—members of the population die out, and the newly adapted ones survive. This is Darwin's "survival of the fittest," but at a different rate of speed.

Extinction, the mysterious and powerful converse of evolution, can sometimes be explained in Darwinian terms. If there is no adaptation to compensate for an environmental change, then the creature will be unable to function effectively in its modified environment and in time will die out. It does not require overspecialization, global warming or cooling, a deadly infectious disease, or an extraterrestrial impact to eliminate a species (although these can certainly speed up the process), but only the inability of the creature to adapt to its changing world. Those animals whose environment remained stable, or where they could engineer minor changes that enabled them to survive for hundreds of millions of years—think of the horseshoe crabs, crocodiles, coelacanths, chambered nautilus—have been awarded the honorary appellation of "living fossils."

Niles Eldredge (1991) explains punctuated equilibrium—or "Punk Eek"—as follows:

> The long period of stasis that characterizes the vast bulk of the history of most species constitutes an "equilibrial" condition. The "punctuation" comes in when a new daughter species buds off from the original parent species, a process that takes thousands of years, rather than the much longer period of stasis, which typically takes *millions* of years. . . . Thus punctuated equilibria offers an additional explanation for why there are usually "gaps" even between closely related species in the fossil record: the transition from parent to daughter species takes place too quickly, and in such a limited part of the parental species' range, that there is little chance that the event will be recorded at all in the fossil record.

Darwin's central idea, that today's animals and plants are descended from earlier animals and plants, is unquestionably valid. That there are millions of life forms on Earth today unequivocally confirms this. We just haven't discovered what it is in evolution that causes actual change, or extinction. There were lots of prehistoric animal and plant forms, most of which died out for unknown reasons, and now there are lots of different forms. The connection between the dead—for which read "extinct"—forms and the living ones is not at all obvious. Everyone now agrees that earlier forms did not simply morph into later forms that were better adapted to living in slightly changed circumstances, and that the less well adapted forms lay down and died, or starved to death, or froze. Paleontologists and evolutionary biologists can look at the evidence before them, and formulate elaborate theories about heredity and genetic modification, but the fact remains that one of the few elements of evolutionary theory that everyone agrees with is that some species have become extinct because humans killed them off. What happened to all the extinct species described in this book is unknown, and what caused some early tetrapods to leave the water, or grow feet, or become salamanders, crocodiles or whales, is also unanswered. But figuring out how it happened is quite different from determining that it did happen. Evolution is a fact of life; it has to have happened. In his book about the evolution of horses (1951), George Gaylord Simpson said:

> Many other histories, more or less complete, have, however, now been compiled from the fossil record. These involve an extremely wide variety, samples from all parts of the plant and animal kingdom—ferns, grasses, oaks, microscopic one-celled animals, clams, snails, fishes, dinosaurs, camels, elephants, and many others. Paleontological demonstration of the truth of evolution can now be based on any of a hundred or more different families as on the horse family. . . . There is really no point nowadays in continuing to collect and study fossils simply to determine whether or not evolution is a fact. The question has been decisively answered in the affirmative.

Except as a specific response to a change in circumstances, there is no known way for a species to modify itself and thus change its genetic structure so that, in future generations, it is transformed into a different animal. If minor modifications are passed along that enable a species to survive while those without the modification cannot compete successfully, we are looking at the answer to the discredited Lamarckian concept of the inheritance of acquired characteristics. Your lifting weights will not make your children stronger, and cutting the tails off generations of rats will never produce tailless offspring. But if a rise in temperature, a change of climate, the appearance of a new predator, or any one of a million minor alterations can amplify a minuscule genetic modification that can be passed along, then the mechanism for evolutionary change will have been found, and we will not have to depend on the wholly inadequate (and mathematically impossible) idea that random changes over vast stretches of time will cause a fish to turn into a reptile, a reptile into a mammal, or a dinosaur into a duck.

The study of early life forms, including most of the vanished species in this book, is rife with unresolved questions. Paleontology, which relies so much on the hardness of rocks, is among the "softest" of sciences, since there is little that can be stated unambiguously, other than the obvious "Here are these fossils." Their age and origin are often subjects of speculation and contention, and as to what they mean or where they might be inserted into a particular phylogenetic series, few paleontologists agree. In the pages that follow, you will encounter many of these differences of opinion, but every scientist quoted here believes in evolution, as do I. The fossil evidence does not provide ready answers to questions of descent or relationships, but it unequivocally demonstrates the durability and complexity of life's history on Earth. Paleontologists will continue to argue about the first complex life forms, the earliest vertebrates, or the origin of whales, but there is no question but that life evolved. In *The Triumph of Evolution and the Failure of Creationism* (2000), Niles Eldredge wrote:

> This simple prediction—that there is one grand pattern of similarity linking all life—doesn't prove evolution, but only because science proceeds by falsifying—disproving—statements we make about how the universe is structured and how it behaves. But we gain tremendous confidence in our statements if, after hundreds of years, everything we have devised to test an

idea fails to falsify it. And so the failure of scientists to disprove evolution over the past 200 years of biological research means that the fundamental idea that life has evolved really is one of the few grand ideas of biology that has stood the test of time. This basic notion of evolution is thoroughly scientific in the strictest sense of the word, and as such is highly corroborated and at least as powerful as the notion of gravity or the idea that the Earth is round and spins on its axis.

This book is about animals whose ancestors came out of the sea and whose descendants returned to it. Myriad forms existed during the billions of years that life has been known to exist on Earth, and the great majority of them are gone. Now there are new and different life forms, and in time, these will certainly go extinct too. In his 1996 book on fossil fishes, the evolutionary biologist John Maisey said:

> Just as the Renaissance discovery of perspective changed the way reality was represented, so our modern understanding of the fossil record has reduced humankind to a single thread in the tapestry of the living world. The history of *Homo sapiens* is interconnected with the history of every other species, living and extinct, that has rented space on Earth. Uniquely, however, humankind has discovered its own origins. Like all species, we evolved from forgotten ancestors, we exist briefly, and we will be gone. This is perhaps the strongest message sent by the fossil record, and we would do well to listen.

Has there been "progress" in evolution? Are recent plants and animals more advanced than their predecessors or, at least, more complex? Alas, no. Of course it is true that a bacterium is a simpler creature than a cassowary or an elephant, in that it has fewer moving parts, but its DNA is as richly endowed as an elephant's, and it functions as well in its own circumstances as any vertebrate does. It might even be argued that certain bacteria do their job a lot better than any other animals, since they can claim an unbroken line of descent that can be traced for 3 billion years, while every other known life form has already died out or is scheduled to do so. For those who like to cite the vertebrate eye as the epitome of evolutionary complexity, consider the eye of the cephalopod, which evolved earlier and is just as successful; or the compound eyes of trilobites, which, with their hundreds of lenses, were far more complicated than the eyes of any vertebrates, and yet the owners of these marvelous eyes have been extinct for 200 million years. There was nothing particularly simple about, say, *Archaeopteryx,* or *T. rex,* but they flourished millions of years ago and exist now only as prepared fossils in museum collections. It is only our self-importance that has led us to assume that we are more advanced, or better designed, than earlier forms; and extinction, which has proved to be the most relentless force on the planet, will come to *Homo sapiens* the not-so-wise, despite his desperate attempts to understand the universe, to prolong his life, or to modify the planet to suit his own needs.

We are the only creatures in 3 billion years that have consciously altered our environment, and because we are the first to recognize the process of evolution (and its integral converse, the process of extinction), we might be able to forestall it in some way, if only to protect ourselves or our descendants from the inevitability of extinction. The animals that have already become extinct without our help have been estimated at some 99.9 percent of all the species that have ever lived, so extinction—whatever it is and however it is defined—is a ruthlessly efficient process.*

In the fullness of geological time, our contribution to the process has been minimal, but our anomalously large brain has enabled us to acknowledge our participation in the more recent extinctions. The dodo, passenger pigeon, great auk, Carolina parakeet, Labrador duck, Eskimo curlew, ivory-billed woodpecker, moa, Steller's sea cow, Caribbean monk seal, quagga, Tasmanian tiger, Hawaiian honeycreepers, and many other, less charismatic species have been eliminated by the careless, clumsy hand of man. Moreover, the list of potential extinctions—whooping crane, California condor, Chinese river dolphin, Mediterranean monk seal, Patagonian toothfish, mountain gorilla, Siberian tiger, giant panda, and so on—is even longer than the catalog of our "successes," and threatens to get longer with every passing day and every annexed acre of rain forest, meadowland, or wetland. Humans have become a hundred times more numerous than any other large land animal in history. "There is no way," wrote E. O. Wilson (1992), "that we can draw upon the resources of the planet . . . without drastically reducing the state of most other species." Our need for food, water, and space has driven wildlife to the edge of survival in so many places that the previously inconceivable question "Why do we need albatrosses (or lemurs, or sea turtles, or tigers)?" is now, sadly, considered a legitimate one to ask. If our experience with other life forms is any indication, we are remarkably ill-suited for preserving anything, since all we have been able to do so far is record a long list of species that we have driven over the edge of the precipice ourselves.

---

* Curiously, this does not mean that fewer animals are alive today than in the past. "The solution of this paradox is simple," wrote E. O. Wilson (1992). "The life and death of species must have been spread across more than three billion years. If most species last, say, a million years, then it follows that most have expired across that vast stretch of geological time, in the same sense that all the people who have ever lived during the past 10,000 years are dead though the human population is larger than it has ever been. The turnover would have been even greater if the grand pattern were dynastic, with one species giving rise to many species, most or all of which yielded to later ascendant groups."

EARLY DAYS

THE ORIGIN OF LIFE IN THE VENTS

There may be a place on Earth—albeit somewhat inaccessible—where we might be able to see conditions not unlike those that the earliest life forms saw. Where the sea floor pulls apart (a phenomenon known as "sea-floor spreading"), cracks or rifts are created in the crust of the Earth. This activity usually, but not always, takes place along the mid-ocean ridge, a 40,000-mile-long undersea series of mountain ranges that snakes around the planet like the seams of a baseball, and is the largest single geological feature on the Earth's surface. The rifts mark the edges of the lithospheric plates, which bear the continents in their inexorable movement on the first 100 to 250 kilometers of the surface of the planet. The plates such as those of North America and Africa drift over the less rigid athenosphere, much like giant icebergs in the ocean. The formation of the plates occurs at the mid-ocean ridge, and they are consumed at "subduction trenches," where one plate slides under the lip of another. Where plates interact, earthquakes occur, volcanoes erupt, and mountains are pushed up. All three were manifest in November 1963, when the Icelandic island of Surtsey was born in a spectacular cataclysm of fire, lava, and steam. The rifts caused by the separation of the plates fill up with lava that wells up from within the earth, flowing outward from the center and moving across the ocean floor. As described by J. R. Heirtzler and W. B. Bryan (1975), "Bizarre as the idea seemed at first, it was becoming evident that the mid-ocean-ridge system was nothing less than a vast unhealed volcanic wound."

The unique conditions around subterranean hydrothermal vents make them strong candidates for a deep-ocean location of the origin of life. Discovered by scientists aboard the research submersible *Alvin* in 1977 at a depth of 8,200 feet in the Galápagos Rift Zone of the eastern Pacific, hydrothermal vents are cracks in the seafloor at the juncture of two tectonic plates. Volcanic activity beneath the plates releases hot gases and dissolved minerals into the ocean, and heats the water to temperatures of nearly 700° F. At these vent sites—subsequently discovered along many other mid-ocean ridges—minerals are spewed into the water in clouds known as "black smokers" that eventually dissolve and disperse into the water column. In the

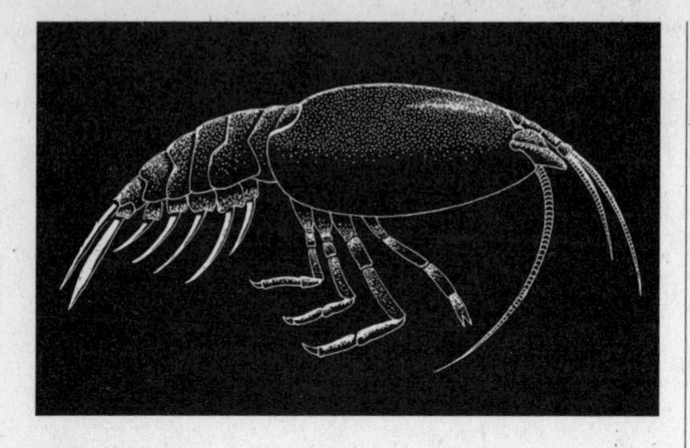

The 2-inch-long rift shrimp (Rimicaris exoculata) *occur in dense schools in the immediate vicinity of hydrothermal vent sites. These shrimps have no eyes (their scientific name can be translated as "rift shrimp without eyes"), but on their dorsal surface there is a pair of organs just below the skin that is light-sensitive.*

vicinity of these vents, a completely unknown fauna was discovered, living not on oxygen, as every other known life form does, but on hydrogen sulfide, a substance that is poisonous to most living creatures. These chemo-synthetic life forms (as opposed to those that are photosynthetic, i.e., able to process sunlight), include 6-foot-long tubeworms with red, feathery plumes but no mouth and no gut; football-sized, snow-white clams with blood-red innards; ghost-white crabs; yellow mussels; floating "dandelions" that are related to the jellyfishes; and eyeless shrimp with light-detecting organs on their backs.

Discovered in 1985 at the 11,000-foot-deep Trans-Atlantic Geotraverse (TAG), the 2-inch-long shrimp *Rimicaris* occur in dense schools in the immediate vicinity of hydrothermal vent sites. They have no eyes (their scientific name can be translated as "rift shrimp without eyes"), but on their dorsal surface there is a pair of organs just below the skin that is light-sensitive. Since they live in total darkness, the ability to "see" is probably unnecessary, but these optical organs may be useful in detecting the faint light emitted by the vents. Like many other hydrothermal vent animals, rift shrimps do not breathe oxygen, but subsist on sulfides dissolved in the water or scraped off the sides of the mineral stacks.

Because the internal digestive tube (trophosome) of the 6-foot-long tubeworm *Riftia pachyptila* has no means of ingesting particulate food matter, their feeding mechanisms baffled scientists until it was discovered that the trophosome was colonized by vast numbers of sulfur-oxidizing bacteria. "We recognized," wrote Childress, Felbeck, and Somero (1987), "that the bacteria and *Riftia* had established what is known as an endosymbiotic relation":

> The *Riftia*-bacteria endosymbiosis is mutualistic. The tube worm receives reduced carbon molecules from the bacteria and in return provides the bacteria with the raw materials needed to fuel its chemolithoautotrophic metabolism: carbon dioxide, oxygen, and hydrogen sulfide. These essential chemicals are absorbed at the plume and transported to the bacteria in the trophosome by the host's circulatory system. The worm's trophosome can be thought of as an internal factory, where the bacteria are line workers producing the reduced carbon compounds and passing them along to the animal host to serve as its food.

Almost everything about the vent tubeworms is special. They must transport poisonous materials through 6 feet of tissue to reach the deeply embedded symbionts; to facilitate carbon dioxide flux from inside to outside they must maintain pressures of dissolved CO_2 approaching that of carbonated beverages. They can grow as much as 3 feet in their first year, a growth rate that surpasses that of any other marine creature. It is not known how tubeworms and other vent animals are able to colonize hot springs, but the larvae might ride the hydrothermal vortices to new sites that they identify by chemical cues.

The rift clam *Calyptogena magnifica* and the mussel *Bathymodiolus thermophilus* also depend on chemosynthetic endosymbiosis, but they have resolved the problem somewhat differently. In the clam, the bacteria are not internal but reside in the gills, where they can obtain oxygen and carbon dioxide from the water. The basic metabolic plan is the same as that of *Riftia*: the bacteria oxidize the sulfide and supply the clam with fixed carbon compounds. "Like other invertebrates harboring sulfur bacteria as endosymbionts *Calyptogena* has a greatly reduced ability to feed on and digest particulate foods" (Childress et al. 1987). Each of the symbiont-containing animals is colonized by a host-specific strain of bacteria; even though the job all these bacteria do is similar, each bacterium has adapted itself to just one species.

These creatures of the hydrothermal vents flourish in a pitch-black, superheated, sulfide-rich environment without any connection whatever with sunlight; they are as far removed from life as we previously understood it as life on another planet. There may have been a fortuitous combination of elements and conditions in the primeval ocean that more or less accidentally created "life." Jack Corliss, one of the discoverers of the Galápagos rift animals (1977), was among the first to suggest that life might have originated in conditions similar to those found in the vents. (Corliss, who was then at Oregon State University, became obsessed with the possibility of the vent systems as the source of life, and left the university to devote himself full time to working on the problem. In 1991, he had moved on to become chief scientist of Biosphere 2, the closed-system environment built in the Arizona desert, and he is now affiliated with the Central European University of Budapest.) Working with Corliss and John Baross, Sarah Hoffman, a graduate student, formulated a theory that life had originated in the Archean Period, about 4.2 billion years ago, on the sea floor, which was probably much more hydrothermically active than it is today. The authors suggested that the water issuing from the Archean vents was so hot that it cracked the molecular bonds of the rocks, and released carbon and carbon compounds such as methane into the solution. Then simple organic molecules formed out of the newly formed chemical elements, and while some of them rose into the water column, some adhered to the rock faces, and formed a clay that provided a safe haven for these molecules, giving them the opportunity to form more complex organic molecules. Out of this jumble of molecules, argued the authors, "biopolymers" could be formed, producing fragmentary nucleic and amino acids, which, in a system far from equilibrium, could organize themselves into new forms, i.e., primitive living cells. In 1985, Baross and Hoffman wrote that the hydrothermal activity "provided the multiple pathways for the abiotic synthesis of chemical compounds, origin, and evolution of 'precells' and 'precell communities,' and ultimately, the evolution of free-living organisms."

In 1988, Stanley Miller and Jeffrey Bada refuted the suggestions of Corliss, Baross, and Hoffman, claiming that "the proposal for a hydrothermal-vent origin of life fails each of the three proposed steps involved in the origin of life." The debate about the origin of life—certainly one of the most intriguing problems in all of science—continues, but it is fascinating to consider at least the possibility that the hydrothermal vents, unknown and unsuspected until the 1970s, might provide some clues. In a 1991 article in the journal *Eos,* John Baross wrote:

> Important new discoveries on the properties of the early earth and atmosphere, including the frequency and size of bolide impacts, have strongly implicated submarine hydrothermal vent systems as the likely habitat for the earliest organisms and ecosystems, while stimulating considerable discussion, hypotheses and experiments related to chemical and biochemical evolution. Some of the key questions regarding the origins of life at submarine hydrothermal vent environments are focussed on the effects of temperature on synthesis and stability of organic compounds and the characteristics of the earliest organisms on Earth. There is strong molecular and physiological evidence from present-day mircoorganisms that the earliest organisms on Earth were capable of growing at high temperatures (about 90°C) and under conditions found in volcanic environments. . . . Further molecular and biochemical characterization of the presently cultured thermophiles, as well as future work with the many species, particularly from subsurface crustal environments, not yet isolated in culture, may help resolve some of the important questions regarding the nature of the first organisms that evolved on Earth.

Coincident with the earliest traces of life on Earth, 3.9 to 3.8 billion years ago, there are indications of a particularly intense bombardment of extraterrestrial bodies here and on the moon. In a 2000 article, David Kring of the University of Arizona wrote that impact events certainly affected life on Earth—an example is the Chicxulub impact 65 million years ago, which signaled the end of the dinosaurs; they also might have "provided the necessary environmental crucibles for prebiotic chemistry and the evolution of life." During a period that is believed to have lasted as long as 200 million years, the moon was hit at least 1,700 times, and "the number of impacts on Earth would have been an order of magnitude larger, implying >10,000 impact events" (Kring 2000). According to Mojzsis and Harrison (2000):

> Recent explorations of the oldest known rocks of marine sedimentary origin from the southwestern coast of Greenland suggest that they preserve a biogeochemical record of early life. On the basis of the age of these rocks, the emergence of the biosphere appears to overlap with a period of intense global

> bombardment. This finding could also be consistent with the evidence from molecular biology that places the ancestry of primitive bacteria living in extreme thermal environments near the last common ancestor of all known life.

"Commonly," wrote David Kring, "this is interpreted to mean that life originated (or survived the impact bombardment) in volcanic hydrothermal systems. However, during the period of bombardment, impact-generated hydrothermal systems were possible more abundant than volcanic ones. The heat source driving these systems is the central uplift and/or pools of impact melt. In the case of a Chicxulub-size event . . . melt pools may have driven a hydrothermal system for 10^5 yr. The dimensions of these systems can extend across the entire diameter of a crater and down in depths in excess of several kilometers." Kring is saying that life either began in hydrothermal vent systems that were generated by impacts, or perhaps life originated in some other fashion, and was best suited to withstand the cumulative effects of this concentrated extraterrestrial bombardment.

Everett Schock of Washington University in St. Louis believes that life began in an environment similar to that of today's hydrothermal vents. In 1992, in the journal *Origin of Life and Evolution of the Biosphere,* he published the article "Chemical environments of submarine hydrothermal systems," in which he noted that the necessary building blocks for life—methane and ammonia, for example—are not available on the Earth's surface, but that the conversion to organic compounds from carbon dioxide and carbon monoxide could have taken place in the sea, especially in the presence of high temperatures. Schock also believes that the early atmosphere of Earth would have been unwelcoming to life because of the constant bombardment by ultraviolet radiation, but deep in the ocean, as hydrogen sulfide spewed from cracks in the sea floor, it mixed with seawater to provide the chemical energy for the synthesis of life.

All cells are surrounded by membranes, which are composed of fat (lipid) molecules. It is difficult to imagine a cell that lacks a membrane, so the question arises: Which came first, membranes or nucleic acids? The answer, says John Howland (2000), is probably "both":

> First, imagine a hot spring on the ancient seafloor. Heated water emerging from the vent would have already percolated through the rocks and clay of the seabed and would be charged with a variety of dissolved substances as well as suspended clay particles. These particles would transport bound molecules that are precursors of life, including amino acids and nucleotides, as well as small polymers formed from them: polypeptides and polynucleotides. In addition, there would probably be lipid molecules present, some of them likely being soluble in hot water, but insoluble in cold. Then, according to the scenario, the hot water would cool on mixing with seawater, and the lipid molecules would precipitate from solution, forming spherical vesicles. If some fraction of those vesicles contained some of the suspended clay parti-

cles inside, with their burden of bound amino acids, nucleotides, and so on, the stage might be really set. The raw materials for protein and nucleic acids would be internal, in close proximity to the clay surfaces, which could catalyze the formation of protein and nucleotide polymers. And a lipid membrane would tidily surround the whole business.

At the 1992 Nobel Symposium, "Early Life on Earth," held in Karlskoga, Sweden, participants presented papers on virtually every aspect of this fundamental and complex subject, which were then collected into a volume published by Columbia University Press (Bengston 1994a).* In "Vitalists and Virulists: A Theory of Self-Expanding Reproduction," Günter Wächtershäuser argued that Pasteur was correct about the impossibility of spontaneous generation in maggots, but wrong when he stated that the generation of a living organism from chemical compounds is impossible. (Wächtershäuser also said that the idea of an "RNA World" requires "an extreme food specialist—more extreme than any heterotroph known today.") He believes that the only possible way for life to have begun is autotrophically, and that "the first organism is a chemoautotroph. It uses the formation of pyrite from hydrogen sulfide as a source of electrons and as its energy source."

With Claudia Huber, Wächtershäuser began testing his idea that a process on the deep ocean floor could transform basic inorganic chemicals into organic chains, the building blocks of life. In 1997 they wrote:

> The origin of life requires the formation of carbon-carbon bonds under primordial conditions. Miller's experiments, in which simulating electrical discharges in a reducing atmosphere of CH_4, N_3 and H_2O produced an aqueous solution of simple carboxylic acids and amino acids, have long been considered as one of the main pillars of the theory of a heterotrophic origin of life in a prebiotic broth. Their prehistoric significance, however, is in question, because it is now thought that the primordial atmosphere consisted mostly of an unproductive mixture of CO_2, N_2, and H_2O, with only traces of molecular hydrogen.

Huber and Wächtershäuser (1998) converted amino acids into their peptides "under anaerobic, aqueous conditions. These results demonstrate that amino acids can be activated under geochemically relevant conditions. *They support a thermophilic origin of life and an early appearance of peptides in the evolution of a primordial metabolism*" (my italics). Metallic ions in sulfides (such as iron pyrites, readily available in seafloor rocks) interact with the carbon- and hydrogen-rich gases that belch from the hydrothermal vents, creating what Huber and Wächtershäuser call

---

* At the conclusion of his introduction, Stefan Bengston wrote, "If you look to this book for final answers about life on Earth, you will be disappointed—it is a time document. The chapters reflect the views, knowledge, and inevitable biases of individual authors or teams of authors, and they are written to convey the experience and excitement of active research projects. . . . You will thus find contradictions in this book, which is as it should be in any vital field of research."

the first organic molecule: acetic acid (CH_3CO_2H), a simple combination of carbon, hydrogen, and oxygen, best known for giving vinegar its pungent odor. The authors believe that the formation of acetic acid is a primary step in metabolism, those chemical reactions that provide the energy for cells to manufacture the necessary biological ingredients for life. They theorize that around the vents, catalytic metallic ions enabled acetic acid to form, and this catalyzed the addition of a carbon molecule to produce the three-carbon pyruvic acid, which reacts with ammonia to form amino acids, which then link up to form proteins.

Wächtershäuser's ideas about the origin of life are among the very few that can be tested in the laboratory, and now George Cody and his colleagues at the Carnegie Institution of Washington's Geophysical Laboratory have done just that. Wächtershäuser calls his theory "the iron sulfur world theory," because he believes that metallic surfaces, particularly those in iron sulfide, "would have been promising facilitators or catalysts that created the precursor chemicals of living cells" (Wade 2000). Cody et al. (2000) placed iron sulfide samples in 24-carat-gold capsules and then subjected the capsules to temperatures of 250°C and pressures equivalent to 2,000 atmospheres. This produced large amounts of pyruvates, in addition to a hydrogen sulfide–like substance that resembled the H_2S produced by volcanoes above and below the sea. Their experiments demonstrated that pyruvates could be synthesized under the proper conditions, and they wrote, "The natural synthesis of such compounds is anticipated in present-day and ancient environments wherever reduced hydrothermal fluids pass through iron sulfide–containing crust. Here, pyruvic acid was synthesized in the presence of such organometallic phases. These compounds could have provided the prebiotic Earth with critical biochemical functionality."

Of Wächtershäuser's theory, the biochemist Michael Adams of the University of Georgia said that he "had developed a very reasonable and testable hypothesis for the origin of organic material relevant to life" (Wade 2000). For the issue of *Science* (25 August 2000) that featured the research paper by Cody et al., Günter Wächtershäuser wrote an essay that he called "Life As We Don't Know It." He said:

> It is occasionally suggested that experiments within the iron-sulfur world theory demonstrate merely yet another source of organics for the prebiotic broth. This is a misconception. The new finding drives this point home. Pyruvate is too unstable to ever be considered as a slowly accumulating component in a prebiotic broth. The prebiotic broth theory and the iron-sulfur world theory are incompatible. The prebiotic broth experiments are parallel experiments that are producing a greater and greater medley of potential broth ingredients. Therefore, the maxim of the prebiotic broth theory is "order out of chaos."

The final chapter of Cindy Van Dover's *The Ecology of Deep Sea Hydrothermal Vents* (2000) is "Vent Systems and the Origin of Life." Van Dover, an *Alvin* pilot from 1989 to '91 and now assistant professor of biology at the College of William and Mary, reviews many of the suggestions presented above, and writes,

> If Woese's (1979) postulate for the origin of life . . . is correct, the earliest stages of life may not only be constructed in our imagination, but may actually be explored in their natal environment. Wächtershäuser provocatively suggests that the progenitor of all forms of life may be tracked down in its earliest habitat. Wächtershäuser asks rhetorically, "[W]here is this original homestead of life?"; his answer: "where hot volcanic exhalations clash with circulating hydrothermal water flow, . . . a place deep-down where a pyrite-forming autocatalyst once gave and is still giving birth to life."

Certainly not the last words on this subject, but some of the most recent came from Birgir Rasmussen, of the University of Western Australia. In *Nature* (June 2000), he reported "the discovery of pyritic filaments, the probable fossil remains of thread-like organisms, in a 3,235-million-year-old [3.2 billion] deep-sea volcanogenic massive sulphide deposit from the Pilbara Craton of Australia." The Pilbara is the site of the Apex Chert, where the oldest fossils in the world have been found (Schopf 1993). It contains one of the most complete sections of Archean volcano-sedimentary rocks, and is one of the two known geologic sequences that can be dated from 3.5 to 3.0 billion years ago. (The other is in Swaziland in South Africa.) Rasmussen interprets the threadlike organisms as nonphotosynthetic organisms that used inorganic matter—in this case sulfur—as an energy source, and lived at temperatures near 100°C, the boiling point of water. If his interpretation of these microfossils is correct, it would be the first evidence of microbial life from an ancient thermal vent system, would push the known range of thermal biota back another 2.7 billion years, and would mean that such environments may have hosted the first living systems on Earth. Rasmussen's suggestion is consistent with the published ideas of Baross et al. (1981) and Huber and Wächtershäuser (1998). Even the National Geographic Society, innocent of the controversy that would follow, named their 1980 film about the discovery of the first hydrothermal vents *Dive to the Edge of Creation.*

WATER, WATER EVERYWHERE

It is a given that all life on Earth depends upon water, but where did all this water come from? Some 4.5 billion years ago, a collision with another body about the size of Mars ejected enough material to form the moon, and boiled away whatever atmosphere the early Earth had, leaving it to hurtle through space as a fiery ball of molten rock. Collisions, however, have proved not to be exclusively destructive; the formation of the moon might be seen as a benefit, but extraterrestrial impacts also gave us an atmosphere and, more important for the development of life, water. Meteorites known as carbonaceous chondrites can contain up to 20 percent water, either in the form of ice or locked up in the crystal structure of minerals, but it is those icy itinerants known as comets that are believed to have been the source of water on the Earth.

In the Oort cloud, beyond the orbit of Pluto, there are believed to be a trillion

(1,000,000,000,000) comets, whose looping itineraries sometimes bring them within range of our solar system. Most comets are composed of ice and frozen carbon dioxide; they are essentially gigantic dirty snowballs. (Halley's comet is a potato-shaped lump about ten miles long with a mass estimated at 10 billion tons, most of which is water ice.) Comets do not collide with Earth very frequently—one may have impacted 65 million years ago, somehow contibuting to the disappearance of the dinosaurs, and one may have vaporized above Siberia in 1908—but it was different in the distant past: "Comets swarmed through the solar system in far greater numbers when the Earth was forming, and would have crossed paths with the planet far more regularly, bringing oceans on their backs" (Ball 1999). Chyba (1990) has estimated that if only 25 percent of the bodies that hit the Earth during the period of heavy bombardment were comets, they could account for all the water on Earth.

(In a 1998 article on the cometary origin of the Earth's water, however, James Kasting identifies "a major snag" in this theory: "Astronomers have found that three comets—Halley, Hyakutake, and Hale-Bopp—have a high percentage of deuterium, a form of hydrogen that contains a neutron as well as a proton in its nucleus. Compared with normal hydrogen, deuterium is twice as abundant in these comets as it is in seawater. . . . If these three comets are representative of those that struck here in the past, then most of the water on Earth must have come from elsewhere.")

Life exists here because Earth is the only planet that we know of with liquid water on its surface, what Kasting calls "the magic elixir required by all living things." But of course the water would not always have been in liquid form; the heat of the planet would have caused it to evaporate almost instantly, and the part of it that was not blasted back into space would have circled the Earth in an atmosphere of steam. As the Earth cooled, condensation caused millennia of rain to fall, and the depressions on the Earth's surface were filled with precipitation. (The Pacific Ocean, the largest and deepest of these depressions, is thought by some to have been formed when the matter that makes up the moon was untimely ripped from the fabric of the Earth, and then caught in the gravitational field of its mother.) In his 1998 article on the origins of water, Kasting pointed out that the creation of oceans requires water and a container in which to hold it. The "containers," of course, are the ocean basins, which were formed as the Earth cooled and solidified. "The mountains of the oceans are nothing like the Alps or the Rockies, which are largely built of folded sediments," wrote the British geophysicist Edward Bullard in 1969. "There is a world-encircling mountain range—the mid-ocean ridge—on the sea bottom, but it is built entirely of igneous rocks, of basalts that have appeared from the interior of the Earth." The oldest sedimentary rocks are about 3.9 billion years old, and because their formation requires liquid water, it is clear that water was upon the Earth less than a billion years after the planet was formed. Life began in the ocean perhaps 3.8 billion years ago, and remained submerged until around 360 million years ago, when the first tetrapods emerged to take up a terrestrial existence, and lead the invasion of the land. For more than 2.5 billion years in the history of life on Earth, all living things on Earth were under water.

The Earth's tectonic plates have been in motion for billions of years, and the shapes on the surface of the Earth—say, 2 billion years ago, or even 50 million years ago—would not be recognizable from space, except in their color scheme: blue for the oceans, and a combination of gray and brown (and eventually, green), with a sometime covering of white clouds. Though it is now accepted that the continents have moved dramatically since their formation, and are still moving, the concept of continental drift was completely unknown before the mid-twentieth century, and early geologists made the not unreasonable assumption that Europe had always been where it was when they were standing on it.

It is obvious to anyone looking at a map of South America and Africa that there is a remarkable similiarity of the east coast of the former and the west coast of the latter, as if the two were parts of a gigantic jigsaw puzzle. Some early geologists believed that the power of the ocean had carved out a great valley between the two continents. In 1908 Frank Bursley Taylor suggested that the moon had once been so close to the Earth, during the Cretaceous Period, that its gravitational pull had dragged the continents apart, and their plowing through the ocean floor threw up the Himalayas and the Alps. Howard Baker presented the idea that Venus had come so close to the Earth that it pulled a great chunk out, forming the moon and setting the continents in motion. Despite the remarkable fit of the coastlines, however, it was not easy to see how they might have come apart. Continents are not supposed to break in half and slide over the Earth's surface like cookies on a greased tin. Weren't the continents rooted in the Earth? How could something that weighed uncountable millions of tons slide around? It was a German explorer named Alfred Wegener who looked at the puzzle and saw the answer.

Wegener had wondered for years about the congruity of the continents. He theorized that there had been a single land mass, which he called Pangaea, during the Carboniferous Period, some 300 million years ago. In 1912 he presented a paper to the Frankfurt Geological Association suggesting that this supercontinent was slowly breaking up. In 1906 he was invited to join a Danish meteorological expedition to Greenland, and then the War to End All Wars broke out and Wegener joined the military. Wounded in Belgium, Wegener spent much of his recuperation developing his theory of continental drift. In *Die Entstehung der Kontinente und Ozeane* (1915, published in 1924 in English as *The Origin of the Oceans and Continents*), he wrote:

> He who examines the opposite coasts of the South Atlantic Ocean must be somewhat struck by the similarity of the shapes of the coastlines of Brazil and Africa. Not only does the great right-angled bend formed by the Brazilian coast at Cape San Roque find its exact counterpart in the re-entrant angle of the African coastline near the Cameroons, but also, south of these two corresponding points, every projection on the Brazilian side corresponds to a similar shaped bay in the African.

He believed that the Mid-Atlantic Ridge, as well as Iceland and the Azores, were composed of material left over when the continents pulled apart. Mountains, he thought, were creases in the Earth's crust caused by the pressure of the moving landmasses, and not, as had been previously believed, "wrinkles" formed by the planet's shrinking like a withered apple. The "shrinking" hypothesis, which had been supported by none other than Isaac Newton, was finally discarded when analysis of radioactive elements (uranium, thorium, and radioactive potassium) showed how much time had transpired since the rock formations had cooled and solidified. Since the core of the Earth is still hot, the Earth could not have cooled enough to produce those towering wrinkles. Furthermore, the Caledonian mountain system in Norway and Scotland has many features in common with the Appalachian system of North America, which indicated to Wegener that these features had once abutted. Summarizing the evidence, he wrote, "It is just as if we were to refit the torn pieces of a newspaper by matching their edges and then check whether the lines of print run smoothly across. If they do, there is nothing left but to conclude that the pieces were in fact joined in this way."

Wegener had postulated that Pangaea began to break up about 180 million years ago, but he was unable to identify the responsible agent. He thought that the continents somehow barged around the ocean floor like ships plowing through pack ice, precipitating earthquakes before them, and leaving the sea floor behind. In 1937, Alexander du Toit, a South African geologist and a lifelong disciple of Wegener, suggested that rather than one continent, there had been two, which he named Gondwanaland and Laurasia. The eminent Cambridge geophysicist Harold Jeffreys was opposed to such heresies, on the grounds that the Earth's crust and mantle were much too rigid for the continents to move around, but F. A. Vening Meinesz countered Jeffreys's argument by suggesting that the mechanism for movement came from thermal convection in the mantle.

Before Wegener, nobody could conceive that the crust of the Earth moved in anything but a contracting manner; the cooling of the planet was thought to cause movement of the Earth's crust. One suggestion held that continents were somehow drowned like the "lost continent" of Atlantis, and another theory was that previously submerged regions had been uplifted and were now part of continents. Both of these ideas were anathema to those who believed that the relative positions of the land and the oceans were immutable, that the continents and the ocean basins were permanent features of the Earth's crust. Indeed, it was not until the middle of the twentieth century, when sea-floor spreading was shown to be the agent that propels the Earth's plates, that most geologists finally came to accept the idea of continental drift. (Those who so vehemently opposed the theory were correct in one respect: the continents themselves do not drift; they move because they are part of the lithosphere, which does move.) Continental drift was eventually employed to explain the appearance of fossils in unexpected and unexplainable places, and later the fossils would be used to verify the mechanism of continental drift, as similar species were found in places that had once been connected—such as southern South America and Antarctica, or eastern North America and western Europe—but had drifted so far apart that they were separated by an entire ocean.

HERE WE COME, READY OR NOT . . .

We don't know whether or not there has ever been water on Mars, or whether there is water under Europa's ice, but we do know that there is water on Earth, and plenty of it. A recent estimate is 326 million cubic miles. The story of the myriad forms of life that colonized the oceans, and later the land, is largely written in stone, in the fossil record.

Whether fast or slow, punctuated or steady, evolution is defined as "change over time," which means that one form has to be somehow modified into another, and this does not seem possible in a single generation. But in *The Triumph of Evolution* Eldredge suggests that it is, and so when we try to find transitional fossils, we are looking for something that barely exists. "Whales first appeared in the Eocene, some 55 million years ago," he wrote. "They are primitive—for whales. But the earliest specimens look like whales, and it is only their general mammalian features that tell us that they must have sprung from another group of terrestrial mammals in the Paleocene." He continues:

> How could we explain these gaps—these quick changes from one set of anatomical features to another? Easily: Gould and I realized that the standard geographic speciation theory as developed especially by Dobzhansky and Mayr was sufficient. Geographic isolation leading to reproductive isolation need not take long to occur: our estimate was from five thousand to fifty thousand years, and some geneticists immediately commented that speciation could go on a lot faster than that. Speciation can be a rapid process—and one *altogether too quick to show up in detail in the fossil record* [author's italics]. The layers of sediment that entomb the fossil record are simply too episodic, too inherently gappy themselves, to record the passage of every successive year—or decade—or century—of geological time. In other words, the naturally occurring gaps between closely related species in the modern fauna and flora, directly caused by the process of fissioning known as speciation, typically happens so quickly that rarely do we catch it in midstream when we scour the fossil record for insights on how evolution occurs.

If I am reading this correctly, Eldredge is saying that the transitions between species happen too fast to show up in the fossil record. If the fossil record were better—less episodic, less "inherently gappy"—then we *would* be able to find the transitional forms. Isn't this what Darwin said in 1859? In *The Origin of Species*, Darwin dedicated an entire chapter to "The Imperfection of the Fossil Record," writing: "The main cause, however, of innumerable intermediate links not occurring everywhere throughout nature, depends on the very process of natural selection, through which new varieties continually take the places of and supplant their parent-forms. But just in proportion of this process of extermination has acted on an enormous scale, so must the number of intermediate varieties, which have formerly existed, be truly enormous. Why then is not every geological stratum full of such intermediate links? Geology assuredly does not reveal any such finely-graduated organic chain; and this, perhaps, is the most serious objection which can be urged against the theory. The explanation lies, as I believe, in the extreme imperfection of the geological record." Eldredge says that the transitional forms went by too fast, like a speeded-up film, for the chance workings of sedimentation to trap each one.

The word "fossil" comes from the Latin *fossilis,* meaning "dug up," and originally referred to any natural object that was dug out of the ground. Fossils' origins were a complete mystery to people of the eighteenth and early nineteenth centuries, but there was no shortage of attempts to explain the mysterious appearance of various animal forms in places where their presence couldn't be explained, such as oysters and fishes on mountaintops. The nineteenth-century English naturalist Philip Gosse maintained that fossils had been strewn around by God as part of the spontaneous creation of a complete world; evidence of a history that had never really occurred. In other words, a God who could make something as intricate as a man would have no trouble with a few stony bones that resembled a fish. Others believed that it was just coincidence that the animal- or plantlike fossils resembled living forms—they had never been alive. Some fundamentalists subscribed to the idea that when Noah's flood submerged the Earth, all animals not on the ark were drowned, and fossils were the remains of those unfortunate animals that missed the boat. By the eighteenth century, most educated people recognized the organic origin of fossils, but the common man often believed in their supernatural origins.

When fossilized shark teeth were first discovered on land, long before any were brought up from the oceans' depths, their origin was a complete enigma. Pliny, the great Roman student of nature, believed that they had fallen from the sky during eclipses of the Moon. They were later thought to be the tongues of serpents that St. Paul had turned to stone while he was visiting the islands of Malta, and in consequence they acquired the name *glossopetrae* ("tongue stones"). They were believed to have magical properties, especially as counteragents against the bites of poisonous snakes, and to that end they were often worn as talismans.

Very few dead animals fossilize, perhaps just one body in many millions. Fossilization requires burial. If a carcass is not buried, the normal processes of scavenging or decay will return the body to the environment—"dust to dust." Animals that lived in or near the water stood a much greater chance of being buried than animals that lived or died on land. The shallower waters of continental margins and inland seas are ideal for the preservation and subsequent exposure of fossils, because the plentiful flow of sediments from the land is the perfect medium for the preservation of bone, sometimes even an outline of a body part, or, in extremely rare instances, the whole body. Another question is how much of the "fossil record" remains buried? Fossils are often found in sedimentary layers that are exposed as cliff faces (Lyme Regis in southern England), or when quarries are excavated (Solnhofen in Germany). In some instances, erosion wears away the surrounding matrix, and the fossils are found sticking out of the ground, or even sitting on top of it. (In Africa, most of the early hominid fossils were found lying on the ground.) Dependence upon the chance quarry or cliff face graced with the appropriate historical conditions necessary for fossilization, or the improbable hope that you will stumble upon a 100-million-year-old jawbone or tooth as you walk, puts the incompleteness of the fossil record into an entirely different light: it is a miracle that we have any record at all of prehistoric life, and those who would reconstruct life styles or phylogenies from this scrappy evidence are to be commended for their creativity, imagination, and fortitude. There is no way of knowing what fossils are entombed hundreds of feet down, in rocks that have not seen the sun for 200 million years, but it is a safe assumption that only a small proportion has so far been revealed. It is the terrestrial version of the story of the oceanographer who responded to a student's remark that "there's an awful lot of water out there" with "And that's only the top of it."

In a review of Henry Gee's *In Search of Deep Time*, Jon Turney of University College, London, wrote (2000):

> Take a complete, illustrated catalogue of London's National Gallery. Shred it into tiny pieces and cast them into the wind from the gallery's steps above Trafalgar Square. Wait a few weeks, then scour the square for surviving scaps of paper. Now try to reconstruct the history of painting from your haul. If you manage to produce a coherent story—schools, styles, genres, named painters and all—you are probably a paleontologist.

The entire science of paleontology is based upon the fossils that have been found, and there can only be speculation in their absence. Although this seems self-evident, it is one of the least-recognized aspects of the study of ancient creatures, for there is a categorical predisposition to draw conclusions on the basis of the remains of animals that have been discovered. Obviously, we can learn nothing from undiscovered animals, but the occasional presence of single specimens—or, at best, a small number of a type of specimen—leads to the inescapable realization that there are vast numbers of creatures that have not been found, and may never be. Paleontologists are continually finding new species and trying to construct an inclusive frame that includes what they have found, but, like the jigsaw puzzle to which paleontology is always compared, most of the pieces are still missing—and are probably destined to remain so.

Except for the inexplicable presence of millions of living species on Earth, the fossils of extinct species are the only evidence of evolution. David Raup, perhaps the foremost extinction theorist of our day, has estimated that there have been on Earth "between 5 [billion] and 50 billion species. And only an estimated 5 [million] to 50 million are alive today." It has been suggested that more than 99 percent of all the species that have ever lived on Earth are extinct, including every single one of our hominid ancestors. Extinction is part of the evolutionary process (or perhaps evolution is part of the extinction process), but how the two intersect is not clearly understood. Indeed, evolution and extinction are equally mysterious, since there is no question that both have been occurring for eons, but how they work, either together or separately, is not known. In a 1997 essay David Archibald wrote, "The only way a steady state of species numbers could be maintained, even while evolution is occurring, is for other species to become extinct. Thus, extinction, rather than being a rare and negative event in human time frames and sensibilities, is actually a very common and positive counterpoint to evolution."

When an animal dies, assuming it is not eaten by scavengers, the bones lose collagen naturally, and become brittle. (Cow bones left in a field are often much lighter than fresh bones.) If they are not buried, the elements will cause them to deteriorate, and they will eventually crumble to dust. If the bone is covered with mud or sand and compacted, the pores will be filled with foreign minerals, a process that varies with the ambient conditions. As layer after layer of sediment presses down on a skeleton, the original form of the animal may be compressed or distorted. It may also be buried, and unless there is some sort of an upheaval, it may remain hidden underground forever. Fossil hunters rely on changes in the Earth's surface to bring their quarry to the surface; most fossils are found by looking down—or up, if the site includes a vertical wall. Digging takes place only after the fossil has been found, to remove it from the ground that has imprisoned it for so long. Even accounting for mountains, canyons, trenches, ridges, mesas, riverbeds, and the great variety of other undulations in the surface of the Earth, the top layer of our planet is mostly horizontal. And since the various stratigraphic layers are usually laid down one atop the other, they are more or less horizontal too. So paleontologists and geologists who would investigate the underlayers have to dig, core, drill, blast, or otherwise invade the Earth's spherical horizontality.

"Theoretical infrastructure," wrote the Edinburgh paleontologist Michael Taylor, "is critical to the interpretation of any fossil." Indeed, without such interpretation, fossils are nothing but bone- or shell- or tooth-shaped hunks of rock. There are two chronologies that apply to the study of fossils: the first is faunal succession, the order in which the animals lived; the second is the order in which they were found. The first—the succession of life forms through time—is essential for the study of evolution, but since fossils represent only a tiny proportion of the actual record and the interpretation of this evidence is the basis of paleontology, the record is not always clear. Part of this succession is the superimposition of one rock layer over another, with lower layers being older than the higher ones, because the lower layers had to be there before the younger ones could be laid down on top of them. Both of these tools are often turned upside down and require more interpretation than the layman might expect. There are, for example, many instances where the faunal succession is not at all obvious, and where more advanced animals seem to precede more primitive forms, leading to the possibility that we might be not looking at a progression at all, but rather two parallel evolutionary tracks.

Perhaps the most unsettling element in paleontology is the randomness of the discoveries. One cannot predict where a particular fossil will fall in a faunal sequence if little is known about that succession, so we are constantly finding things that have to be plugged in *somewhere,* but we often have only a vague idea of where they are supposed to go. Consider Philip Gingerich's 1983 discovery of *Pakicetus,* a new link in the chain of whale evolution. Up to that time, all known ancestral whales had flippers, but Gingerich reconstructed *Pakicetus* with legs, which completely rearranged our ideas about whale evolution. One might assume that the most recent animals would be found closest to the surface, and the older animals deeper in the Earth, but of course, geology is not always so cooperative as to maintain a strict chronology, and upthrustings, earthquakes, reworkings, and other disturbances often rearrange stratigraphic layering so as to make the obvious older-at-the-bottom chronology less reliable.

The fossil record is, almost by definition, incomplete. First of all, it is biased in favor of those animals that leave skeletons, and most of those are destroyed by natural forces long before they get an opportunity to fossilize. If the carcass has come to rest in the bed of a lake, a stream, a river, or the bottom of the ocean, the soft parts are likely to be eaten by scavengers. The remainder of the carcass then has to be buried in sediment—only sedimentary rock contains fossils. Therefore, animals that live on or in the sediment are most likely to be fossilized, whereas creatures that swim above the bottom are less likely, and land animals the most unlikely of all. As sediment piles on top of sediment, the lower layers are compacted, and gradually turn into sedimentary rock. Very small particles, such as those found in clay, will form a rock known as shale. A variety of sandstones are formed by the deposition of various types of sand. Shales and sandstones result from the accumulation of particles that have been brought from elsewhere, usually by the action of the water. Limestones and chalks, on the other hand, are formed by the accumulation of the shells of living creatures such as "forams"—animals whose shells are composed primarily of calcium carbonate that is absorbed from the water by the animal as it builds its shell. Our world dances to the fossil record, but the tunes and the rhythms are often more mysterious and disconcerting than the music our children listen to.

FROM FOSSIL TO SQUID

THE GARDEN OF THE EDIACARA

We have found fossil evidence of ancient life forms that do not conform to any known plan. Some of these soft-bodied "animals" are flat like pancakes; others look vaguely like jellyfishes; some are sort of wormlike; and others resemble nothing even remotely recognizable. They may have floated, they may have lain on the bottom, or they may even have been buried in sediment. Most are quite small, but at least one was more than 3 feet across. They have been dated at between 565 million and 543 million years ago, making them the oldest multicellular life forms so far discovered.

Darwin admitted that he had no idea how life began. In *The Origin of Species* he also confessed that he couldn't understand how complex life seemed to spring up in the Cambrian Period without any evidence of life forms before that. He wrote:

> There is another . . . difficulty, which is much more serious. I allude to the manner in which species belonging to several of the main divisions of the animal kingdom suddenly appear in the lowest known fossiliferous rocks. Most of the arguments which have convinced me that all the existing species of the same group are descended from a single progenitor, apply with equal force to the earliest known species. For instance, it cannot be doubted that all the Cambrian and Silurian trilobites are descended from some one crustacean, which must have lived long before the Cambrian age, and which probably differed from any known animal. . . . To the question why we do not find rich fossiliferous deposits belonging to these assumed earliest periods I can give no satisfactory answer.

He wasn't looking in the right place.

In 1946, an Australian mining geologist named Reginald Sprigg was prospecting north of Adelaide, South Australia, in a section of the North Flinders Ranges known as the Ediacara Hills. The name, pronounced "ee-dee-*ak*-ra," is from an Aboriginal word that means "veinlike spring of water." Sprigg found fossilized imprints of what were apparently soft-bodied organisms, preserved mostly on the undersides of slabs of quartzite and sandstone. Most were round, disc-shaped forms that Sprigg dubbed "medusoids," which means a shape resembling a jellyfish. He published two papers in the *Transactions of the Royal Society of South Australia,* and in both he noted this resemblance. Others of these organisms, however, resembled worms, arthropods, or even stranger things. Initially Sprigg thought that these fossils might be from the Cambrian Period, but later work established that these fossils are in fact late Precambrian. The Ediacaran fauna is sometimes referred to as the Vendian fauna, a reference to creatures of the same age—565 million to 535 million years ago—that were found along the coast of the White Sea in Russia, an area where the Vends, a Slavic people, once lived. Although their identification remains problematical, the Ediacaran fossils are the oldest record of animals in the world.

The Ediacaran fauna comes in all sorts of strange shapes and sizes, and its organisms are as easily defined by what they lack as by what they have. They have no heads or tails; no fronts or backs; no circulatory, nervous, or digestive systems. The ones found by Sprigg, the most common, are still referred to as medusoids, but they do not resemble any known jellyfishes. One type (*Cyclomedusa*) looks like a small fried egg; it was the type found by Mark McMenamin in Sonora, Mexico, in 1995 and dated as "the world's oldest fossil." (There are actually fossils older than McMenamin's 600-million-year-old *Cyclomedusa,* but they are cyanobacteria.) *Cyclomedusa* is probably the most common and widespread Ediacaran fossil. It also has one of the largest size ranges, from a few millimeters to about a meter in diameter. Formerly thought to represent a floating jellyfish of some sort, *Cyclomedusa* is now considered by some to have been a benthic (bottom-dwelling) polyp, somewhat like a sea anemone.

Other forms include the animal that left the mysterious imprint that McMenamin claimed was partially responsible for his interest in the Ediacaran fauna, the picture that appeared in the Time-Life Nature Library book *The Sea* in 1960, with this caption:

> MYSTERY FOSSIL, first of its rare kind ever found, has no known relationship with any other living creature living or dead. It comes from pre-Cambrian rock strata in South Australia.

It has since been identified as *Tribrachidium,* which means that it has triradiate (three-part) symmetry, otherwise unknown in the animal world. (Certain well-known plants, like clover, are triradially symmetrical, but no animals are.) Others, found in the English Midlands, are shaped something like ferns, and their classification as plants or animals is still in question. There are Ediacarans shaped like garlic cloves (*Inaria*), and one (*Arkarua*) that looks like a tiny hubcap with fivefold (pentameral) symmetry, which has led James Gehling (1987) to describe it as "the earliest known echinoderm." (*Arkarua* comes from an Aboriginal name for a mythical giant snake of South Australia, where the fossil was discovered.) *Arkarua* occurs alongside *Dickinsonia, Tribrachidium, Cyclomedusa,* and other familiar Ediacaran animals as well as many new and as yet undescribed species. The fossils preserve what appears to be a five-lobed central region that is interpreted as five ambulacral grooves, which are characteristic of echinoderms (spiny-skinned animals, such as starfishes, sea urchins, and sand dollars). Named for Reginald Sprigg, *Spriggina* looks something like a soft-bodied trilobite, mostly because it has a crescentic structure at the "head" and a segmented "body." (Some paleontologists, including Adolf Seilacher, believe that the fossil is more like a frond, and the headlike "cepahalon" is actually a kind of holdfast.)

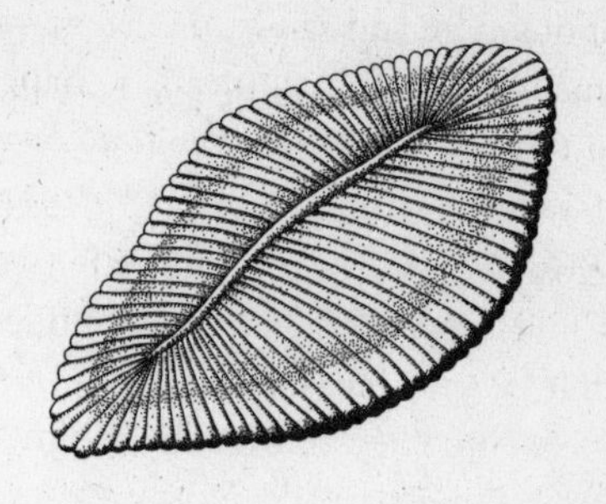

Dickinsonia *"has the distinction of being the only fossil to be described as a jellyfish, a coral, a sea anemone, an annelid worm, a polychaete worm, an arthropod [joint-legged animal], a bacterium, a protozoan, a member of a new phylum, a member of a new kingdom, and even a creature from outer space" (Mark McMenamin).*

Probably the best-known but least understood of all the Ediacaran fossils is *Dickinsonia,* which, as McMenamin wrote (1998), "promises to provide a tremendous amount of information concerning the paleobiology of these organisms. If only we knew what it was. It has the distinction of being the only fossil to be described as a jellyfish, a coral, a sea anemone, an annelid worm, a polychaete worm, an arthropod, a bacterium, a protozoan, a member of a new phylum, a member of a new kingdom, and even a creature from outer space." It is roughly oval in shape, with a midrib that bisects it on the long axis. Radiating from the central rib are tubular partitions that become wider as they approach the perimeter. At one end, the distance from the midrib to the perimeter is considerably longer than at the other. There is no way of determining which—if either—end is the head and which is the tail. The largest known specimen of *Dickinsonia* is 16 inches long, about the size and shape of a beaver's tail. *Dickinsonia* is known from Vendian rocks of South Australia and from the Winter Coast of the White Sea, in northern Russia.

Like many of the fossils of the better-known Burgess Shale, in the western Canadian Rockies, and the Chengjiang sites in Yunnan province, the Ediacaran animals—if indeed they are animals—defy classification. This is largely because there are almost no characteristics apart from classes of symmetry that are demonstrably present in any living organism. For example, *Kimberella* has recently been confidently reinterpreted as a mollusclike animal rather than a box (cubozoan) jellyfish, but neither interpretation is supported by a single morphological feature that is unequivocally shared by the fossils and the group in question. The original interpretation was based on a small number of specimens from South Australia that looked like four-part box jellyfishes lying on their sides. Recently, many well-preserved and large specimens over an inch across from the Winter Coast of the White Sea region of Russia were found by Mikhail Fedonkin, of the Russian Academy of Sciences, and Ben Waggoner, now of the University of Arkansas. Fedonkin and Waggoner (1997) have shown that *Kimberella* was a bilaterally symmetrical animal that had rigid parts. They reasoned that *Kimberella* probably had a tough shell-like covering that rigidly stood up into the sediment when the animals were buried. Thus, *Kimberella* appears to be somewhat like a mollusk—or, as they wrote, "We conclude that *Kimberella* is a bilaterian metazoan, more complex than a flatworm, more like a mollusc than any other metazoan, and plausibly bearing molluscan synapomorphies such as a shell and a foot." Nevertheless, it is still uncertain which group of modern animals is most closely related to this interesting animal.

Gregory Retallack (1994) doesn't believe that the Ediacarans were animals at all, but lichens. He suggests that if they were really soft-bodied animals, they would have been squashed flat by the weight of the sediments. Instead, Retallack writes, they were almost totally resistant to compaction, and their structure would have been more similar to that of lichens, which are made of sturdy molecules like chitin. Moreover, Retallack says, their large size, some of them up to 3 feet across, is much more in keeping with a sessile (permanently attached) organism that gained its nutrition via symbiosis with photosynthetic organisms. If they were animals, there is no plausible explanation for how they might have obtained their nutrition. However they did so, they are among the first complex living things to occupy the Earth, and because the planet was covered with water during their ancient and ill-documented existence, their history is important in our search for the earliest marine life forms.

Australian paleontologist Martin Glaessner believed that the Ediacaran fauna could be assigned to known phyla such as Cnidaria (jellyfishes), thought to be among the most primitive of animal forms. He therefore viewed (1958) the Ediacarans as the missing links between the small, simple ancestors of today's jellyfishes and the anemones. The paleontologist Adolf Seilacher—who has been called by Stephen Jay Gould (1989) "the finest paleontological observer now active"—does not believe that the Ediacaran fauna is related to any living forms, and that all of them died out. He proposed (1988) that these extinct forms, which he christened vendozoans, were built rather like air mattresses—he described them as "quilted pneu structures"—that lay flat on the sand; their resemblance to jellyfishes and sea pens is only coincidental. He wrote,

> There are many possible reasons why vendozoans should have been flat. Since they fail to show unequivocal signs of intestinal tracts (axial structures were interpreted as a gut in *Dickinsonia,* but as a stem in pinnate forms), it is possible that they were gutless. In this case, all metabolic processes (nutrient intake, respiration, excretion) had to take place through the body surface, which should have been maximized accordingly.

Seilacher also answered the question about how animals without hard parts might have left such clear impressions, by postulating a thick bacterial mat that covered the ocean floor; when the vendozoans were covered by sand, they were pushed into the bacterial mat, and their outlines were preserved, like batter in a waffle iron. Mark McMenamin agrees with him (and also with Gould in calling Seilacher "the best paleontologist of his generation") and suggests that the Ediacaran animals could photosynthesize because they had symbiotic algae in their tissues, rather like corals today. Other workers had also suggested that the Ediacarans might have been photo-symbiotic, like plants, capturing light for energy. Mikhail Fedonkin published a paper in 1983 in which he wrote that "extant cnidarians, tubellarians, and representatives of other invertebrate groups are known to have symbiotic algae within their tissues. These algae participate in oxygen exchange with, and provide much of the food requirements for, their animal hosts. It is possible that this type of symbiosis is one of the most ancient, and furthermore it is possible that it was significantly more widespread in the Vendian than it is today." Virtually every Ediacaran fossil imprint is whole; there is no sign of any bite marks or missing pieces. The absence of predators, and not their plantlike orientation, is the reason that McMenamin named the faunal assemblages the "Garden of Ediacara" (1986). He wrote, "Could we be dealing with a predator-free ecosystem? A peaceful seafloor garden, of course! Ediacara." Seilacher opined that they were "an evolutionary experiment that failed with the coming of macrophagous predators."

The most abundant assemblages of Ediacaran fauna are those from the Flinders Ranges of Australia, and the White Sea region of Russia, accounting for 60 percent of the described taxa (Martin et al. 2000). To date, the best fossils have been found along the shoreline cliffs of Zimnie Gory, on the White Sea, just north of Arkhangelsk. They include *Cyclomedusa, Charnia, Tribrachidium, Dickinsonia, Kimberella,* and several other types, buried in layers of coarse siltstone, claystone, and sandstone. (The earliest known Ediacaran fossils were found in the Mackenzie Mountains of northwest Canada, from approximately 600 million years ago.) As described by Fedonkin and Waggoner (1997), *Kimberella* was probably mollusklike, and Seilacher (1999) suggested that the radula scratches on the South Australian rocks were made by an animal similar to *Kimberella,* and similar scratches were found below the dated ash at Zimnie Gory. "One of these creatures," wrote Richard Kerr (2000), "may have been the long-sought animal whose descendants split into the two great lineages of modern life: the protostomes—mollusks, annelids, and arthropods—and the deuterostomes—the echinoderms and chordates."

Adding an Ediacaran fauna to another location, Hagadorn and Waggoner (2000) found "the frondlike *Swartpuntia,* and tubular fossils . . . similar to *Archaeichnium, Cloudina, Corumbella,* and *Onuphionella*" in the Wood Canyon Formation of Nevada. Some of these forms can be dated up to and into the Cambrian, unusual for the Ediacarans, but the authors found that the limited Nevada fauna suggested a paleogeographic link between the southwestern United States and southern Africa in the Vendian period. In the Late Proterozoic (Precambrian), the supercontinent Rodinia that rifted apart to form Laurentia (North America), Baltica (Europe), Asia, Australia, Antarctica, India, and Greenland showed that the pieces that would become southern Africa and northern Laurentia were connected. And in 2000, Hagadorn, Waggoner, and Christopher Fedo published a paper, "Early Cambrian Ediacaran-Type Fossils from California," describing fossils from just over the Nevada border, and part of the same neoprotozoan and Cambrian strata of the Great Basin. In three locations in California (the White-Inyo region, Death Valley, and the Mojave Desert), they found "discoidal structures" and the now-common *Swartpuntia,* which is similar to *Spriggina.*

Mark McMenamin of the University of Massachusetts has devoted virtually his entire professional career to the study of Ediacarans. In 1986 he wrote an article for the journal *Palaios* in which he coined the term "Garden of Ediacara," and after studying these enigmatic life forms for another twelve years, he wrote a book with the same enchanting title (1998). McMenamin worked on Ediacarans in Namibia with Adolf Seilacher and with Lynn Margulis at the University of Massachusetts. (Margulis's son and frequent coauthor, Dorion Sagan, wrote the foreword for McMenamin's book.) In a chapter entitled "A Family Tree," McMenamin discusses the imminent publication of his book, which was about to go to press without, as he puts it, a "satisfactory explanation of what the Ediacarans *were*." This obviously would not do, so after much reflection and examination of specimens, particularly *Pteridinium* from Namibia, he decided that the cells of the Ediacarans divided not in the normal manner, where a cell divides and then each resulting cell divides, forming four cells, but in a unique fashion, in which cell families reproduce into cognate families. "This process," he wrote,

> explains the morphology of what I will call the one-cell, two-cell, three-cell, four-cell, and five-cell Ediacaran specimens. These can be further categorized as to whether the founder cell families give rise to offspring (cognate families of cells). In other words, if there are one-member cell families, then the total number of cell families in the mature Ediacaran organism equals the original number of cell families. If the cell families reproduce, however, then the number of cell families increases with time as successive iterations of cell families are grown. These Ediacaran forms are the ones with long axes because the iterating cell families tend to stack up.

He then classifies the known Ediacarans into categories according to the number of cell families, where *Cyclomedusa* is a one- or sometimes two-celled family; *Tribrachidium* (the "mystery fossil" from the Time-Life book) is descended from a three-cell iteration, and thus trilobate; *Spriggina* is the result of four-cell iterating families, etc. He then draws up a cladogram (a treelike diagram) of Ediacarans, indicating that they branched off from the protist ancestor in one direction, and all other animal groups took the other direction. He concludes that "the creatures of the phylum Vendozoa . . . shared a common ancestor with them from a protist ancestor, but developed a highly unusual approach to multi-cellularity." At the conclusion of the "family tree" chapter in his 1998 book, McMenamin writes, "I have cracked the code of the Ediacaran puzzle."

Not everyone agrees. Simon Conway Morris (1993) and Jim Gehling (1999) are two paleontologists who believe that the Ediacarans were early forms of known animals, not anomalous creatures that redefined cell division; others, like Seilacher, believe they were unique and became extinct before the Cambrian Period began, 540 million years ago, leaving no descendants. The answer to the mystery of the Ediacarans has not been found, nor has their place in the early history of life been ascertained. Indeed, there may not be a single answer, but parts of everybody's ideas may have to be blended together to produce some sort of universal Ediacaran phylogeny. In *Wonderful Life* (1989), Stephen Jay Gould wonders how the world would look today if all those weird creatures from the Burgess Shale had not died out, and also speculates on what might have happened "if the flat quilts of the Ediacara had won on their second attempt. . . . Could life have ever moved to consciousness along this alternate pathway of Ediacara anatomy? Probably not. . . . The developmental program of Ediacara creatures might have foreclosed the evolution of internal organs, and life would have remained permanently in the rut of sheets and pancakes—a most unpropitious shape for self-conscious complexity as we know it." In a 1998 article, Ben Waggoner acknowledged that there was no clear answer to the question "What are the Ediacaran organisms?" In any event, wrote Peter Ward and Donald Brownlee in *Rare Earth* (2000),

> The Ediacarans do make for good theater—enigmatic, mysterious, and the first on the stage. They were a hard act to follow, but a great wave of diversification was occurring at the end of their time in the limelight, a wave that continues on planet Earth still. With the second act of the Cambrian Explosion, undoubted animals appear on the scene.

Recent fossils found in Namibia show us that members of the Ediacara biota populated the oceans up until the end of the Precambrian, 545 million years ago. With the close of the Precambrian, much of the Ediacara biota became extinct. Many of them seem to be "failed experiments" in the evolution of life on Earth that became extinct just before the Cambrian Explosion of life forms. Other Precambrian fossils do appear to represent the ancestors of modern animals. It is generally agreed that simple burrows and trace fossils (fossils of tracks), such as those made by *Helminthopsis,* found in Upper Precambrian rocks were made by primitive worms. These worms, and some other members of the Ediacara biota, survived the extinction event that killed off most of the Ediacarans and took part in the greatest evolutionary event in Earth's history: the Cambrian Explosion. Within 35 million years of the end of the Precambrian, representatives of essentially all modern phyla were present in the Cambrian seas.

WONDERS OF THE BURGESS SHALE

Is the history of life on Earth somehow preordained? Is there a force at work that controls the way living things evolve or become extinct? What would life be like today if some primitive forms had died out before they could endow their descendants with certain characteristics?

When the railroad was being built through the Canadian Rockies in British Columbia, some of the workmen reported finding fossils, which encouraged a visit from a geologist in 1884. Millions of trilobite-bearing slabs were found scattered all over the ground, just above the treeline. Trilobites are extinct marine invertebrates that had a flattened oval body, divided into three sections by a pair of lengthwise furrows (see p. 48). The area was named the Ogygopsis Shale, for a particularly abundant trilobite found there. But it was not until 1909, with the arrival of Charles Doolittle Walcott, director of the Smithsonian Institution and one of America's premier paleontologists, that the area was recognized as one of the most important paleontological sites in the world, especially the adjacent Burgess Shale site, which was only 10 feet deep and a couple of hundred feet long. By 1912, Walcott had published a whole series of papers describing the fantastic fauna he had found, but he was becoming more and more involved in the administration of the Smithsonian and had less time for paleontology. He died in 1927, and although the material he had collected was available for study in the Smithsonian and the Museum of Comparative Zoology at Harvard, it was almost totally ignored until the 1960s, when James Aitken of the Geological Survey of Canada took an interest in the fossils, and assigned Harry B. Whittington, an authority on trilobites, to head a project to analyze the fossils of the Burgess Shale.

As Whittington began to examine specimens of a spiny, arthropod-like creature that Walcott had named *Marrella splendens,* he soon realized that he was facing a task of incredible complexity and richness. (*Marrella,* which was only about three quarters of an inch long, was the most abundant animal in the Burgess Shale, suggesting that it might have aggregated in shoals, and been a staple of the diet of many predators, rather like krill today.) It was around 1970 that two Cambridge University students, Derek Briggs and Simon Conway Morris, signed on to complete their doctoral research on species of the Burgess Shale fauna, Briggs on the crablike *Canadaspis perfecta,* and Conway Morris on the worm originally named *Canadia sparsa* but quickly rechristened *Hallucinogenia* because of its unlikely appearance. In *Crucible of Creation* (1998) Conway Morris wrote, "The apparent weirdness of *Hallucinogenia* began to be taken as an exemplar of the Burgess Shale. Life in this community came to be regarded as a Cambrian bestiary with an organic exuberance that completely overshadowed the dull ordinariness of typical Cambrian faunas." As originally reconstructed, *Hallucinogenia* had seven pairs of stilt like "legs" and a series of tentacles on its back, but in 1990, Richard Cowen, in his textbook *History of Life,* flipped the organism over, suggesting that "Conway Morris thought it was a scavenger tottering around on its spines [but] I thought it was a mud-grubber in the sediment with the spines protecting it."

Found in the Burgess Shale was a primitive onychophore (small worm) known as *Aysheaia,* a creature with 20 wrinkled legs that ended in claws, and rings around its body. It was close enough to modern onychophores to cause these to be classified as living fossils, and it was hoped it might give us a clue as to the transitional form between segmented worms and trilobites.* In his monograph on *Aysheaia,* Whittington (1978) contended that this little wormlike creature was *not* ancestral to the ony-

---

* The invertebrate class Onychophora, commonly known as "velvet worms" or "walking worms," are small invertebrates that live in damp, terrestrial environments on every continent except Europe. They have a simple, three-segmented head equipped with tiny eyes, paired antennae like those of a slug, and a segmented trunk that

chophores, but Stephen Jay Gould—whose whole discussion of the Burgess Shale animals centered around the concept that few of them were ancestral to living forms—wrote (1989), "I believe that, on the balance of the evidence, *Aysheaia* should be retained among the Onychophora." Also, there are those who regard the mysterious Ediacaran *Spriggina* as a forerunner of arthropods and trilobites. Others think it might have been a sessile, fern-shaped animal that clung to the bottom with the horseshoe-shaped element where the cephalon (head-shield) of the trilobites is located.

Among the most enigmatic fossils found in the Burgess Shale, and elsewhere, were sclerites, tiny scalelike plates that were characterized by a mesh of hexagonal cells with a round hole in each cell. Nobody knew quite what to make of these mysterious structures, which were obviously part of *something,* but what part of what? In the Chengjiang region of south-central China, the paleontologists Chen, Hou, and Lu (1985) found the answer in an enlongated, wormlike creature with scalelike plates spaced along its sides, each plate positioned in conjunction with a leg. The discovery of the complete fossil of *Microdictyon sinicum* did not answer the question of what purpose the sclerites served, but at least it located them on a particular animal. Surprisingly, the position of the legs on *Microdictyon* gave researchers a clue to the mystery of *Hallucinogenia,* because they could now see that the spikes that were originally thought to have been legs were spines, analogs of the plates on the flanks of *Microdictyon.*

Unknown as a source of fossils until 1984, the area of Chengjiang in China's southwestern border province of Yunnan has achieved a reputation that is equal to that of the Burgess Shale. Between 1984 and 1996, thousands of fossils have been unearthed from the Early Cambrian strata at the base of Maotianshan Mountain. Like the Burgess Shale and Solnhofen in Germany (the source of the *Archaeopteryx* fossils), the finely grained sandstones of Chengjiang retain exquisitely detailed impressions of muscles and tissues, making it possible to get a much better idea of what a particular animal actually looked like, instead of having to guess and extrapolate from the often fragmented shelly fossils. In their 1991 article about another "armored lobopod"* like *Microdictyon,* Lars Ramsköld and Hou Xianguang wrote of the Chengjiang Formation, "It is recognized as the most important early Phanerozoic *Lagerstätte* [deposit] besides the Burgess Shale. . . . Over 3,000 remarkably preserved 'soft-bodied' specimens collected to date have revealed a diverse fauna, now partly under study in a Chinese-Swedish project."

In his 1998 discussion of *Hallucinogenia,* Conway Morris wrote, "My original reconstruction had one small problem: the animal was upside-down." Ramsköld and Hou wrote, "The Burgess Shale problematicum *Hallucinogenia* has been regarded as

---

is not divided into major sections. Like the trilobites, however, and unlike most worms, they have about 20 pairs of legs, each of which ends in a flexible knob with two claws on it. Unlike arthropods, the walking worms (generic name *Peripatus*) do not develop a rigid cuticle, but they do periodically shed their exterior skin.

* A lobopod is the soft walking appendage of the lobopodians, a group of primitive arthropods that flourished in the Cambrian such as *Hallucinogenia* and *Microdictyon* and are represented today by the onychophores (Conway Morris 1998).

representing either a separate phylum or the detached part of some larger organism. A general resemblance to *Microdictyon* was noted recently. In the light of the above, *Hallucinogenia* can be reinterpreted as an 'armored lobopod.' The structures originally described as legs are here regarded as spines of paired plates. . . . The 'dorsal tentacles' of *Hallucinogenia* are reinterpreted as legs. Only a single row is visible." In *Wonderful Life,* Stephen Jay Gould wrote: "Yet *Hallucinogenia* is so peculiar, so hard to imagine as an efficiently working beast, that we must entertain the possibility of a very different solution. Perhaps *Hallucinogenia* is not a complete animal, but a complex appendage of a larger creature, still undiscovered."

Ramsköld and Hou described *Xenusion* and *Luolishania,* which resembled *Microdictyon* and *Hallucinogenia* in that they were "caterpillar-like animals . . . with 11 pairs of legs, above which are rounded, outwardly-facing plates." (Now that it has been turned over, *Hallucinogenia* has spikes, not plates, but they are in the same dorsal location.) The Chengjiang animals are close enough to *Aysheaia* to place both species in the Onychophora, which means that the fauna of the Burgess Shale, along with that of Chengjiang, can be assigned to known groups, and these animals are not so far removed from known groups after all. In his 1991 review of the Ramsköld and Hou article, Stefan Bengston said,

> Shall we then never understand life forms that do not have living relatives presenting similar anatomy? No doubt we shall, but the revelation that some of the more "bizarre" Cambrian creatures now look more like onychophorans with shoulder pads is a healthy reminder that any fossil—whether an isolated sclerite, a preserved soft body, or something else—presents but a fraction of its body for inspection.

The scourge of the Cambrian seas was Anomalocaris, *a yard-long swimmer with huge eyes, grasping appendages, and a pineapple-ring mouth. It must have preyed upon trilobites, because they existed in great profusion at that time; furthermore, there is fossil evidence of a trilobite with a bite taken out of it by an* Anomalocaris.

Probably the weirdest of all the Ogygopsis Shale creatures was the one that J. F. Whiteaves had originally named *Anomalocaris canadensis* ("strange crab from Canada") in 1892 because he thought it was the headless trunk of a shrimplike arthropod. (Derek Briggs later concluded that, because the fossil part resembled the *leg* of an arthropod, *Anomalocaris* was a very large, multilegged creature not unlike today's crabs and lobsters.) In the Burgess Shale, a jellyfishlike fossil, named *Peytoia* by Walcott, was found that had a hole in the middle of the disk and toothlike projections around the hole. Neither a hole in the middle nor teeth are present in any known jellyfishes, fossil or recent, but because it resembled nothing else that he knew of, Walcott persisted in identifying it as a jellyfish. (It actually looked like a pineapple slice with teeth around the hole in the middle.) Walcott had also described a holothurian (sea cucumber) he called *Laggania,* with a body "so completely flattened that the tube feet are obscured, the outline of the ventral sole lost, and the concentric bands almost obliterated." While examining a *Laggania* specimen, Conway Morris found a structure similar to Walcott's *Peytoia,* and concluded, not unreasonably, that the jellyfishlike *Peytoia* had come to rest on the holothurian and the two creatures, unrelated except in death, had been preserved together.

Remember the old story about blind men examining an elephant? Each of them arrived at a different conclusion about the nature of the beast from the part they examined: the man who examined the trunk said that an elephant was like a snake; the one who felt the flanks said it was like a wall, and so on. The "elephant" of the Burgess Shale was found in pieces. In a 1985 publication, Whittington and Briggs described additional *Anomalocaris* specimens, erecting two distinct species, *Anomalocaris canadensis* and *A. nathorsti*, and showed that *Anomalocaris* (the original "leg"), *Peytoia,* and *Laggania* were not different animals at all, but parts of *Anomalocaris.* It was a unique predator with a pair of large anterior grasping appendages, the "legs" that Briggs thought were repeated along its length, and a rounded mouth between them (the "jellyfish" of Walcott), and a flattened body, Walcott's "holothurian." *Anomalocaris* had large eyes on stalks, and appeared to swim by undulating the overlapping plates along the sides. It probably swam around until it located its prey, which it picked up with its feeding appendages and brought to its circular mouth. Since the feeding organ (originally believed to be an entire animal) was seven inches long, the entire creature was probably more than three feet in length, by far the largest of the Burgess Shale animals, and was probably the apex predator of the Cambrian seas.

In 1996, Desmond Collins summarized the "evolution" of *Anomalocaris.* Though we have no idea what evolutionary trail might have led to this bizarre Cambrian predator, we can follow its journey through the literature, as it metamorphoses from a shrimp to a jellyfish to the bug-eyed monster that graces the cover of Conway Morris's 1998 *Crucible of Creation.* Beginning with Whiteave's 1892 discovery of the "crustacean," we can track it through Walcott's additions of the "jellyfishes" (*Peytoia*) that were actually the mouth, and spotlight Derek Briggs's 1979 realization that the disparate pieces could actually be combined into a single animal. In 1980, in confirmation of Briggs's proposal, Harry Whittington examined a fossil from the Burgess Shale that had the original *"Anomalocaris"* attached to the head, and the "jellyfish" where the jaws would be. Whittington and Briggs also showed that *Laggania cambria,* which Walcott had described as a holothurian, was actually a kind of anomalocarid too. Since 1975, Collins has led expeditions to Yoho National Park, British Columbia, the site of the Burgess Shale, and has collected numerous fossils that are referrable to *Anomalocaris,* and has been able to reconstruct its appearance quite accurately. Collins has placed *Laggania* in the order Radionta (the name derives from

The 2-inch-long Wiwaxia *was covered in flattened scales that overlapped more like a bird's feathers than an invertebrate's shell. It was not a snail, it was not an arthropod, it was not a worm, and, as with so many of the animals of the Burgess Shale, nothing like it has ever been seen again.*

the radiating teeth of the jaws) and the class Dinocardia ("terrible crab," from its role as the largest Cambrian predator). *Laggania* is different enough from *Anomalocaris* to warrant its own genus, but similar enough to be placed in the same order.

When Walcott examined a fossilized flattened oval creature about an inch long, he decided to classify it with the polychaete worms because he thought it showed evidence of segmentation. Walcott named it *Wiwaxia,* a name he derived from the Indian word *wiwaxy,* meaning "windy." Like many of the creatures of the Burgess Shale, it was not reminiscent of anything else; in fact, it turned out to be not even remotely related to the worms. As reconstructed by Simon Conway Morris, in a monograph that Gould calls "a thing of beauty," *Wiwaxia* was covered in flattened scales that overlapped more like a bird's feathers than an invertebrate's shell. Arising from two longitudinal axes are raised rows of spines, presumably, but far from certainly, for protection. It was not a snail, it was not an arthropod, it was not a worm. Like so many of the animals of the Burgess Shale, it was sui generis, and nothing like it has ever been seen again. As of now, some Burgess Shale forms are seen as ancestral to or at least related to some known forms, but others have no known affinities, and seem to have achieved some very weird shapes indeed and then died out. A review of the fossil biota of the Burgess Shale reads like nothing more than a guidebook to the animal life of another planet. *The Fossils of the Burgess Shale* by Briggs, Erwin, and Collier (1994) includes a photograph of each of the known forms that have been uncovered. These photographs, viewed alongside detailed line drawings of the complete animal, demonstrate clearly why these animals have such a special place in Earth history.

While the ancestors of insects, dinosaurs, snails, starfishes, sharks, elephants, trout, hummingbirds, and human beings were beginning their long journey toward biodiversity, many of the strange creatures found in the Burgess Shale didn't make it. Some of the latter are *Marrella, Wiwaxia,* and *Opabinia,* a three-inch-long, five-eyed predator that roamed the ocean floors, grabbing its victims with a flexible nozzle that had clawed pincers at the end; and of course our old friends *Anomalocaris* and *Hallucinogenia.* According to Gould's revisionist interpretation, the creatures of the Burgess Shale turned the "tree of life" on its head; instead of widening into an almost infinite variety of forms, animals like *Opabinia, Anomalocaris, Hallucinogenia,* and all the other five-eyed, spike-backed, or shrimp-footed creatures did not evolve into current forms but simply vanished from the record, demonstrating that the variety of animals on earth was decreasing, a process he called "decimation."

This so narrowed the possibilities that only a handful of forms survived to populate the world today.

Gould's most controversial conclusion, expressed in *Wonderful Life,* was that the present state of animal life was not inevitable, and that a "replaying of the tape" would not necessarily produce the results that we now experience. If, for example, the Ediacaran fauna had not become extinct and the metazoans had, we would have ended up with a completely different world, one that would not necessarily include *Homo sapiens.* (One is reminded of Ray Bradbury's story of time travel "The Sound of Thunder," where people go back in time and accidentally step on a butterfly, so when they return to the present, the world is changed.) Although Gould's book celebrates Conway Morris, Derek Briggs, and Harry Whittington as the heroes of the Burgess Shale story, Conway Morris came to disagree so strongly with Gould's conclusions that he wrote *Crucible of Creation* largely to refute Gould. In his preface, Conway Morris wrote: "One of my purposes in this book is to discuss the basis upon which Gould builds many of his conclusions, especially concerning the roles of historical contingency and the evolutionary explanations for the apparently remarkable range of morphological types we see in the Cambrian. . . . As will be clear from this book I believe, however, that a different message can be read from the Burgess Shale than that promoted by Gould."

Either something happened around 540 million years ago and spawned a host of new life forms, or nothing special happened, and some of the older life forms died out, while others developed into some of the more familiar creatures we see today. These possibilities (with of course, any number of variations), have divided the invertebrate paleontologists into two camps: those who believe, with Gould and Seilacher, that most of the Ediacaran and Burgess Shale fauna was too weird to be placed anywhere in the phylogeny of known animals and disappeared; and those who believe, with Conway Morris, Derek Briggs, Richard Fortey, and others, that the Ediacaran and Burgess Shale fossils for the most part can be assigned positions in the evolutionary development of early invertebrates, and that their descendants are still with us. How they got here (or didn't) is the much-debated subject of current evolutionary theory, and the lines have been drawn. The differences in interpretation and the conclusions are often completely in opposition, and some of the participants have already "slugged it out," to use one of Fortey's fortuitous phrases. As recounted, Simon Conway Morris, in *Crucible of Creation,* presented his views and specifically criticized those of Gould. That led to the publication of a head-to-head confrontation in *Natural History* magazine between Simon Conway Morris and Stephen Jay Gould, entitled "Showdown on the Burgess Shale"—which was about as vitriolic and nasty as science writing can get.

In the December 1998–January 1999 issue of *Natural History,* the editors introduced the "showdown" with this paragraph:

> Almost a decade ago, Harvard paleontologist and *Natural History* columnist Stephen J. Gould published *Wonderful Life: The Burgess Shale and the Nature of History.* In addition to chronicling ongoing work on the Burgess creatures,

> Gould used these fascinating fossils to exemplify his view of evolution. A few months ago, in *The Crucible of Creation: The Burgess Shale and the Rise of Animals,* invertebrate paleontologist Simon Conway Morris, of Cambridge University, a key player in Burgess research, challenged Gould's interpretations. We invited Conway Morris to summarize his argument, which we publish here, along with Gould's reply.

Conway Morris argues that Gould misrepresented Charles Walcott as inept because, according to Gould, he attempted to "shoehorn a range of previously unknown creatures into a few familiar categories to fit his preconceptions." Morris also maintains that Gould's idea of rerunning the tape would not, as Gould suggests, produce a world that contained no humans or anything like us. "Surely," says Conway Morris, "this whole argument, focusing on the implausibility of humans as an evolutionary end product, misses the point. It is based on a basic confusion concerning the destiny of a given lineage, be it of a human family or a phylum, versus the likelihood that a particular biological property or feature will sooner or later manifest itself as part of the evolutionary process. The point is that while the former, say the evolution of the whales, is from the perspective of the Cambrian explosion no more likely than hundreds of other end points, the evolution of some sort of fast, ocean-going animal that sieves sea water for food is probably very likely and probably almost inevitable."

The curious thing about this feud is Conway Morris's turnaround. In his earlier opinions, quoted by Gould in *Wonderful Life,* he held that Walcott was probably wrong, and that the Burgess fauna really did *not* belong in any known phyla. Gould so admired the brilliant work done by the Cambridge University workers on the Burgess Shale animals that he wrote, "Paleontology has no Nobel prizes—though I would unhesitatingly award the first to Whittington, Briggs, and Conway Morris as a trio." In his "Showdown" reply, Gould writes,

> Conway Morris has chosen, less in this article than in his book, to be imperiously dismissive of my ideas, as if no sensible person could ever advocate such prejudiced nonsense. But he never tells us that *Wonderful Life* treats him, in his radical days as a graduate student, as an intellectual hero. I developed my views on contingency and the expanded range of Burgess diversity directly from Conway Morris's work and explicit claims, and I both acknowledged my debt and praised him unstintingly in my book.

In his book, Conway Morris takes Gould to task over and over again, criticizing mostly his "rerunning the tape" concept (Gould's notion that neither humans nor intelligence would have emerged if there had been but the smallest change in the evolutionary events of Precambrian times), but also many of his other opinions as well. Here is an excerpt from Conway Morris's first chapter, about which Richard Fortey says, in *Trilobite!*, "I have never encountered such spleen in a book by a professional":

> Again and again Gould has been seen to charge into battle, sometimes hardly visible in the struggling mass. Strangely immune to seemingly lethal lunges he finally re-emerges. Eventually the dust and the confusion die down. Gould announces to awestruck onlookers that our present understanding of evolutionary processes is dangerously deficient and the theory is perhaps in its death throes. . . . One source of unease in Gould's writing is what appears to some people as the fine line between argument and rhetoric.

Certainly the differences between Gould and Conway Morris cannot be easily resolved. As with so many things paleontological, differences of opinion and interpretation abound. In a 1998 review of Conway Morris's book, historian Peter Bowler wrote:

> To a historian of science such as myself, the books by Gould and Conway Morris seem themselves like a rerunning of the tape of history, but in this case there is a loop that was first played in the late 19th century and is now repeating itself almost exactly. For all the new discoveries and the modern apparatus of cladistic analysis, the alternative visions of the nature of history are as clear today as they were to the biologists who tried to defend their belief in a purposeful universe against the assault of Darwin's *Origin of Species.* Gould himself once wrote about the "eternal metaphors" of paleontology, and on that point, Conway Morris has merely confirmed his claim that rival versions of nature are still in play.

When it was originally found, the fauna of the Burgess Shale was considered unique. Since then, similar fossils have been found in China and Greenland, but the specimens of British Columbia are the best preserved, most numerous, and most diverse. But it is not only its preservation or diversity that makes this collection of fossils so significant: they represent one of the earliest appearances, a half a billion years ago, of complex animals on Earth. The Burgess Shale corresponds in time to the so-called Cambrian Explosion, the era when simple animals were succeeded by a host of complex ones: the first animals with limbs, segmented bodies, shells, jaws, claws and teeth. As far as we know, all of these were marine forms; animals did not emerge from the ocean for another hundred million years. Yet in less than 10 million years—the wink of a geological eye—the ancestors of all multicelled animals that walk, fly, swim, or crawl began their long journey toward their present forms. Included in this catalog was *Pikaia,* the first chordate, and therefore our earliest known ancestor, but many of the creatures of the Cambrian oceans were invertebrates that lacked bones that could easily fossilize. Some of them, however, had composite shells that fossilized beautifully, and presented us with an encyclopedia in stone in which we can read detailed entries about the multifarious blossoming of the evolution of animal life.

THE CAMBRIAN EXPLOSION

The Devonian Period was named for Devon, in England, where layers rich in fish fossils had been found. Until the 1830s, when Roderick Murchison, a geologist, began looking for something that preceded it, the Devonian was the oldest known fossiliferous rock formation. In Wales, Murchison found a layer that was beneath the Devonian rocks, and he named it Silurian, after the Silures, an ancient Wesh tribe that had successfully fended off the Romans. Adam Sedgwick, Murchison's rival in geology, then discovered an even older layer, which he named Cambrian, from Cambria, the ancient name of Wales. The opposing geologists were unable to agree on the boundaries of their two layers, but when Murchison was appointed director of the National Geological Survey, he simply ordered that "Cambrian" be deleted from all official books and maps. Regardless of Murchison's high-handed resolution of the controversy, "Cambrian" is the term most commonly used nowadays to refer to the period 540 million to 510 million years ago. It often appears in discussions of the Cambrian Explosion, an event in evolutionary history that neither Murchison nor Sedgwick could have imagined in their wildest geological fantasies.

About 200 million years ago, violent earthquakes and volcanoes accompanied the breakup of the protocontinent Gondwanaland, and massive mountain ranges were thrust up from below the sea. Some events connected with these seismic upheavals prompted an unprecedented surge of growth in animal life on Earth, which is known as the Cambrian Explosion, or life's "Big Bang." For the nearly 4 billion years that preceded this mysterious and momentous event, life had consisted of bacteria, multicelled algae, and single-celled plankton. Yet in the geological instant of 10 million years, the ancestors of all animals appeared. There were no inhabitants of dry land; all Cambrian life forms lived in the sea, most of them on the muddy sea floor. This period saw the abrupt appearance of animals with limbs, segmented bodies, shells, jaws, claws, tentacles, and teeth. The first of these new life forms appeared in the (slightly Precambrian) Ediacaran fauna of Australia; in the Burgess Shale of Canada, the Cambrian animals took the form of worms, shellfish, and arthropods, none of which had ever been seen before, and many of which would never be seen again. In *Life: A Natural History of the First Four Billion Years of Life on Earth* (1998), paleontologist Richard Fortey discusses the evolutionary "explosion":

> It is probably the most significant threshold to be crossed since the first organic molecules joined together to polymerize the first molecular chains. This is where leisureliness disappeared from our story. The animals that evolved in the Cambrian would have crawled over one another to be first to mate, evading the attention of predators on the way. Competition was introduced into ecology: those animals led exciting lives, vying with one another, sculpted for fitness. By contrast, the mat-formers which popularized the endless stretches of Precambrian time, and even the Ediacara fauna, may have led lives almost devoid of incident. When this changed, it changed forever.

Until the period known as the Tommotian, which immediately followed the disappearance of the Ediacaran (Vendian) fauna, no living things had skeletons. Richard Cowen (2000) wrote that "one of the most important events in the history of life was the evolution of mineralized hard parts in animals. . . . Beginning rather suddenly, the fossil record contains skeletons: shells and other pieces of mineral that were formed biochemically by animals." The evolution of hard body parts marks the beginning of the Paleozoic Era, and its oldest subdivision, the Cambrian. The Cambrian Explosion was heralded by the worldwide presence of what have been dubbed—for want of a better name—the "small shelly faunas" (SSFs). Some, like the archaeocyathids ("ancient cups"), were strange little cup-shaped things that seemed to grow from the ocean floor, and may have been related to sponges. They were often formed of one cup nested in another, the two held apart by narrow, radial walls. (None of the soft parts of any of these animals have been preserved.) Other "small shellies" were rather more mollusklike, with a caplike shell, such as *Latouchella,* and the enigmatic *Yochelcionella,* which consisted of a little cap and with an "extraordinary 'snorkel' beneath the curled tip [that] was probably the 'exhaust pipe' for water passing over the gills, which lay under the mantle and shell" (Norman 1994). *Tommotia,* the index fossil of the period, is preserved as an angular, ribbed shell that might have been occupied by some sort of a primitive gastropod or cephalopod.

Large numbers of angular plates have also been found from this time, which are now believed to have been the body armor of halkieriids. These creatures, which Conway Morris (1998) describes as "armoured slugs," could glide over the sediment on a soft undersurface, but the slug's upper body was covered in as many as 2,000 scaly armor plates, called sclerites. At Sirius Passet in northern Greenland in 1990, Conway Morris and J. S. Peel found a fossil of *Halkiera evangelista* that had a brachiopod-like shell at either end, which might at first look like a coincidence, but Conway Morris suggests that when the animals were very small, the shells were almost juxtaposed, and if it developed the ability to swing one shell beneath the other, perhaps as a defensive action when threatened, "Could this represent the first step towards becoming a brachiopod?" *Wiwaxia* (of the Burgess Shale) also had sclerites, and Conway Morris believes that the halkieriids could have given rise to *Wiwaxia,* which then might have been an ancestor of the polychaete worms. (Remember, Conway Morris believes the Burgess Shale animals can mostly be placed in existing phylogenies, whereas Gould believes they cannot.)

The Cambrian Explosion has attracted the interest of scientists since Darwin (1859) wrote that "the sudden manner in which several groups of species first appear in our European formations;—the almost entire absence, as at present known, of formations rich in fossils beneath the Cambrian strata—are undoubtedly of the most serious nature." In his 1987 article "The Emergence of Animals" (published as a book in 1991, with coauthor Dianna McMenamin), Mark McMenamin wrote, "Perhaps the most astonishing aspect of the Cambrian faunas is that so many radically different animals appeared in such a short interval. What gave rise to this sudden diversification?" He answers, "A key element in the establishment of animal communities is predation which establishes a hierarchic chain of food transfer," but beyond that, he is able to offer little in the way of explanation. In *Trilobite!* (2000), Fortey says that Mark and Dianna McMenamin, in *The Emergence of Animals,* "claimed up to a hundred animal phyla 'exploding' into life in the Cambrian; most of them are also claimed to have died out, leaving no progeny. They popped up like so many jack-in-the-boxes and then auto-destructed, in the manner of some post-Dada extravagance designed to outrage. . . . The extraordinary thing to an objective reader is that there is no attempt to justify why these hundred or so 'Cambrian phyla' should be recognized as one of the great divisions of the animal kingdom—how, exactly do they differ from each other so much that they 'deserve' to be called a phylum?"

Now it appears that evidence from the Moon may help to explain the Cambrian Explosion. Geologists and geophysicists from California have found that there was an increase in the bombardment by asteroids and meteorites of the Earth and its satellite about 570 million years ago, and this may have contributed to the rapid increase of life on Earth. Continental drift, erosion, and the large percentage of the surface of the Earth that is covered in water means that the record of extraterrestrial hits is sparse, but none of these factors apply to the Moon. Culler, Becker, Muller, and Renne (2000) examined soil samples brought from the Moon by *Apollo 14*. Using radiogenic isotope analysis to examine tiny glass droplets that form when the lunar soil is impacted by an asteroid, they were able to date them within a range of 30 million years. They found that the number of impacts decreased over the past 3.5 billion years to a low point between 400 million and 600 million years ago. A subsequent fourfold increase in impacts suggested to the researchers that some life forms on Earth may have been eradicated, making room for the development of new forms. (The asteroid that collided with the Earth 65 million years ago off Yucatán at Chicxulub is believed by many to have led to conditions that wiped out the nonavian dinosaurs.) Of course, most of the smaller asteroids are burned up by the Earth's atmosphere, so it is difficult to compare the conditions on the airless Moon and on the Earth.

In their 1997 article on the evolution of animal body plans, Erwin, Valentine, and Jablonski wrote:

> For its first 3 billion years on Earth, life was no larger or more sophisticated than a single cell. All of that changed when almost 600 million years ago new, multicellular life forms appear in the fossil record. Starting with simple soft-bodied creatures this evolutionary innovation culminates in the "Cambrian explosion," a burst of biological creativity unprecedented in Earth's history. The Cambrian epoch saw the creation of hundreds of life forms stranger than anything humanly imaginable. Many of these animals are now extinct, but those that remained established all of the basic body plans of all animals that evolved since.

Only in the past few years have enough data on the molecular biology of development and Neoproterozoic-Cambrian environmental change accumulated for scientists to begin critical evaluation of this remarkable interval of evolutionary change. At Harvard, Knoll and his colleagues (1996) have contributed to the development of chemostratigraphic methods that provide a means of correlating the Precambrian-Cambrian (PC-C) boundary successions independent of the animals they wish to evaluate. They have also participated in stratigraphic and radiometric studies that have sharply constrained the timing of these events; produced geochemical data that support the hypothesis that atmospheric oxygen levels increased just before macroscopic animals evolved; and demonstrated that phytoplankton show evolutionary dynamics much like those of PC-C animals, indicating that ecology played an important role in a Cambrian explosion that extends across kingdoms.

What do we know about the beginning of animal life? In 1998, Adolf Seilacher, Pradip Bose, and Friedrich Pflüger published their findings, in *Science,* of evidence of animal fossils in central Indian sandstones that were 1.1 billion years old. The stratigraphic sequences of limestones, sandstones, and shales of the Vindhyan Formation, where the fossils were found, was calculated by the potassium-argon method and its age calculated accordingly. They weren't particularly impressive fossils, just worm burrows (otherwise known as "trace fossils," because they are not fossils of animals, but fossils of what the animals *did*), preserved in stone. But worm burrows needed worms to make them, and since the earliest known animals were supposed to have appeared some 540 million years ago, this discovery added another 460 million years to the history of animal life on Earth—a prodigious leap backward.

Moreover, the ability to make burrows at all requires muscles of some sort, and Seilacher and his colleagues suggested that these billion-year-old animals moved by rhythmic muscle contractions, or peristalsis. Seilacher et al. believed that these burrows had been preserved because the animals that made them had tunneled under microbial mats, and the mats, covered with thin layers of sand, preserved the traces. (This idea is not unlike Seilacher's interpretation of the Ediacarans, which, he said, left clear impressions because a thick bacterial mat covered the ocean floor, and when the Vendozoans were covered by sand, they were pushed into the bacterial mat, and their outlines were preserved.) These "new" fossils were "too irregular to be . . . cracks, too sharply delineated to be wrinkles, and too large to be attributed to protists or fungal rhizoids." Some of the burrows were almost a quarter of an inch in diameter, which qualifies them as "macrofossils."

No sooner had the Seilacher et al. *Science* paper appeared than the dissenters chimed in. In a later article, a *Science* editor, Richard Kerr, quoted Andrew Knoll as saying, "If you see centimeter-scale coelomate organisms and then don't see another for 400 million years, you have a lot to explain." (A coelomate organism is one with a coelum, a lined cavity between the gut and the body wall.) And then the Indian paleontologist Rafat Azmi (1998) said, in a letter published in *Science*, that Seilacher et al. were wrong, because he had just published a report in which he identified "small shelly fossils" from a layer of the same Lower Vindhyan sandstones that were 545 million years old, but that were *below* the trace fossils of Seilacher et al. Then Simon Conway Morris, Sören Jensen, and Nicholas Butterfield responded to Azmi's letter by saying that his "announcement of small shelly fossils" is probably incorrect, since "there is little reason to accept [his] identification of the Vindhyan material as organic, let alone being any sort of small shelly fossil."

Azmi's "small shelly fossils" were examined by a select team of Indian scientists, who reported "that the identification of fossils by R. J. Azmi is far from convincing, and that more detailed work [would be] necessary before the authenticity of the find is accepted." Despite these findings, published in *The Journal of the Geological Society of India*, Azmi plans to persevere with his claim, and says that "there is no question of any contamination or misrepresentation" (Bagla 2000).

Unlike Conway Morris and his colleagues, Marin Brasier (1998) believes that Azmi's "small shelly fossils" are "well-preserved and typically early Cambrian skeletal fauna," and therefore, that the Seilacher fossils "may be little more than 540 million years old, close to the beginning of the Cambrian explosion." After the Ediacaran "animals" disappeared, except for some sponges, there were no animals longer than a few millimeters. Then came the trilobites.

WHAT HAPPENED TO THE TRILOBITES?

Trilobites are—or were—joint-legged, hard-shelled creatures with multifaceted eyes, that endured for 300 million years and died out before the dinosaurs even made their debut. "Who are we johnny-come-latelies," asks Richard Fortey (2000a), "to label them as either 'primitive' or 'unsuccessful'? Men have so far survived half a per cent as long." There is no typical trilobite. Most of them crawled along the bottom of Silurian and Ordovician seas, but some were swimmers. There were thousands of kinds, ranging in size from ladybug to serving platter. Here they come, in what Fortey calls "as odd a parade as any carnival could offer: . . . Some smooth as eggs; others spiky as mines; giants and dwarfs; goggling popeyed popinjays; blind grovellers; many flat as pancakes, yet others puffy as profiteroles." How these fascinating creatures began is one of the great mysteries of evolution. The first known trilobites, from the Early Cambrian, 540 million years ago, were already fully formed and had a sophisticated visual system—the first known eyes.

During the Cambrian Period, the continents and seas were arranged very differently than they are today. There were four major continents, Gondwanaland, Angara, and two sections of Euramerica, which were inundated by a rising sea level, causing thick sedimentary deposits to accumulate. This sedimentary rock (conglomerate sandstone, shale, and limestone) was formed in shallow seas that covered large areas of present-day Europe, Asia, and North America. The Cambrian rock layers are the first to contain easily recognizable fossils, virtually all of which were marine invertebrates. The dominant forms were trilobites, but snails, brachiopods, and sponges are also present. More than half of the known fossils from the Cambrian Period are trilobites, and they were the dominant life form on Earth for much of the 325 million years that we call the Paleozoic Era.

Some 570 million years ago, the newly developed process of photosynthesis made oxygen available to creatures that lived in the sea, and in the relatively short geologic space of a few million years, all sorts of invertebrate animals arose to share in what was to become the Cambrian Explosion. Trilobites appear fully formed in the fossil record, with their incredibly complex and efficient compound eyes; we have few clues as to what preceded them. Trilobites appeared roughly at the same time as the Ediacaran fauna, first discovered in Canada and later in Chengjiang in southwestern China. Most of the Ediacarans vanished, but the trilobites dominated the seas for another 300 million years, representing one of the great success stories in evolutionary history. The fact that they are now extinct and have been for 200 million years is by no means an indication of failure. Extinction is an integral part of the evolutionary process, and sooner or later, every species reaches a point where it can no longer function efficiently and dies out, because of replacement by a better-adapted form, overspecialization, predation, climatic change, a catastrophic event, or a process of species senescence that we do not understand.

Richard Fortey is a senior paleontologist at London's Natural History Museum. His book *Trilobite!* is a loving exposition of these enigmatic animals, and brilliantly covers almost every aspect of their history, paleontology, and biology. He introduces us to Thomas Hardy's description of a trilobite in his first chapter and discusses the work of his predecessors and contemporaries and his own not insignificant contributions, but the book is mostly about the wonderful world of the trilobites and their importance in tracing the history of life on Earth. In his introductory chapter, he invites us to

> come and see the world as it once was through the crystal eyes of the trilobite. We shall find out how trilobites tell us the pattern of evolution, and how it can be read from the rocks. We shall discover how faith in trilobites not only moves mountains but shifts whole continents. We shall see how cast-off shells can be reanimated into living animals. We shall understand something of the richness of the animal kingdom. Through trilobites, we shall take possession of the geological past.

The irregular appearance of fossil trilobites in eastern North America, western Canada, northern Europe, and Australia can only be explained by the movement of the continents 500 million years ago, when they were shoved around by the movement of the Earth's tectonic plates. Before that, no landmass was situated where we now find it. Trilobites have provided some of the best-known and best-preserved of all marine fossils. More than 15,000 species have been described, ranging in size from a quarter of an inch to almost three feet.* The fossils are the exoskeletons, so most of what we have to go on is the hard parts, but some died and were buried during the molting process, so there are also some fossils of the "soft-shell" phase. They

* In 2000, Canadian paleontologists discovered the world's largest recorded complete trilobite fossil on the shore of Hudson Bay in Churchill, Manitoba. The 445-million-year-old fossil is over 28 inches (72 cm) in length, nearly a foot longer than the previous record holder. The giant trilobite was found and recovered during a long-term field project led by Graham Young, associate curator of geology and paleontology at the Manitoba Museum, and Bob Elias of the University of Manitoba.

are now tentatively classified as arthropods (the name means "jointed legs," and the phylum includes crabs and insects), but they probably differ enough to require a phylum of their own.

The ancestry of the trilobites is poorly known; obviously they did not arise fully developed from the Precambrian ooze, but were descended from a primitive ancestral arthropod that lived before the Cambrian. The first fossilized trilobite was described from Welsh limestones in 1698 by the Reverend Edward Lhwyd, custodian of the Ashmolean Museum at Oxford, who called it "the skeleton of some Flat-Fish." (This specimen has now been identified as the Ordovician trilobite *Ogygiocarella debuchii.*) The word "trilobites" used to denote a distinct group of animals first appeared around 1780, in the work of M. T. Brunnich. The American paleontologist Charles Doolittle Walcott (1850–1927), discoverer and namer of many of the fossils of the Burgess Shale, was a dedicated trilobitologist (if there isn't such a word, there ought to be) and was the first to discover the structure and function of trilobite legs. Harry Whittington, who followed Walcott into the Burgess Shale a hundred years later, described the limbs in great detail and considered that the outer filamentous branches on the legs were probably gill branches. We have already met Walcott and Whittington in the story of the Burgess Shale, and as Fortey (a student of Whittington's) modestly says, "Now it is my turn to enter the story." Nobody could understand how a common Ordovician trilobite found in New York State, called *Triarthus,* could have survived on the sea floor if the rocks in which the fossils are found are "so sulphurous that they stank of rotten eggs when you broke them open with your geological hammer." In Sweden they were even called "stinkstones." Only when the chemosynthetic animals of the hydrothermal vents were discovered living where sulfur abounds did Fortey realize that *Triarthus* and the other olenid trilobites, which lived from the Late Cambrian to the Early Ordovician (505 million to 445 million years ago) in what is now Scandinavia, had very likely developed a method of extracting energy from the bacteria's biochemical reactions, and thus might have been "the first animals to live symbiotically with sulfur bacteria."

In the hydrothermal vents are worms, shrimps, and clams that have adapted to the superheated, sulfurous conditions by becoming functional symbionts with sulfur bacteria, which provide a source of energy in a low-oxygen environment where sulfide accumulates. In a 2000 paper, Richard Fortey presented the idea that the olenid trilobites, which "were tolerant of oxygen-poor, sulfur-rich sea floor conditions . . . were chemoautotrophic symbionts." The olenids were characterized by a wide, multisegmented thorax, a thin cuticle, and the occasional development of brood pouches. In other species, the mouth parts (hypostome) functioned in the manipulation of food, but they were atrophied in some members of this group. The earliest olenids, like *Olenus,* have normal hypostomes, but in later species, like *Peltura* and *Porterfieldia*, it is degenerated and atrophied. To Fortey this strongly suggests "indirect ingestion of bacteria," and he finds this atrophication "hard to explain any other way." Fortey believes that the "spectacular" brood pouches of some olenids might have been used to maintain "a steady level of oxygenation for larval trilobites until they too could begin a symbiotic existence." If his analysis is correct, it would represent the earliest occurrence of chemosymbiosis in the fossil record. Inspired by Lynn Margulis, McMenamin (1998) discusses the possibility that some of the 600-million-year-old Ediacaran animals were photosynthetic symbionts, but there is nothing in the Ediacaran fossils that actually suggests symbiosis.

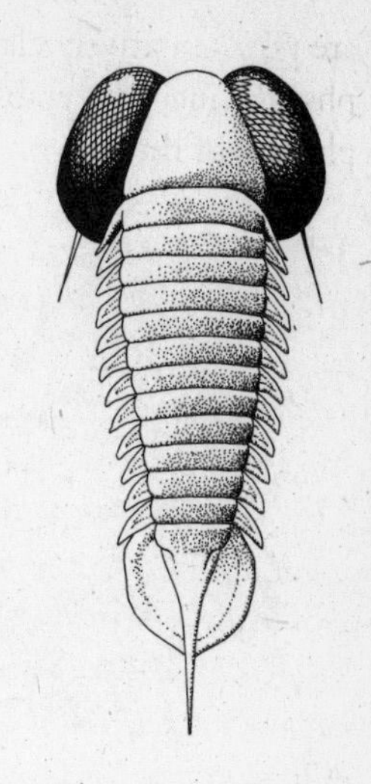

A very few trilobites were free-swimming, a determination made by paleontologists who realized that the eyes of species like Opipeuter *descended below the plane of the shell, and would have dragged in the mud if the animal had crawled on the bottom.*

All trilobites are characterized by a flattened, oval body with a pair of longitudinal furrows that divided the shell into three sections—hence the name trilobite, which means "three-lobed." The animals were also divided into three sections latitudinally—the head (cephalon), body (thorax), and tail (pygidium)—but it is the longitudinal divisions that are responsible for the popular name. The upper shell, which was usually segmented, was thicker than the lower, and is therefore better represented in the fossil record. But among the abundant trilobite fossils are some that show the underside of the complete animal and therefore the legs, and others where death and subsequent fossilization occurred before the new shell had had the opportunity to harden. From these fossils we have concluded that these marine creatures remained close to or on the bottom, and some may have buried themselves in the sand with only their eyes protruding. A very few species were free-swimming, a determination made by paleontologists who realized that the eyes of species like *Opipeuter* descended below the plane of the shell, and would have dragged in the mud if the animal crawled on the bottom. (Richard Fortey named *Opipeuter inconnivus*; the term means "one who gazes without sleeping.") Food particles were probably stirred up by the legs and passed forward to the mouth. Since the mouth had no large mandibles and no teeth, we can infer that trilobites were not usually predatory and were restricted to soft or small food items. After the Cambrian, many trilobites developed the ability to roll up, rather like a sowbug, to protect their softer underside against predators, and in the great lexicon of fossil trilobites, there are some that have been preserved in the "enrolled" position.

Despite a vast variety of designs, all of them conformed to the basic body plan. The tribolite thorax was segmented into as many as 40 elements, many of which ended in backward-directed spines. The posterior portion of the thorax of *Ogygopsis* (for which the section of the Burgess Shale called Ogygopsis Shale is named) contained many elements fused to form a single plate. Most trilobite fossils were formed from molted exoskeletons, and so the legs are usually missing, but enough legs have been found to allow paleontologists to deduce that trilobites had paired legs attached to each segment of the cephalon, thorax, and pygidium. Each leg was divided into a lower segment for walking and an upper segment equipped with feathery gills for breathing. The legs of some species may have had flattened lobes—like those of some

One of the earliest known trilobites is Olenellus, *first found in New York State and later in Scotland. Even the earliest trilobites had well-developed multifaceted eyes, the first complex eyes in history.*

crabs—that were used for swimming, but it is believed that most trilobites walked along the bottom, or burrowed into the sediment, much the way horseshoe crabs do today. The periodic molting of trilobites and the fossilization of the molted outer shell creates a situation where, "theoretically, at least, it is possible to find the same trilobite animal—in various stages of growth—twenty times or more in the fossil record" (Eldredge 1976). The fossil record indicates that the trilobites dominated the Cambrian fauna for another 320 million years—they were the dominant life form on Earth. Among the host of weird and unusual animals found in the Burgess Shale was the trilobite *Olenoides,* which, compared to such enigmas as *Marrella, Anomalocaris,* and *Hallucinogenia,* looks like a familiar friend.

One of the earliest known trilobites is *Olenellus,* first found in New York State and later in Scotland. Fifteen million years later than *Olenellus* we have the relative giant *Paradoxides,* with spines from the cephalic region "extending back like a pair of swords," and the tanklike *Illaenus,* "an armadillo among trilobites," which could roll up into a nearly perfect sphere. The palm-sized *Calymene* is one of the better-known species; the somewhat smaller *Odontopleura* is covered with spines, and smaller still is the big-eyed Devonian trilobite *Phacops,* about an inch and a quarter long, usually found in dense aggregations and looking like nothing so much as a collection of flattened, fossilized screws. Fortey describes more participants in the parade passing through history:

> Here is *Cheirurus* with a tail like a cat's claw carrying a few large curved spines; then *Dalmanites* with its single spine extending back for the pygidium, like a marlinspike. Now comes *Scutellum,* flat as a flounder, but with a huge pygidium shaped like a ribbed fan—one flip and it is past us. A gigantic *Lichas* is nearly as flat, but with the glabella blown up, like so many bubbles, with a fluted pygidium with a coarse saw-edge. Nearby there are some tiny trilobites buried in the soft mud, probably diminutive blind grubbers called *Shumardia.* There follow many more spiny monstrosities—*Comura,* with so many vertical spines that it looks like fakir's nightmare.

Trilobites had eyes, and these were the most ancient visual system on the planet. Ricardo Levi-Setti is a professor of physics and the director of the Enrico Fermi Institute at the University of Chicago. Trained as a physicist, he has a passionate interest in trilobites and has studied them intensively. He is the author of *Trilobites* (University of Chicago Press, 1993), one of the definitive works on these fascinating creatures. Levi-Setti wrote an article called "Ancient and Wonderful Eyes," published in the first and only issue of a magazine called *Fossils*. Its opening paragraph:

> More than 500 million years ago, trilobites began to develop some of the most extraordinary eyes that have ever existed on this Earth. They were among the first eyes ever created, and they are the very earliest eyes that have been preserved in the fossil record. And yet, despite their ancient origin, these eyes were remarkably efficient and elegant in their design—capable of seeing far more and far better than you would perhaps think possible.

The Moroccan trilobite *Fallotaspis*, dating from 540 million years ago in the Early Cambrian, had large, well-developed eyes, so there must have been eyed creatures long before that. We may never know what creature first developed some sort of primitive visual apparatus, but it could have been as much as 200 million years before *Fallotaspis*. Trilobite eyes were multifaceted, like those of crustaceans and insects, and were covered by a single transparent corneal membrane. Along with every other element of the exoskeleton, this membrane was shed when the animal molted. In the earliest species the number of lenses was not large, but later types had as many as 12,000 lenses, all of them densely packed and arranged in specific patterns. Some genera had huge compound eyes, while others had mere dots or granules. The eyes of trilobites differed in size, shape, and number of lenses, but they were all similar in one respect: they were made of calcium carbonate ($CaCO_3$) or calcite, a common, colorless mineral that is transparent in its purest form. (All the limestones, marbles, and chalks are crystallized calcites, colored by impurities.)

Some trilobites could see in all directions at once; in front and behind, above and below. They could see close and distant objects without refocusing. Niles Eldredge's Ph.D. thesis was on the big-eyed trilobite genus *Phacops*. In his 1976 discussion of trilobites, which appeared in the same issue of the short-lived *Fossils* magazine, Eldredge wrote, "These lenses—technically termed aspherical, aplanatic lenses—optimize both light collection and image formation better than any [other] lens ever conceived. We can be justifiably amazed that these trilobites, very early in the history of life on Earth, hit upon the best possible lens design that optical physics has ever been able to formulate!"*

The great majority of trilobites had holochroal eyes, with numerous lenses closely packed; each lens is in direct contact with its neighbors. Each of the lenses is a

---

* Researchers at Cornell University have recently discovered that *Xeno peckii*, a tiny insect parasitic on paper wasps, has eyes that they describe as "a modern counterpart to the structural plan proposed for the eyes of some trilobites." Buschbeck et al. (1999) described an eye that appears "as a cluster of large convex lenses, giving it a raspberry-like appearance." They deduced that each unit has its own retina, and that each of these units is image-forming, unlike the compound insect eye, which has many more, but smaller, facets and decomposes the image into tiny points of light.

single calcite crystal, and the lenses are arranged on the curved surface of kidney-shaped eyes, some of which were sessile (stalked). In some pelagic forms, holochroal eyes were enormous, suggesting that their owners were "diurnal migrants, inhabiting deep waters of low light intensity during the day and coming up to the surface at night" (Horváth, Clarkson and Pix, 1997). "Phacopid trilobites were generally bottom-dwellers," wrote Horváth et al., "and as far as is known, none were pelagic." These phacopid eyes, named for their owners, are usually larger and have fewer lenses than the holochroal. The lenses consist of two optically homogeneous units of different refractive indices characterized as a "doublet." The optical properties of these amazing organs, also called schizocroal eyes, had been guessed at, but Levi-Setti and Euan Clarkson were the first to recognize that certain minute impurities in the calcite lenses of the phacopid (schizocroal) eyes bent the light just enough to compensate for the spherical aberration inherent in the basic design of the lens, and transmitted the light in a true path. Using a fossilized trilobite eye as a substitute camera lens, Ken Towe of the Smithsonian Institution was able to show through photographs exactly how accurate the eyes of trilobites actually were.

In the Cambrian seas, was there any animal capable of attacking or eating the tough-shelled trilobites? Indeed there was, and we have already met this formidable creature: it was *Anomalocaris*, the yardlong swimmer with huge grasping appendages and a pineapple-ring mouth, which so many people had identified *not* as the sum of its parts. If the latest interpretation is correct, then the parts add up to the dominant predator of those ancient seas, which would have to have preyed upon trilobites, because they existed in such profusion—and besides, there is fossil evidence of a trilobite with a bite taken out of it by *Anomalocaris*. It is also possible that the largely upward-looking eyes of the trilobites were useful in detecting *Anomalocaris* and other predators that hovered above them. One of the suggestions that has been forwarded for the demise of the trilobites is that they originally developed in the absence of predators, and were later unable to modify themselves sufficiently to avoid those that arrived on the scene. Even their armor, spines, and enrollment might have been inadequate to protect the trilobites from *Anomalocaris*, the scourge of the seas.

At the end of the Ordovician, 510 million years ago, a major mass extinction occurred, which took out large proportions of the living species of graptolites (tubular, colonial animals with feathery gills that could leave their tubes), conodonts, nautiloids, brachiopods, and trilobites, which had "a bewildering array of morphologies and presumably life styles" (Hallam and Wignall 1997). Most, if not all, of the swimming trilobites died off. "The Devonian," wrote Hallam and Wignall, "was not a pleasant time for trilobites, and the group was progressively whittled down by both ordinary ("background") and mass extinctions." Only those species related to the small genus *Proetus* survived. Yet once again they bounced back, speciating anew and repopulating the Carboniferous seas where their relatives had been eliminated. Still, the trilobite clock was winding down.

Fortey writes that his "hope has faded that, when today's mid-ocean ridges were explored by bathyscape, in some dimly-known abyss there might still dwell a solitary trilobite to bring Paleozoic virtues into the age of the soundbite. . . . Three hundred million years was course enough." The exit of the trilobites from the stage of life puts one in mind of Franz Joseph Haydn's Symphony no. 45, written in 1772 and known as the "Farewell" Symphony. Instead of the usual fast, cheerful final movement, Haydn wrote a slow movement, and directed the players to leave the stage one by one, blowing out the candles of their music stands as they left the stage. As the oboes played their final notes, for example, they got up, put their instruments in their cases, and left. Then the cellos, the trumpets, the basses, and the clarinets walked off the platform. The full orchestra became a small orchestra, then a chamber group, then a quartet. The symphony ends with just the first and second violins in a plaintive diminuendo, as the symphony winds down to its conclusion. For *Philippsia*, the last—and therefore the "youngest"—of the trilobites, which has been found in rocks dated from the Permian, 260 million years ago, there was no second violin, and when the sea-floor stage went dark, the trilobites were gone forever.

"LIVING FOSSILS"

Evolution, as we understand it, modifies various animal forms over time until they metamorphose into other forms or vanish altogether, becoming extinct. To deny this is to deny the very essence of evolution. So the recognizable ancestors of most of the living animals today did not look very much like their living descendants, except in those particulars that enable us to identify the early versions. But there are some creatures, known by the oxymoronic term "living fossils" (fossils, by definition, cannot be alive), that have somehow managed to avoid the evolutionary process altogether, and have come through the eons in almost the same form they acquired 100 million or 200 million years ago.

The concept of a "living fossil" was first introduced in 1859 by Darwin in *The Origin of Species*. He wrote, "As we here and there see a thin straggling branch springing from a fork low down on the tree, and which by some chance has been favoured and is still alive at its summit, so we occasionally see an animal like *Ornithorhynchus* [the platypus] or *Lepidosiren* [the lungfish], which in some small degree connects by its affinities two large branches of life, and which has apparently been saved from fatal competition by having inhabited a protected station."

Somewhere in the Triassic Period, about 200 million years ago, arose a type of creature with a two-part body consisting of the prosoma, the largest segment, which includes the head, and the opisthosoma, the middle segment, which consists of the heavy shell. The head and thorax are fused together to make the cephalothorax, and there is a spiky tail, also known as the telson. These animals have compound lateral eyes, each consisting of about 1,200 separate photoreceptive units called ommatidia, covered by a single thick cornea. There are also two simple eyes, or ocelli, in the center of the head. The five pairs of legs are used for locomotion, for capturing prey, and for feeding. These legs all have claws, except for the last pair. The mouth has no appendages of its own, and opens at the base of the legs to form a sort of chewing mill. Five sets of overlapping plates, called book gills, are used for respiration. Females are somewhat larger than males, and can reach a length of 2 feet, including the tail. This is a horseshoe crab, though it is not a crab at all, and bears only a faint resemblance to the horseshoe. In the past, it was also known as the king crab. Its scientific name is *Limulus polyphemus*; *limulus* is derived from the Greek and means "oblique," and probably refers to the tail; and *Polyphemus,* the name of the one-eyed Cyclops slain by Hercules, is probably a reference to the eyes set high on the shell. Horseshoe crabs are officially arthropods (joint-legged animals), but they differ enough from the crustaceans, spiders, insects, centipedes, and millipedes that they have been given their own class, the Merostomata, which also includes the living scorpions and the extinct sea scorpions (Eurypteridae). They are further classified in the order Xiphosura, from *xiphos*, Greek for "sword," and *uros*, "tail."

In the spring, as the coastal waters of the Atlantic and Gulf coasts of the United States warm up, horseshoe crabs swim into shallow estuaries and wait for the highest tides, caused by the full moon, to come ashore to breed. The male rides piggy-back upon the female, and as she deposits as many as 20,000 small, greenish eggs in a shallow pit dug in the sand, he releases his sperm over the eggs. The billions of eggs laid each spring on protected shady shores like Delaware Bay provide vital nourishment for hundreds of thousands of migrating shorebirds. The tide has usually receded by the time mating has been completed, so the stranded horseshoe crabs—still piggy-back—dig into the sand and wait for the flood tide so they can return to the sea. Two weeks later, as the spring tides again wash over the eggs, they hatch, and the little crabs head for the moonlit sea, where they will feed on organic debris. Horseshoe crabs can swim, but they do it upside down and awkwardly, and spend most of their time grubbing through bottom sand for the worms and mollusks that are their main food.

Because there are 350-million-year-old fossils of creatures that closely resemble today's horseshoe crabs, *Limulus* has long been considered the paradigm of the "living fossil": an animal that seems to have avoided evolutionary modification, and looks the way it did millions of years ago. As Daniel Fisher (1984) wrote,

> The Xiphosurida, or horseshoe crabs, are often cited as a classic example of arrested evolution. They have been so consistently associated with this concept, in both professional and popular literature, that their reputation for extreme conservatism in form and behavior is probably more widely known than any other single aspect of their biology or history.

"Arrested evolution," technically known as bradytely (*brady* means "slow"), is a term used to refer to an unusually slow evolution within a single lineage. But *Limulus polyphemus* has not come down unchanged for 300 million years, nor is it the only horseshoe crab ever to have been found in the fossil record. *Mesolimulus* was a 3-inch-long type found in the Solnhofen shales of Germany; *Paleolimnus* was found in the Permian of Kansas; and *Aglapsis* had a head shield and segmented shell, not unlike a trilobite. Moreover, there are other, similar species of horseshoe crabs that are not *Limulus* and are alive today, such as *Tachypleus tridentatus* and *Carcinoscorpius rotundicauda*, found in southern and eastern Asian waters from Japan to Indonesia and India. Niles Eldredge writes (1984) that in only a few cases is the same name used for an ancient species and a current one, and this standard might be useful in the definition of a living fossil. On the other hand, since just about everything that ever lived is extinct—there is nothing alive today that looks like an ammonite, a trilobite, or *Anomalocaris*—we can probably forgive use of the term when it is used to designate an animal whose lineage has remained unbroken for 200 million years.

The question of why some species perdure while most others die is difficult to answer, but in *Fossils: The Evolution and Extinction of Species* (1991), Eldredge writes that he believes he has an answer, at least for *Limulus*:

> Horseshoe crab behavior and ecology can help us understand the reasons for their extreme evolutionary conservatism. They are often the very last species to be driven from an estuary as human pollution transforms the environment far beyond the tolerance limits of most species. And that is precisely the point. Horseshoe crabs are ecological *generalists*: they can withstand a wide range of conditions, natural as well as human-induced. Marine creatures, they can tolerate wide fluctuations of salinity, both in the briny conditions of heightened salt content, as well as the brackish conditions typical of estuaries, where fresh riverine waters mix in with the salty sea. . . . And that hardiness, that jack-of-all-trades nature that stamps horseshoe crabs so clearly, holds the clue to why they have remained so stable, so evolutionarily non-changeable for hundreds of years

Horseshoe crabs can survive just about anything—except the ultimate predator, *Homo sapiens*. A news article in the *New York Times* for August 9, 2000, indicated that crab fishermen in Virginia, who had been collecting 355,000 horseshoe crabs per year, were threatening the very existence of the species. Horseshoe crabs are collected for bait and also for their blood, which contains special cells that swarm to the site of a wound and kill certain kinds of bacteria. The pale blue blood contains a substance known as LAL, for limulus amoebocyte lysate, that is now used as a fast, effective way of testing drugs to make sure they are free of harmful bacteria before they are given to people. "Ten years ago," wrote the *Times* reporter Francis X. Clines, "an estimated 1.24 million crabs spawned on the beaches of Delaware Bay, laying trillions of eggs vital to the spring hemispheric bird migration. But scientists who favor the sanctuary say a variety of research has shown a decline of as much as 90 percent in horseshoe crabs across the last decade, a result of overfishing."

HERE COME THE MOLLUSKS

If they could read genealogical charts, the earliest mollusks could trace their ancestry as far back as the Early Cambrian, about 500 million years ago. David Norman (1994) describes what these ancestors looked like:

> The archetypal mollusc is thought to have had a large, flat muscular foot by which the animal crept along the sea bed. At the front of the foot was a head with some sort of sense organs (tentacles) and a mouth housing a peculiar feeding device known as a radula. This was a strip of tissue with a mat of tiny teeth on the surface, which could be pushed out through the mouth and scraped over the sea bed. Above the foot, the body was hunched up to form a cavity for the main body organs, covered by a sheet of tissue known as the mantle. The mantle was important because the shell of the mollusc grew from it. The shell and mantle overlapped the sides of the foot and enclosed a narrow cavity along the sides of the body into which hung a series of leaf-shaped gills. The foot was anchored to the shell by several pairs of large, powerful muscles which often left distinctive marks on the inside of the shell.

Variations on this theme of a covering shell include gastropods (slugs, snails, and limpets), bivalves (clams, oysters, mussels, and cockles), and a class known as Monoplacophora, or "one plate-bearers," a reference to their single shell. Their shells were pointed, cap-shaped affairs, about an inch long, and one of the Cambrian forms was known as *Scenella*, from the Greek *skene*, "tent." From the scars on the interior surface of fossil shells, we know that the early monoplacophorans had segmented muscles, suggesting an ancient connection with segmented worms, or maybe even arthropods. The monoplacophorans arose during the Cambrian radiation, some 500 million years ago, and were believed to have gone extinct by the Middle Devonian, 150 million years later.

On May 6, 1952, aboard the Danish research vessel *Galathea*, in a trawl pulled up from 11,878 feet in the Mexican Pacific, the Danish malacologist (mollusk-specialist) Henning Lemche found ten specimens of a segmented, limpetlike mollusk with several pairs of gills. He did not actually examine the little creatures until 1956, and published the first description of *Neopilina* the following year. In a popular account (1957), Lemche wrote, "One of the specimens had gotten separated from its shell, and I could see on its upper side small, whitish, shining streaks, which I recognized as muscles. Three pairs of these near the head were close together. Farther back were five other pairs. This is the same number of muscles that is found in chitons, but chitons have eight shell plates, one behind the other, and this animal had only one shell, as snails do." Moreover, snail shells show a varying amount of twisting (torsion), whereas the new animals did not. It was not a chiton, and it was not a snail, but rather an animal that had been recorded only from fossils of the Cambrian Period, 350 million years ago.

The discovery of living fossils permits the study of the biology of an almost extinct group of organisms in ways that would be impossible from the hard parts preserved as fossils alone. They provide living links to now generally extinct groups. As previously mentioned, in the fossil monoplacophorans a unique serial repetition of paired muscle scars occurs on the inner surface of the shell. Interestingly, although originally a main characteristic of the monoplacophorans, these muscle scars do not occur in the living representatives of the group.

Members of the genus *Neopilina* have eight pairs of serially arranged pedal (foot) retractor muscles. Not only are the muscles serially arranged, but there is a serial repetition of paired nerve connectives, nephridia (kidneys), gills, and, to a lesser extent, gonads and auricles. (The condition of repeated body parts is known as metamerism.) Mollusks are not ordinarily considered to be a segmented group, but the metameristic succession of structures suggested to scientists describing *Neopilina* that it might belong to a segmented group of mollusks. This had considerable significance in the evolution paradigm because it made *Neopilina* a potential "missing link" between the unsegmented mollusks and the segmented annelids (earthworms, etc.) and arthropods (insects, spiders, and crabs). The discovery of *Neopilina* has contributed to new theories about the evolution of mollusks. The gastropods, bivalves, scaphopods (*Dentalium*) and even the cephalopods (which are mollusks), may have evolved from an ancestor similar to *Neopilina*.

Since the initial discovery of *Neopilina galatheae* in 1952, at least seven more species have been described, all of which have been collected in water ranging from about 6,000 to almost 20,000 feet deep. Three additional species of *Neopilina* (*N. adenensis*, *N. bruuni*, and *N. oligotropa*) and four of *Vema*, another new and closely related genus, have been collected in the eastern and central Pacific, the South Atlantic, and the western Indian oceans (Batten 1984). Found almost 20,000 feet down off the coast of northern Peru, *Vema ewingi*, which has six pairs of gills to *Neopilina*'s five, was named in honor of the research vessel *Vema*, owned by Columbia University's Lamont Geology Laboratory, in whose nets the second genus of living monoplacophorans was brought up, and Maurice Ewing, its chief scientist. Although reasonably widespread, these primitive mollusks have gone undetected until relatively recently, but their existence in a deep water habitat did not support the previously popular notion that more ancient forms of life would be found in the oceans' depths. Further sampling of the deep-sea fauna has so far not revealed more ancient life forms, but there is a lot of ocean and ocean floor still to be examined.

BRACHIOPODS

Brachiopods are relatively rare animals today, but they were much more abundant in ancient seas, when they probably filled the niche that is occupied by today's clams and oysters, and they too are often called living fossils. The brachiopod shell is composed of two halves, with each half having a slightly different shape. Where the paired shells of clams are designated left and right, those of the brachiopods are designated dorsal and ventral, and the ventral is usually larger. They are sometimes known as lampshells because of their resemblance to a Roman oil lamp. Like the trilobites, they flourished for a long time, but then diminished to ecological insignificance. (The trilobites took the concept of insignificance to new heights, becoming utterly and irrevocably extinct.)

Fossil brachiopods from the Cambrian onward are so numerous (more than 30,000 species have been identified) that they are used as "index fossils," which can be used to date a particular stratum. Current brachiopods range in size from 1 to 6 inches, but some fossils are 14 inches in length. There are now some 300 species of living brachiopods; their range today is worldwide, and they are particularly numerous in the Antarctic. Most of the space inside the brachiopod shell is occupied by an organ called the lophophore, which is composed of rows of hollow tentacles that pump water through the animal, maintaining a steady flow of oxygen and filtering tiny food particles from seawater. Brachiopods do not move very much; most are held to the bottom by a stalk. (The name, which can be translated as "arm-foot," came about because an early investigator mistook the coiled lophophores for arms with which the animal pulled itself along.) They have been divided into two classes: the *Articulata*, with shells that are hinged with interlocking "teeth," and the *Inarticulata*, in which the shells are held together by muscles only. A representative living species is *Lingula*, which is almost identical in structure to some ancient brachiopods. ("Lingula" is a diminutive of *lingua,* "tongue" in Latin, which is a good description of the shape of the animal's shell.) *Lingula* is usually found buried in the sand in Hawaiian waters, at depths down to 140 feet.

When Charles Walcott first examined the fossils of the Burgess Shale, he was inclined to affiliate some of the stranger specimens with known phyla. Of course, many of these animals have no such affiliations, but—to use Gould's phrase—Walcott "shoehorned" them into known categories. Thus, faced with the shrimplike *Yohoia*, with its multisegmented body and "great appendages" at the head end, Walcott placed it among the brachiopods. He reasoned that the appendages were claspers used by the male to hold the female during mating—a brachiopod feature—but later examination revealed that both males and females had them. (Eventually, *Yohoia* was placed in a group known as arachnomorphs, which includes the horseshoe crabs and trilobites.) Even stranger was his designation of the five-eyed *Opabinia* as a brachiopod; but at least it had a "nozzle" on the anterior that resembled the stalk of the traditional brachiopods. *Opabinia* is not a brachiopod, but it is not easy to say exactly what it was. The "nozzle" had claws on the end, and was probably used to grasp worms or other prey items. With its flaplike structures along the trunk and its tail fan, it reminds us of *Anomalocaris*, the scourge of the Cambrian seas, and was probably related to it.

Yohoia and *Opabinia* are no longer classified as branchiopods, but there are several species of inarticulate branchiopods associated with the Burgess Shale. They include *Acrothyra gregaria*, one of the commonest fossils of the Burgess Shale fauna; *Lin-*

gulella, with a small fragile shell; *Micomitra*, which had a delicately incised shell and setae (bristles) that extended far beyond the margin of the shell; and *Paterina*, which looked like *Micomitra* but was black and lacked the setae. The articulate brachiopods were represented by *Diraphora* and *Nisusia*, which were characterized by fine radiating ribs, not unlike today's cockles.

CEPHALOPODS

Many of the early soft-bodied creatures have not left their fossil remains for paleontologists to examine, but those that had shells have contributed much to our understanding of evolution. The soft parts have long since disappeared, but their shells have endured, giving us a picture of the ancestors of today's clams, oysters, and snails. Most octopuses, squid, cuttlefishes, and nautiloids (known collectively as cephalopods) leave little hard evidence behind, but many of the early forms were equipped with shells, giving us a picture of prehistoric sea floors inhabited by creatures with elaborately curled and twisted shells, sometimes 10 feet in length. As these shells grew, they were partitioned off into a series of chambers connected by a channel, the siphuncle, through which ran a strand of tissue well-supplied with blood. Of the ancient cephalopods, the ammonites and belemnites are extinct, there are two surviving nautiloid species, and the squids have come to populate the world's oceans.

The class Cephalopoda contains two subclasses, the Nautiloidea and the Coleoidea. Nautiloidea are defined as "firm outer shell with mother-of-pearl layer secreted by mantle present. Arms multiple (several dozens), slender, suckerless, in two circles. Head and arm bases dorsally covered by a fleshy hood closing aperture when animal retracts into shell. Funnel not fused ventrally. Two pairs of gills." Coleoidea: "Shell internal or absent . . . 8 or 10 appendages, disposed in a single circle, always armed with suckers or hooks. Head and arms not covered dorsally with hood. Funnel fused into conical tube. One pair of gills" (Nesis 1982). Of course, information as to the numbers of arms and gills, the form of the funnel, and all other soft parts, are largely useless for identifying fossils.

NAUTILOIDS

Because there are nautiloids alive today, we tend to think that the earlier ones looked like the current versions, with a chubby, curved shell marked with brownish stripes, out of which a many-tentacled animal peeks with keyhole eyes. But this is far from the case. Paleozoic nautiluses came in all sorts of shapes and sizes, only a few of which would look familiar to the nonpaleontologist. The coiled shell of the living nautilus was only one of several arrangements. For example, some had a straight shell, but the shell of the Late Cambrian *Plectronoceras* was curved and had a deep living chamber. (The names of many fossil nautiloids end in *ceras,* the Greek word for "horn," because many of the shells were horn-shaped.) There are hardly any examples of fossilized soft parts of early nautiloids, but from the living species, we assume that the tentacles had no suckers.

The earliest known cephalopod was the cap-shaped *Plectronoceras,* which emerged in the Cambrian, 200 million years ago. Found in the Fengshan Formation of northeast China, it was a nautiloid distinguished by a shell that enclosed an air space to provide buoyancy, evidenced by the presence of septa (smoothly curved interior walls) and a siphuncle that allowed for the extraction of water from the chambers. Early nautiloids lived in a many-chambered shell and had numerous arms—sometimes as many as 90, in contrast to squids, which have 10, and octopuses, which have 8. Nautiloids populated the world's oceans—which were not necessarily where they are now—for almost 500 million years, more than enough time to develop many different shapes and lifestyles. *Gonioceras* looked kind of like a flattened cone-shaped paper cup with a cephalopod peeking out of the wide end; the elongated *Michelinoceras* had the same shell arrangement, but with a longer cup that was not flattened. The shell of *Phragmoceras* looked remarkably like the stubby horn of a goat, flattened on the bottom—from where, we assume, the animal emerged. Like gastropods today, some nautiloids developed spikes, spines, or wings on their shells, but as with the snails, the function of these adornments is unknown. The eyes of the primitive nautiloids probably weren't very good, and they probably couldn't see these decorations—or much else, for that matter. Most of the early nautiloids crawled or crept along the bottom, and most were of moderate size, but some reached incredible proportions: the Precambrian genus known as *Endoceras* reached lengths of 10 to 12 feet. Its very long, straight shell would have made steering a problem in open water, so we assume that it crawled along the bottom. With the water jet at one end and the center of buoyancy somewhere near the middle, the animal would have rocked like a see-saw if it had not dwelled on the sea floor. *Cyrtoceras* was a little creature that crawled like a snail, with a perky, vertical shell like a crescent moon. (*Cyrto* is Greek for "curved.") The shell of all nautiloids was divided into chambers by simple septae, which differentiated them from the ammonites, which had a similar outer shell but a far more complex interior.

Maybe because its shell is uncommonly lovely, with its graceful shape and wavy sepia stripes, or perhaps because it is a "living fossil" that affords us an unprecedented glimpse of life hundreds of millions of years ago, the chambered nautilus has been an object of fascination for scientists and nonscientists for ages. Renaissance goldsmiths turned the shells into jewel-encrusted chalices. It has given its name to submarines, from Captain Nemo's in *Twenty Thousand Leagues Under the Sea* to Robert Fulton's 1797 model to the first American nuclear sub, the one that passed under the

The chambered nautilus (Nautilus pompilius) *is the only survivor of an ancient group of cephalopods that lived 100 million years ago. Nowadays, they are being collected for their beautifully striped shells, and in many parts of the South Pacific they are considered endangered.*

North Pole in 1958. The French named a submersible *Nautile*. It was celebrated by Oliver Wendell Holmes in his 1858 poem "Nautilus," which ends with well-known lines that are, however, biologically incorrect, as the *Nautilus* cannot leave its shell:

Build thee more stately mansions, O my soul,
As the swift seasons roll!
Leave thy low-vaulted past!
Let each new temple, nobler than the last,
Shut thee from heaven with a dome more vast,
Till thou at length art free,
Leaving thine outgrown shell by life's unresting sea!

Until 1958, when the first ones were shown at the Nouméa Aquarium in New Caledonia, no nautiloids had ever been exhibited alive in an aquarium. Then followed Monaco and Tokyo, and in 1976, the Waikiki Aquarium in Honolulu, where the first *Nautilus* eggs, collected from the wild, were successfully incubated and hatched in 1990.

Nautilus pompilius and its immediate relatives are the only living cephalopods with an external shell; they are the last survivors of a group of animals that flourished long before there were any bony fishes. (Living squid and cuttlefishes have only a vestigial reminder of the exterior shell, a "gladius" or "pen" that is an interior support element. The chalky "cuttlebone" that is used in birdcages is the pen of a cuttlefish.) The shell of nautiloids, which serves for protection and also as a flotation device, is composed of a series of chambers connected by the siphuncle, by means of which the animal can regulate its buoyancy by adjusting the amount of gas in the chambers. A nautiloid never leaves its shell, but if it is removed, it resembles a squid with an overabundance of tentacles. These tentacles, numbering as many as 90, can be retracted into sleevelike sheaths and are used for capturing prey and for sensing. The nautilus can retract into its shell and close it off with a leathery hood. It propels itself slowly by taking water into the body and ejecting it through a funnel, which is not a closed tube as in other cephalopods. The eyes, which are lensless pinholes open to the sea, are not very efficient, and because the nautilus can only differentiate light and darkness, it has to depend on other senses, particularly smell, to find food and/or a mate.* Nautiluses feed at night, making nightly migrations up the face of the reef in search of food. They are believed to be mostly scavengers. The eggs, more than an inch in diameter, are the largest of any invertebrate's.

Most of the nautiloids were wiped out millions of years ago, and the only living representative of their line was thought to be *N. pompilius* and its immediate rela-

* Although it was long assumed that nautilus found its prey by smell, only recently did anyone think to find out what they smelled *with*. Jennifer Basil, Roger Hanlon, and their colleagues (2000) experimented with captive nautilus at the Marine Biological Laboratory in Woods Hole, and discovered that when they plugged the rhinopores, fleshy papillae located just under the eye and open to the exterior, with Vaseline, the cephalopods could not locate an odor plume. When the Vaseline plugs were removed, the nautilus could locate the odor from distances of up to 30 feet.

tives. In *In Search of Nautilus* (1988), Peter Ward wrote that all of its relatives disappeared by the end of the Miocene, some 5 million years ago, and there were no fossils after that. "Nautilus was certainly living," he wrote, "but it was no fossil. In fact, it had no fossil record at all." Later, he learned that there was indeed a fossil record of *Nautilus,* and the nautiloid fossils that Richard Squires (1988) of California State University at Northridge had collected from 50-million-year-old rocks in Washington State were the oldest chambered nautilus fossils ever found. In a 1998 article, Ward wrote:

> The living fossil we call nautilus survived the great cosmic collision that killed the dinosaurs, and many other changes in the 65 million years since that catastrophe. Its presence on Earth through all this time gives us one more small peek into the workings of evolution, which we are learning can tick slowly as well as at more staccato rates. The long voyages of the nautilus, from the depths into the shallows each night, are thus a perfect metaphor for its evolutionary history, which comes up to our world unchanged from the great depths of time.

One aspect of Ward's research into the biology of *Nautilus* and its place in cephalopod history took him to Papua New Guinea to examine the first "king nautilus" (*Nautilus scrobiculatus*) ever captured alive; it was brought up in a trap by Ward's fellow cephalopaleontologist Bruce Saunders of Bryn Mawr College. The previous specimen was known only from a shell that had washed ashore around 1900. When Ward and Saunders (1997) examined the specimen, they realized it was so different from *N. pompilius* that they placed it in an entirely new genus, which they called *Allonautilus* ("other nautilus"). The hood—which in *N. pompilius* is reddish-brown, the same color as the stripes on the shell, and sparsely spotted with white—was black and covered with fleshy white tubercles. The shell was a different shape; most surprising, it was "covered with a thick, shaggy orange fur" (Ward 1998). This was the periostracum, often observed on the shells of gastropods but never before seen in nautiloids. Further investigation revealed that the gills of the king nautilus were only about half the size of those of a chambered nautilus. With David Woodruff of the University of California at San Diego, Ward and Saunders (1997) sequenced the DNA of specimens collected from numerous locations in the South Pacific and concluded that there were indeed two distinct branches of living nautiloids, *Nautilus* and *Allonautilus.* With fossil evidence uncovered by Neil Landman of the American Museum of Natural History in New York, Ward and Saunders were able to show that *N. pompilius* was extremely primitive, and "may even be the ancestor of most of the nautiloids present on our planet for the past 75 to 100 million years" (Ward and Saunders 1997). As for *Allonautilus,* it is much younger, and is probably an evolutionary offshoot of the older species. (In addition to *Allonautilus scrobiculatus,* there are four currently recognized species in the genus *Nautilus: N. pompilius* from various locations in the southwest Pacific; *N. belauensis* from Palau; *N. macromphalus* from New Caledonia; and *N. stenomphalus* from eastern Australia.)

When Schipp, Chung, and Arnold (1990) examined the digestive organ, or coelum, of *Nautilus,* they discovered abundant colonies of a bacterium that they tentatively identified as "probably from the *Pseudomonas* group." The protective and nutritive advantages to the bacterium are obvious, but what the nautilus gained from this symbiotic relationship was not so obvious. And as with many other bacterial symbionts, the method of infection was not apparent. The authors asked, "If they are essential to the *Nautilus*, are they acquired from the environmental seawater in early life stages, or are they somehow derived during the formation of the egg capsule by the adult?"

When the cephalopod specialists Martin Wells and Joyce Wells of Cambridge University and Ron O'Dor of Dalhousie University in Halifax, Nova Scotia, examined specimens of living nautiluses in Papua New Guinea, they were surprised to discover that the animals could survive for several days in water with very little oxygen and could even survive for hours when there was no oxygen in the water at all. In response to oxygen starvation, *Nautilus* pulls into its shell and slows its heart beat to one or two beats per minute, recovering quickly when placed in oxygenated water. They can probably draw on the large amount of oxygen in their blood, or even the oxygen in the buoyancy chambers of the shell. (*Allonautilus*, with its much smaller gills, cannot remain out of the water for more than a couple of minutes, suggesting that *Nautilus* came first and that the king nautilus was a recent arrival.)

All nautiloid cephalopods, including the living representative of the group, had an external shell divided by smoothly curved interior walls, the septa, into chambers that are connected by the siphuncle. The seams where the septa meet the shell wall (called sutures) of nautiloids, which make the exterior pattern on the shell, are simple zigzag patterns. Of the numerous lines of nautiloids, only the various *Nautilus* species and *Allonautilus* are alive today, but the group was well represented in the Paleozoic. Although not necessarily ancestral to the other cephalopods, the nautiloids are certainly the oldest, indeed the *only,* survivors of an extraordinarily diverse group of shelled cephalopods that has been recorded from a fossil record of more than 500 million years.

AMMONITES

Ammonites were also early cephalopods, but unlike the nautiloids, where a couple of survivors afford us some information on how their ancestors lived, there are no ammonites left, so we can only speculate. Judged by their abundance and variety, however, as well as the duration of their life on Earth, ammonites are among the most successful animals in history. Most ammonite shells are similar in general appearance to that of the nautilus, but there are also bizarre variations on the theme. During the Jurassic and Cretaceous periods, ammonites evolved more streamlined shells for swimming and the structure of the shell became stronger. Different shell shapes emerged as well, such as snail-like or uncoiled. (The name comes from the Egyptian god Ammon, who was often shown with ram's horns.)

Ammonites first appeared during the Silurian Period (438 million to 408 million years ago), and proliferated into the Carboniferous (355 million to 290 million years

In this reconstruction of a living ammonite, the animal is shown resembling a nautilus, although there is nothing in the fossil record to support such an assumption.

ago). Like so many other living creatures, they took an enormous hit in the Permian Extinction, 250 million years ago, when more than 90 percent of all species in the oceans were eliminated. (On land, more than two thirds of the reptiles and amphibians vanished too.) Although reduced to a single family by the Permian "mother of mass extinctions," the ammonites bounced back and proliferated into the Mesozoic (175 million to 65 million years ago), when, with more than 80 known families, they dominated the marine invertebrate fauna. As with the trilobites, which dominated the seas for hundreds of millions of years, ammonite extinction was a repeated phenomenon and occurred throughout their collective history. The last of the ammonites died out 65 million years ago, in the K-T (Cretaceous-Tertiary) Extinction that also took out the dinosaurs

Relative to body size, their jaws were the largest of any known cephalopod's, and, like most other cephalopods except the nautiloids, they had excellent vision. In water that was rarely more than 300 feet deep, the larger ammonites, all of which seem to have been predators like modern squid, fed on largish animals including mollusks, echinoderms, other ammonites, and crustaceans. The smaller ones fed on much smaller animals, including plankton and small benthic creatures like foramaniferans. It is not clear whether, like *Nautilus*, they fed on carrion.

Although many fed on the ocean floor, others may have caught their prey while floating or swimming. Their shells were divided into a series of air chambers that provided buoyancy, and like other cephalopods, ammonites moved by jet propulsion, expelling water through a funnel. The fertilized eggs developed into juvenile ammonites known as larval protoconches, which were tiny compared to the eggs of the nautiloids and drifted freely as plankton. As the ammonite grew and extended its coiled shell, it moved forward within the shell, sealing off chambers behind it. Within the coil was a series of progressively larger chambers divided by thin septa (singular: septum). Only the last and largest chamber was occupied by the living animal. Passing through the septa was the siphuncle, which extended from the ammonite's body into the empty shell chambers. The ammonite secreted gas into the shell chambers, enabling it to control the buoyancy of the shell. Estimates of their life span are exceedingly variable, ranging from 1 to 20 years.

From the fossilized shells, which are usually casts or molds of the infilling of the chambers, we have a bounteous record of their shape and structure, if not their lifestyle. Like nautiloids, ammonites had septa dividing the chambers, but the ammonites developed complex saddles, lobes, and sutures, which looked as if someone had tried to cram too-large pieces inside the shell and had to fold or otherwise distort them to make them fit. Found in sedimentary rock layers of limestone and clay and in clay mineral iron ore, ammonite shells are important index fossils; they often link the rock layer in which they are found to specific geological time periods. The shales of the Western Interior Seaway (also known as the Pierre Seaway), which covered much of midwestern North America in the Cretaceous, from 130 to 65 million years ago, are particularly rich in ammonite fossils, which are unusual in that the original shell is often preserved along with the inner nacreous lining. The predominant ammonites there are the numerous species of the genus *Baculites*, which can be used to date the various regressions and transgressions of the seaway: 80 million years ago *Baculites obtusus* was the predominant form; 7 million years later, it was *B. compressus'* time, and by 69 million years ago, *B. clinolobatus* dominated. The members of the family Baculitidae were characterized by a long, narrow, tapering shell that ended in a tiny spiral turnup, known as the amonitella. Most were less than a foot long, not counting the animal, but the shell of *Baculites grandis* reached a length of more than 3 feet.

In other cases of fossilization, the original shell is replaced by minerals such as iron pyrite. The visible seams, called suture lines, are well preserved in fossils. (The ammonite suture lines are zigzagged whereas those of the nautilus are straight.) Many ammonites were closed spirals, like the familiar nautilus shell, but they also achieved fantastic variety, including shells that looked like a normal coil in profile but were hubcap-thin when viewed head-on; and several species that looked like elongated dunce caps, ridged saxophones, slinkys, paper clips, smoking pipes, or drip-castles.

All ammonite shells grew from the base outward, so the original shaft and subsequent turns retained their form as the animal added new material at the very place where it was in permanent contact with the growing shell. The same growth stages are attributable to the tightly coiled forms like *Eubostychoceras*, the kudu horn–shaped *Hyphantoceras*, and probably the strangest and most irregular of all, *Nipponites mirabilis*, which grows in an open spiral, then changes the growth plane and soon envelops its shell in its own coils, like a snake trying to tie itself into a knot. Takashi Okamoto of Ehime University in Japan, who plotted the growth of *Nipponites* as a computer program, likened its rolled form to a pile of dog or human feces, but Neale Monks said "it is actually very regular—a sine wave folded back on itself through 90 degrees with each half wave" (personal communication).

Heteromorph ammonites are those with a shell whose coils were not touching, but formed open spirals like a ram's horn. These shells are similar to no living creature's, and we can only speculate about how such bizarre animals might have lived. Peter Ward (1979) wrote: "Previous investigators have speculated about the mode of life of these torticonic cephalopods, with the favored interpretation using the obvious convergence with gastropod shape as an indication of a vagile [wandering] benthonic existence." In other words, he is saying that these stable but not particularly maneuverable creatures crawled along the bottom with their shells above them, like snails. But studies of the internal structures of the shell, by means of which the animal con-

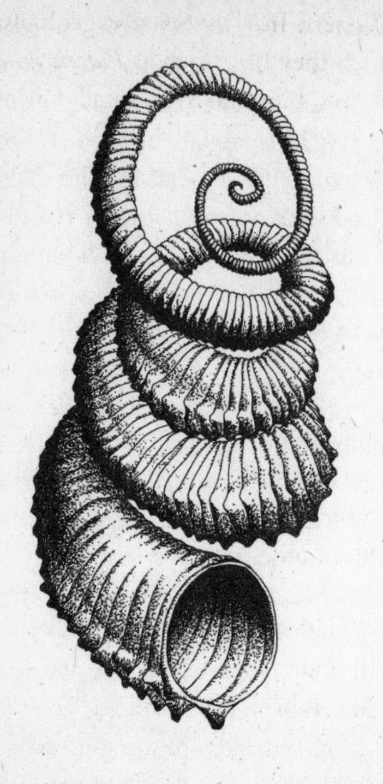

The fossilized shell of a heteromorph ammonite, Didymoceras. *It is not at all obvious how the animal might have moved, or how such a fragile shell structure was protected.*

trolled its buoyancy, have led to numerous different interpretations. At the conclusion of his 1979 paper, Ward wrote, "The later Cretaceous . . . torticones, however, would have had a much more difficult time in this mode of life [shell aperture facing downward] for the majority of these genera developed a 'U'-shaped body chamber after an initial tortoconic stage, [which] must have forced the body to be positioned upward or facing the surface of the sea." He suggests that these ammonites might have lived a mesopelagic (midwater) existence, in habitats similar to those "occupied by the poorly streamlined but tremendously abundant cranchid and spirulid squids."

Despite the general similarities in the shape of the shells of nautiloids and some ammonites, it may be that the biology of ammonites was closer to that of squids. In 1993, David Jacobs and Neil Landman argued that the living *Nautilus* was a poor model for the function and behavior of ammonites. Comparisons with *Nautilus* have led, say the authors, to the unwarranted assumption that ammonites stayed largely in their shells, but this might not have been the case at all, and ammonites with long body chambers might have extended the mantle well out of the shell. Furthermore, *Nautilus*, with the smallest mantle volume relative to its size of any modern cephalopod, has the least effective jet propulsion, which consists of a siphon (the hyponome) that can be aimed backward to shoot out water when the animal moves in the direction of its tentacles—which is most of the time. If ammonites extended their bodies out of the shell—which *Nautilus* does not do—they might have been capable of faster swimming than the poky *Nautilus*. If the ammonites were propelled by water taken in and expelled by mantle contractions, the method employed by squids, it is "likely that the immediate shared ancestors of coleoids and ammonites had a coleoid-like mantle." It has been too easy to compare ammonites with nautiloids, but "the inference that a *Nautilus*-like hood, [large] tentacle number, or pinhole eye was present in ammonoids is hard to support. . . . In addition to these soft-part features, the reproductive behavior of ammonoids may have been similar to that of coleoids, where migration, congregation, and mass mortality occur as part of breeding behavior." Also, the small size of coleoid and ammonite embryos, as contrasted with the huge eggs of *Nautilus*, suggest a further connection.

How do we know that ammonites sometimes congregated in large numbers?

Near the town of Kremmling, Colorado, in the Western Interior Seaway, geologists discovered a concentration of fossil ammonites which they identified as *Placenticeras meeki*. The largest ones were the size of automobile tires, but most were smaller. Most had been enclosed in sedimentary concretions that were formed when the shelled carcasses died and drifted to the bottom, where they were covered with mud. Eventually, rock-hard concretions were formed, the surrounding sediment was flattened into shale, and the ammonites fossilized. They remained buried deep within the earth until the Rockies rose through the shales and lifted the fossils to the surface. They are now buried on a grassy slope that covers several square miles, waiting to make their first appearance in 70 million years. Kirk Johnson of the Denver Museum of Natural History has recovered many of the ammonite fossils and transported them to the museum, where they will be catalogued and some of them placed on exhibit. It is estimated that there are thousands of ammonites at the Kremmling site, and with a few exceptions, most of them are females, identifiable by their larger shells. Johnson (1999) believes that ammonites, like squid, reproduced once and then died:

> Perhaps the hill of giant fossil shells in the Colorado Rockies was the site of a Cretaceous night of mass spawning. Phases of the moon cued the ammonites to congregate; after the males had fulfilled their role, they left the scene. Imagine the sea by the light of the moon and the muffled clatter of thousands of forty-pound female shellfish as they release the next generation in the form of planktonic embryonic ammonites. Then, with their life cycle complete, they shudder, sink to the seafloor, and begin their trip into the fossil record.

In Klaus Ebel's 1992 study of the mode of life and soft body shape of heteromorph ammonites, he proposed that the animals led a benthic existence, crawling or walking along the bottom, rather like snails. At one time, the bizarre shapes of these ammonites suggested to some that this was evolution gone mad, an uncontrolled frenzy of useless design that soon led to extinction, but Ebel wrote that "heteromorphs are considered as entirely vital, yet specialized members of the ammonitic faunal spectrum that were not condemned to become extinct solely because of their strange shell forms." Ebel posited the crawling position as the only possible explanation for heteromorph forms, where the animal, unable to draw the soft parts into its shell, carried the shell on its back. He illustrates the various growth stages of *Subptychoceras* (the "paper-clip" ammonite), where the shell starts out as a straight shaft, which then turns down sharply, makes a 180-degree turn, and as it continues to grow, keeps doubling back. The largest of the paper-clip ammonites is the genus *Diplomoceras*, described from an almost complete specimen found by William Zinsmeister and A. E. Oleinik (1995), which resembled a clothesdryer vent hose bent into the shape of a paper clip. The specimen clearly shows that the ammonite does not shed any of the early parts of the shell and keeps everything intact throughout its life. Six feet long and 6 inches in diameter, the shell of *Diplomoceras maximum* would have been more than 13 feet long if straightened out.

In a 1998 study that appeared in the online journal *Palaeontologica Electronica,* Neale Monks and Jeremy Young of the Natural History Museum in London analyzed the functional morphology of heteromorph ammonites and suggested that unlike the nautiloids, where the soft body parts fill the entire living chamber, the body of these ammonites was much smaller, and the animal could move in and out of the shell, but could not leave it. On the basis only of the appearance of the shells, previous workers had assumed that the ammonites fed on plankton and floated with the opening upward, but Monks and Young believe that the animals fed with the opening downward, and they therefore "can be visualised as having been rather like a small octopus with a mobile burrow or cave. When feeding, the aperture would have been angled toward the sediment allowing the animal to forage in the sediment, pulling itself along with its arms, digging up the sediment, digging up small molluscs, worms and crustaceans." The ram's-horn shells of such ammonites as *Hamites* are robust, with constrictions, spines, and ribs, which might have made them less vulnerable to predators as they floated away. The authors recognize that their interpretations are "circumstantial at best," but they are closer to the lifestyle of living cephalopods which makes them somewhat more likely.

Mature ammonites typically fell into two distinct size classes: macroconchs, the larger, female forms, and microconchs, the smaller males. They varied in size from the minuscule *Ponteixites*, less than three fourths of an inch across, to the unbelievable *Parapuzosia*, from Montana and Wyoming, whose coiled shell could be 6 feet across. Another giant was *Pachydiscus catarinae*, which was 5 feet across. (One exceptional specimen was found in a Late Cretaceous deposit at Point Loma, San Diego County, California, on a rock and boulder beach within a concretion that weighed over 500 pounds. Because removal of the specimen required the use of a helicopter, it was dubbed "the flying ammonite" by the collectors and those who prepared it for exhibition.)

The ammonites' complex, wrinkled septa produce angular or convoluted sutures on the outside of the shell. These extinct cephalopods have been divided into three groups on the basis of the type of sutures: goniatids have relatively simple and undulating sutures; ceratids have sutures with smooth, large "hills" alternating with saw-toothed valleys; and ammonitids have very complex, almost fernlike branching and convoluted sutures, almost fractal in their complexity. The septa are also different in the various ammonites. The reason for the intricate folding is not known, but it was long regarded as an adaptation to deep water, where the pressure is so much greater than near the surface. However, in a 1997 study, Daniel, Helmuth, Saunders, and Ward wrote that "any departure from a hemispherical shape [such as that seen in the shell of *Nautilus*] actually yields higher, not lower, stresses in the septal surface." If they are correct, it means that the conventional explanation has to be rejected, and the ammonoids with the most complex septa did not necessarily inhabit the deepest waters.

There is no "goal" in evolution, and certainly the process is not programmed toward the development of more complex forms, but this is one thing that seems to happen anyway. Of course, a move in the direction of complexity does not always occur; think of the microbial forms that have not changed in 3 billion years. But there are some cases where the earlier forms are simpler and the later ones more complex, and the ammonoids present a textbook case of evolved complexity.* In a 1999 study, Saunders, Work, and Nikolaeva analyzed the sutures of 588 genera of fossil ammonoids and found that descendants were more than twice as likely to be more complex than their ancestors. Since it has been argued that more complex sutures are *less* effective at buttressing the shell, and that the most complex ammonoids inhabited shallow waters—where predation was considerably more effective—the drive toward complexity appears to be a trend in ammonite evolution, even after three mass extinctions threatened to wipe them out altogether. And one finally did, regardless of their complexity.

From the locations where fossil ammonites have been found, something of their lives and deaths can be inferred. They were common in deeper waters, and rarely found in the shallows. (There are not now, nor have there ever been, fresh-water cephalopods.) Because so many have been found in huge aggregations, it has been assumed that many ammonite species were social animals. In 1960, Erle Kauffman and Robert Kesling published a paper entitled "An Upper Cretaceous ammonite bitten by a mosasaur," in which they went into incredible detail to show that "the shell was bitten repeatedly and bears dramatic evidence of the fatal encounter" with a mosasaur, a marine reptile that flourished in the seas of the Cretaceous Period, from 90 million to 65 million years ago. They plotted the tooth patterns of the mosasaur against the holes, and published nine pages of photographs to accompany the 45-page paper.

There are those, however, who believe that this dramatic event, long a staple of mosasaur and ammonite lore, did not occur at all—at least not as a bite. In his 1998 book *Time Machines*, Peter Ward devotes an entire chapter, "The Bite of a Mosasaur," to the subject of how to eat a nautilus. He recounts his observations of a sea turtle attack on a nautilus in captivity in New Caledonia, after which the nautilus's shell was fragmented, like a porcelain plate hit with a hammer. Then he tells of his examination of a collection of *Placenticeras* shells, where the holes "seemed to conform to the sizes and shapes produced by limpets." He includes a discussion of

---

* In *The Evolution of Complexity by Means of Natural Selection* (1988), John Tyler Bonner wrote, "There is an interesting blind spot among biologists. While we readily admit that the first organisms were bacteria-like and that the most complex organism of all is our own kind, it is considered bad form to take this as any kind of progression. In the first place, to put ourselves at the pinnacle seems to show the kind of egocentricity that has been a plague to science and fostered such ideas as [that] the earth is the center of the universe or that man was specially created. There is a subconscious desire among us to be democratic even about our position in the great scale of being. From Aristotle onwards there was an idea that there really was a progression towards perfection from plants and lowly worms to human beings. The basis of this scale of being was not evolutionary, but reflected the degree of difference from ourselves, and therefore is unacceptable to us today. In an early notebook Darwin cautioned that we should never use the terms 'lower' and 'higher' (although he was sensible enough not to follow his own advice), and I have been reprimanded in the past for doing just this. It is quite permissible for the paleontologist to refer to strata as upper and lower, for they are literally above and below each other, and even geological time periods have these adjectives—for example, upper or lower Silurian to mean more recent and more ancient, respectively. But these fossil organisms in the lower strata will, in general, be more primitive in structure as well as belong to a fauna and flora of earlier times, so in this sense 'lower' and 'higher' are quite acceptable terms."

Erica Roux, a graduate student who tried to make round holes in nautilus shells, which are a close approximation of unfossilized ammonite shells:

> Erica constructed an artificial mosasaur jaw. It did not look much like the real thing, being fabricated of metal with series of teeth made out of nails and screws, but nevertheless it approximated the real thing in many ways. The "teeth" descended onto the shell surface just as a mosasaur jaw would have, and a gauge attached to the jaw showed the amount of pressure needed to produce a break. A nautilus shell was put between the jaws, and the type of damage inflicted on the shell was observed. For several days [the lab] reverberated with cracks and snaps, as shell after shell fell victim to those jaws of death. An army of mosasaurs could not have had so much fun. And in all the carnage that ensued, not once was a round circular hole approximating the size of a mosasaur's tooth ever produced.

Ward believes that mosasurs did indeed eat ammonites; they just crushed the shells to get at the soft edible bits. But in a 1999 paper Adolf Seilacher wrote, "The claimed 'mosasaur bites' are probably all caused by limpets rasping on necroplanktonic *Placenticeras* shells, compactual puncturing of the pits and diagnostic beveling of the rims." In other words, Seilacher believed that the ammonites were already dead and lying on the bottom when they were colonized by the limpets. The story of ammonite-crunching mosasaurs dies hard, and there are many who refuse to accept the prosaic limpet explanation. They do not believe that the limpets could possibly have organized themselves into the straight lines or V-shaped formations that reflect the tooth arrangement in the mosasaur's jaws on both sides of the ammonites as in Kauffman and Kesling's paper.

The ammonites dominated the Ordovician and Silurian seas, but, wrote Ward in 1988, "The response to the Devonian challenge of the fishes led to major evolutionary breakthrough among the nautiloids: they produced a creature capable of competing with the fishes by tackling the three great problems of the chambered cephalopod design: growth rate, shell strength, and reproductive strategy." So they grew faster, to avoid being eaten by fishes when small, the shell was redesigned, and the number of eggs was increased. Still, the ammonites flourished for 100 million years, with numerous species appearing and then disappearing for reasons that cannot be explained. Each of three distinct crises reduced the ammonite population by an incredible 90 percent, so that by the end of the Permian, only a few varieties survived.

The Permian Extinction, which occurred about 225 million years ago, was probably the most catastrophic in Earth's history. It might have been connected to the impact of an extraterrestrial body, changes in sea level or climate, or the convergence of the continents into a single landmass; but whatever it was, as many as 90 percent of all terrestrial and marine species became extinct. Most of what preceded this moment would never be seen alive again; their extinctions marked the end of the Paleozoic Era. In *On Methuselah's Trail* (1992), Ward describes how he went looking for ammonite fossils at Zumaya on the Bay of Biscay in northern Spain. He found some close to the K-T boundary, but he could not find a single instance of an ammonite species that had survived. But the ammonites were already in decline during the Turonian Period, at least 20 million years before the K-T extinction. Whether it was a comet, an asteroid, volcanic eruptions, or a combination of these that did in the ammonites of the Cretaceous is not known, and all the remaining forms suffered the effects of a terminal extinction 65 million years ago.

The ammonites so dominate the Mesozoic fossil fauna that some paleontologists call this period from 265 million to 65 million years ago the Age of Ammonites. There are some areas where the number of individuals from a single species runs into the hundreds of millions. By the Late Cretaceous, just before the ammonites vanished, there were 246 known ammonite genera, compared to 14 nautiloid genera. What happened to them? Perhaps the ammonites relied too heavily on a single food source, such as some planktonic species that was destroyed by the catastrophic event, whatever it was, while the nautiloids, with their great depth range, were able to eat a variety of food. Today's nautiluses pass through 500 vertical feet daily, rising from the sea floor to eat carrion, discarded lobster shells, and hermit crabs, which they locate by their sense of smell. If an asteroid's impact caused sunlight to be cut off, the phytoplankton that lived by photosynthesis would have died, and the creatures that depended upon them would have starved. If, as some have proposed, 90 percent of the calcareous plankton became extinct at the close of the Cretaceous, the ammonites might have lost their primary food source. But if the ammonite embryos, some of them no larger than one millimeter in diameter, were part of the plankton that was destroyed, the ammonites might have ridden this catastrophe right into extinction. In the Cretaceous rocks of Montana, thousands of preserved embryonic ammonites have been found, which may be the smoking gun that provides the solution to the mystery of the disappearing ammonites.

BELEMNITES

The belemnites, order Belemnoideas, were a group of marine cephalopods, similar to modern squids and cuttlefishes in that they were dartlike, rapid swimmers, and could move tailward or tentacleward with equal facility. "Belemnite" comes from the Greek *belemnon,* meaning "dart" or "javelin." The belemnites, like living squid and octopuses, have two gills in the mantle cavity (the ammonites and nautiloids have four). In those rare fossils where the soft tissue has been preserved, it can be seen that belemnites had ten tentacles of equal length, set with rows of little hooks rather than suckers, and, like the modern cephalopods, they had an ink sac.

The belemnites' squidlike body enclosed a cone-shaped internal shell, called the phragmocone, which terminated at the tail end in a solid, pointed element, known as the rostrum or guard. Composed of solid calcite ($CaCo_3$) and with a depression at the tip known as the alveolus, this is the part most commonly found as a fossil. (The phragmocone and the pro-ostracum, a bladelike forward extension, are rarely preserved.) The phragmocone, which was used for buoyancy control, incorporated an internal shell consisting of gas-filled chambers connected by a siphuncle, as in the shell of the living nautiloid and the pro-ostracum.

The belemnite phragmocone resembles a straightened nautiloid shell, and the pro-ostracum corresponds to the calcified pen or gladius of living squids and cuttlefishes, but no living cephalopod has a solid guard at the posterior end, so the function of this bullet-shaped element can only be guessed at. In an attempt to answer the question, Monks, Hardwick, and Gale (1996) reconstructed a belemnite known as *Cylindroteuthis puzosiana* and found that the guard added to the model's stability. They then outfitted the model with cardboard fins of various shapes, and subjected it to wind-tunnel testing. Cuttlefish maintain stability by emptying and flooding some of the chambers, but the belemnite phragmocone was not suited for such activity, and they suggested that once emptied, it remained so, and that the belemnite controlled its buoyancy by active swimming. The solid guard counteracts the buoyancy of the phragmocone and "provides a rigid structure from which the fins could generate lift and turning forces."

Gut contents of fossil marine reptiles—ichthyosaurs and crocodiles, for example—often contain numerous cephalopod hooklets. Well preserved belemnites found in Oxford Clay in England show phragmocone and pro-ostracum damage, indicating the fleshy body parts had been bitten off by a predator, probably a narrow-snouted ichthyosaur. Many Jurassic fishes have teeth capable of crushing hard objects, and in one shark fossil from the Jurassic of Germany the gut region is filled with belemnite guards. These durable fossils are often found in huge concentrations, giving rise to the term "belemnite battlefields." Accumulations of this sort may have occurred as a result of death after mating, a phenomenon in many living species of squid; following some unknown catastrophic mass-mortality event, such as an underwater avalanche; or possibly as a result of predation accumulations, which might have occurred when a single animal fed on large numbers of belemnites and died as a result. A fossil shark from Holzmaden, Germany, contains some 250 belemnite guards crowded into its gut; more recently, sperm whales have been found with the beaks of 14,000 squid in their stomachs (Berzin 1972).

In some parts of the Oxford Clay of southern England, fossil belemnites are so numerous that they pose considerable problems for brick manufacturers. The most common British genera are *Cylindroteuthis* and *Lagonibelus*, which averaged around 20 inches in total length, and the largest known belemnite was *Megateuthis*, of the Jurassic of Europe and Asia, which measured more than 11 *feet* from tentacle tips to tail tip. The belemnites were especially abundant in Jurassic and Cretaceous seas 200 million to 140 million years ago, and they went extinct 65 million years ago at the time of the K-T event, along with the nonavian dinosaurs and many other life forms. The belemnite *Belemnitella americana* has been chosen as the state fossil of Delaware because it is especially common in the Mount Laurel Formation along the banks of Chesapeake Bay and the Delaware Canal in Delaware.

The belemnites' shape and pattern have inspired many images and tales throughout history. In the Middle Ages, the bullet-shaped belemnite fossils were called thunderbolts, thunder-arrows, or sometimes devil's fingers, because some people believed that they were hurled to the Earth during thunderstorms when lightning flashed in the sky. A person struck by lightning would be therefore killed by a thunderbolt. Folklore healers believed that stomachaches could be cured by scraping off and swallowing a little of the "thunderbolt," and in Scotland, water in which belemnites had been soaked was supposed to cure horses of worms. According to Kenneth Hsü, "Belemnites are called pen-rocks in Chinese because their cylindrical shells have a stubby and a pointed end like a Chinese writing pen." When Harold Urey wanted to test fossils to see if their isotopic composition (particularly the presence of oxygen 18) could be used to determine the temperature of ancient oceans, he chose belemnites because their original skeletons are preserved, and are not replaced by minerals as is often the case with other fossils.

SQUID

The ancient cephalopods with shells—ammonites, belemnites, and nautiloids—left abundant fossil evidence of their existence, but what of the ancestors of today's soft-bodied squids and octopuses? Some of them have hard parts too, but they are not as obvious and sturdy as "thunderbolts," corrugated ram's horns, or 10-foot-long cones. The only hard part of an octopus is its beak, and many species of living squids have hooks or claws on their tentacles, as well as a bony gladius or pen, which is a vestigial reminder that teuthids are descended from shelled ancestors. In a 1977 review of the evolution of dibranchiate (one pair of gills), cephalopods, Desmond Donovan wrote,

> The fossil record of many coleoid groups is very bad. Except for belemnites and aulacocerids [primitive belemnites], they are rare fossils, and the living fauna includes many families that are not known from the fossil record. Furthermore, there exist puzzling fossils which do not easily fit into any coherent story. Much of the descriptive literature is very old, and needs revision, and there have been few recent attempts to construct phylogenies. . . . The phylogeny of the living coleoids has to be compiled from hopelessly inadequate paleontological evidence.

In light of the dearth of fossils of squid, comparisons of fossil taxa with recent organisms can be useful. This obviously works best where the group being studied is fairly tight and coherent, and the fossil can be slotted in neatly. It works less well when most members of the group being studied are extinct, as is the case with hominid phylogeny, where all but us are gone. The absence of hard parts in fossil squids is what led Desmond Donovan to make his comment about "hopelessly inadequate paleontological evidence."

There are basically three existing views of coleoid phylogeny. The traditional view has a nautiloid as an ancestor. Naef (1921), and Morton (1958) have divided the early coleoids into two groups, the belemnites and cuttlefish on one side, and spirula, squid, and octopuses on the other. Jeletzky (1966) has the coleoids as basically a single lineage with various groups branching off sequentially: belemnites, squids, spirula, cuttlefish. He didn't know where to put the octopuses. Theo Engesser (1998) has produced a third phylogeny, which has the ammonites and coleoids as sister groups; the belemnites branched off early on and didn't give rise to anything else. The vampyromorph and octopus group branches off early on, followed by the oegopsid group. The remaining coleoids are divided into a spirula group and a myopsid group.

The Oxford Clay of England is especially rich in cephalopod fossils, with ammonites predominating and belemnites in large numbers but of more limited variety. *Belemnoteuthis,* which was probably a belemnite, was also preserved with "complete phragmocones with pro-ostraca [and] many specimens had mineralized soft parts, including the mantle, head, arms, and even ink sacs" (Page and Doyle, 1991). Other localities in the Oxford Clay yielded the probable vampyromorph *Mastigophora* and a cuttlefish called *Trachyteuthis.* Also found are the arm-hooks of *Belemnoteuthis,* sometimes indicative of the diet of Jurassic mosasaurs, ichthyosaurs, plesiosaurs, and fishes. Cephalopod fossils of the lithographic shales in Holzmaden and Solnhofen, Germany, also show many details of cephalopod anatomy, including mantle, arms, hooks, and ink sacs.

The fossils of the Loligosepiina (e.g., *Mastigophora*) bear a certain resemblance to the living *Vampyroteuthis,* a deepwater, exotic cephalopod that has eight arms and two long filaments that correspond to the feeding tentacles of squid. Usually classified between the squids and the octopuses, *Vampyroteuthis infernalis* ("vampire squid from hell") is a brownish-colored, gelatinous cephalopod with enormous blue eyes that are proportionately the largest eyes of any living animal. It has two tongue-shaped fins in the posterior portion of the mantle, and behind them, two large composite light organs. There are also numerous small light organs scattered on the underside of the mantle. "Jurassic Loligosepiina were large animals," wrote Donovan, "some of them probably a metre or more in length, which lived in shallow water. *Vampyroteuthis* is small and shows adaptations to a deep water existence. It may thus be an example of a once-successful animal surviving in the deep sea after it has been superseded elsewhere by more highly evolved forms." In her classic (1949) report on *Vampyroteuthis,* Grace Pickford wrote, "It [the gladius] most nearly resembles that of Belemnosepiidae, but cannot be placed with these forms because it is not calcified. However, there can be no doubt that it is a teuthoidean type; there is no trace of a chambered phragmocone, only a cup shaped conus, and the origin of the mantle muscle follows the margin of the shell-sac both laterally and ventrally." Because *Vampyroteuthis* has several octopod characteristics (not least, of course, its eight arms) it might suggest that the Octopoda diverged from the vampyromorphs (the Loligosepiina) (Donovan 1977).

During the middle of the Cretaceous Period, some 85 million years ago, the Western Interior Seaway stretched from the Gulf of Mexico to the Arctic Circle, including all of Saskatchewan, North and South Dakota, Kansas, Nebraska, Oklahoma, and most of Texas. A particularly rich fossiliferous area is the 600-foot-thick formation known as the Niobrara Chalk, which extends from southwestern Kansas to south-central Manitoba, but it is best exposed in northwestern Kansas, where badlands have been cut along the bluffs of the Smoky Hill River and its tributaries. The waters of the Western Interior Seaway were not particularly deep, rarely exceeding 1,000 feet, but they nevertheless provided a habitat for large squids. Many ichthyosaur fossils have been found with cephalopod hooks in their gut, suggesting that they, like most modern marine predators, fed on squid. (Not all modern squids have hooks, so their absence in fossil predator gut contents does not mean that the mosasaurs did not eat squids; it only means that there is very little hard evidence to show that they did.) It has been shown that some marine reptiles could get the bends (Rothschild and Martin, 1988), which indicates that they were deep divers, but the fishes and sharks that they fed on did not live in the depths, and certainly the diving bird *Hesperornis* (one was found in the gut of the mosasaur *Tylosaurus*) was a near-surface inhabitant. "We know," wrote Rothschild and Martin, "that the modern sperm whale has an affinity for giant squids, and similar but unrelated large squids are fairly common fossils in the same deposits that produce most of the mosasaur remains." Even in the relatively shallow Western Interior Seaway, the major deep-water prey item would probably have been cephalopods.

The fossil evidence indicates that large squids were fairly common in the waters of the Western Interior Seaway. The remains of a large squid known as *Niobrarateuthis walker* was first described by Richard Green in 1977 from a specimen that was found in Kansas and was named for the Niobrara Chalk Formation and *teuthis,* which is Greek for "squid." Several other large genera of squid have been identified from the shales of the Western Interior Seaway, including *Enchoteuthis* and *Kansasteuthis* (Miller and Walker 1968). A specimen of *Tusoteuthis* was found in the stomach of the Cretaceous fish *Cimolichthys* in the Pierre Shale of Kansas, and Nicholls and Isaak (1987) described six specimens of the same species that were found in the Pembina shales of Manitoba, dating from approximately 79 million years ago. The preserved parts are restricted to the pen, which consists of two parts: the conus, a flattened, leaflike shape; and the stemlike rhachis, which is attached to the conus like the handle of a shovel. Together the parts look very much like a short-handled paddle or broad-bladed spear. The gladius, which is also known as the pen, is all that remains of the shell of the ancestral teuthids. This chitinous structure lies on the median line of the dorsal side of squids and cuttlefishes. In fast-swimming oceanic squids the pen is a straight, nonflexible, slender rod, but in shallow-water coastal species the stem is short and the conus is spatulate and fairly wide.

The longest of the gladii of the six Manitoba *Tusoteuthis* specimens is almost 4 feet, suggesting a very large squid indeed. The gladius of *Architeuthis,* the largest living squid, is feather-shaped, and although not as sturdy as that of *Tusoteuthis,* it is almost as large (Aldrich 1991; Förch 1998). Because the gladii in *Tusoteuthis* and *Architeuthis* have both been measured at approximately 4 feet, it seems not unreasonable to postulate a similar overall body size, even though the fossils have left not the remotest suggestion of tentacle length. The largest specimen of *Architeuthis* ever measured was 57 feet long, including tentacles (Kirk 1888). But though they may be the same size, *Architeuthis* and *Tusoteuthis* are very different squid, because they inhabit very different environments. The living giant squid is known to inhabit depths of

5,000 feet, off New Zealand, for example, whereas *Tusoteuthis* could not have lived in water that was deeper than 1,000 feet, because that was the maximum depth of the Western Interior Seaway.

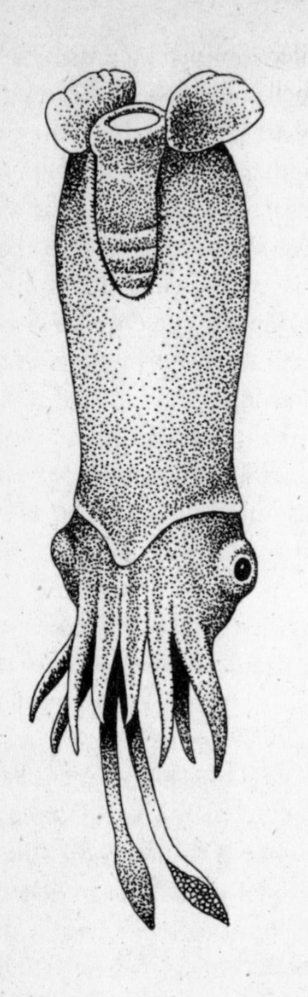

Spirula *is a small cephalopod with a coiled interior shell. The animal swims head down, and has a light organ that aims upward.*

One of the most remarkable events in the British oceanographic research vessel *Challenger*'s three-year (1873–76) collecting expedition was the unexpected appearance of *Spirula spirula,* a primitive cephalopod that was believed to have died out 50 million years earlier. Like the belemnites, it had a chambered coiled shell inside the body, but where the belemnite's phragmocone was straight, that of *Spirula* was an open planispiral, shaped like a horn coiled in a single plane without the coils touching one another. Of the living cephalopods, only *Nautilus* has a chambered, coiled shell, and it is on the outside. Almost every group of marine creatures has a "skeleton in its closet," a living fossil that existed tens or even hundreds of millions of years ago, and exists today, usually in a somewhat modified form. For the elasmobranchs (sharks), it is the six-gill and seven-gill sharks; for the fishes we have the lungfishes and the coelacanths; the crocodiles have a lineage that can be traced back unbroken, but not unmodified, for 200 million years; and in the cephalopods, there is *Spirula.*

Although Thomas Huxley was the first scientist to describe *Spirula,* most of the work on living specimens was done by Johannes Schmidt, a Danish zoologist who found the breeding grounds of eels in the Sargasso Sea. His nets, designed to catch tiny eel larvae, were also effective in capturing *Spirula.* In his 1922 announcement, Schmidt said, "Few animals have been of more interest to zoologists than the little cuttle-fish *Spirula.* Related to the extinct belemnites, and characterized by having an interior, chambered shell, it occupies an isolated position among recent species." *Spirula* is a small, muscular species that reaches a length of about 3 inches, half of which is tentacles. It is found in mesopelagic waters of tropical open oceans, where it makes daily migrations from depths of about 1,600 feet to within 300 feet of the surface (Clarke 1970). It has eight arms, and two longer, partially retractile tentacles; in life, it is whitish with reddish-brown markings. The paired fins are unusual in that they are oriented in a vertical, transverse position, affixed to the posterior of the little cephalopod like the rear tires of a car; the fins of all other cephalopods are oriented horizontally, like the wings of an airplane. Its narrow arm crown, bulging eyes, the peculiar orientation of the fins, and the presence of a coiled, internal shell make this species very different from all other cephalopods. The shell is divided into 25 to 37 gas-filled chambers and is found at the posterior end of the animal (the end away from the tentacles); its buoyancy causes the animal to orient in a tentacle-down position. Scientists on the *Galathea*'s 1950–52 round-the-world oceanographic expedition "caught many of them, and one evening in the Indian ocean we took some live ones, which we were able to watch in the aquarium in their characteristic position, bottom up" (Kramp 1956). In addition to the shell, *Spirula* is equipped with a bead-like light organ that glows with a steady yellowish-green light, making it visible from above. Upon the death of its owner, the buoyant shell floats to the surface and often washes ashore, sometimes appearing in great quantities. Collectors know these shells as "little post horns," for their resemblance to the horns used on old-fashioned horse coaches in Europe. The calcareous shell retains the phragmocone and siphuncle of its distant ancestors—whatever they might have been.

Spirula is not a belemnite, even though Schmidt said it was "related to the belemnites" and Idyll (1965) called it "one lively ghost of the belemnites." Is it an ammonite? Not according to Neale Monks (1999): "*Spirula* is a coleoid, and probably related most closely to the cuttlefish (though that is not certain). Fossil spirulids are known from the Pleistocene of the Canary Islands, the Pliocene of New Zealand, and the Miocene of Japan. The probable ancestors of *Spirula* are known as far back as the Eocene, which begin coiled like *Spirula* but straighten out later. A key difference between *Spirula* and ammonites is the way they coil. In *Spirula* the coiling is endogastric—under the animal. Ammonites, in contrast, have shells which are exogastrically coiled—that is, the shell coils over the back of the animal."

Today, cephalopods serve as an important link in the food chain for many oceanic vertebrates, and are therefore a fundamental component of life in the ocean. Phylogenetic analyses have provided some insight into their evolution, but the rise of these fascinating creatures—particularly squids—to a predominant place among the oceans' representative megafauna is still largely an enigma.

THE VERTEBRATES

Geological processes strongly influence the relative likelihood of preservation of animals from different environments. Shallow margins of the ocean and other large bodies of water, deltas, and coastal plains receive copious sediments and are subject to subsequent tectonic activity. These processes contribute to a rich fossil record. In contrast, the oceanic depths and terrestrial environments beyond the margins of large water bodies provide few vertebrate fossils. The relative position of the continents, the extent of shallow inland seas, and the amount of tectonic activity have varied throughout geological history and influence the probability of preservation and recovery of fossils from different environments during different periods of time. The fossil has the potential for documenting the rate and pattern of evolution, but its irregularities have made these data difficult to interpret.

—ROBERT CARROLL, 1988

The first vertebrates—in a sense the ancestors of all fishes, amphibians, birds, and mammals—have proved to be most elusive. In a 1997 essay entitled "They must have come from somewhere!" Keith Thomson wrote, "The problem with vertebrate origins is that the ancestors we seek lived so long ago and were so soft bodied that their remains may never be found. This leaves us with a motley collection of living tunicates, hemichordates, and cephalochordates (plus a few fossils) to sort through. As for the known invertebrate groups, it is patently fallacious to derive 'any one highly differentiated animal type from another of comparable complexity,' but that has not stopped many from trying."

Where ammonites, nautiloids, belemnites, and trilobites sometimes left calcified shells for paleontologists to find, early marine reptiles, fishes, and mammals have bones that can fossilize, and sharks leave great quantities of teeth behind, so that there is often calcified or lithified evidence of their existence. Very rarely, fossilized soft parts are found to give us a hint as to what the inside of the animal looked like. In addition, living fishes, sharks, and marine mammals allow us to learn something about the lifestyles of their ancestors by observation. The only living marine reptiles are turtles, crocodilians, a few snakes, and one lizard, the marine iguana, so our understanding of the extinct ichthyosaurs, plesiosaurs, and mosasaurs is confined to our often fragmented deductions based on their almost always fragmented fossil remains. But like virtually all other early animals, the early vertebrates left very little behind for paleontological theorists to work with.

WHAT DID IT MEAN TO HAVE A BACKBONE?

Conodonts were first described by Christian Heinrich Pander, a Russian paleontologist with a German name, who made the first report of tiny toothlike fossils in 1856 and named them conodonts, or "cone-teeth." (These first fossils, whether or not they were actually teeth, were originally called conodonts, but today the animal itself is called a conodont and the teeth are called conodont teeth.) After publication of Pander's report, paleontologists continued to find numerous tiny toothlike objects in marine limestones, but no one had any idea of what the original owner looked like or, for that matter, whether the objects were actually teeth. Whatever they were—or were called—they were very useful as "index fossils," because fossil conodonts from one rock sample differed from those of another and therefore could be used to date stratigraphic layers. Ideas about their origin ranged from the bizarre to the ludicrous and included fishes like lampreys or hagfishes, worms, parts of arthropods, pharyngeal structures of annelid worms, copulatory organs of nematodes, gill-rakers of extinct fishes, the radulae of gastropods, fossil algae, and a floating plant imaginatively named *Conodonotophyta chattanoogae*. In 1953, nearly a century after the objects were first described, paleontologist George Gaylord Simpson wrote that they "were not really teeth or jaws [and were] of unknown origin," which seemed to sum up the conodont controversy.

Though these fossils are abundant in Ordovician rocks, we know they were primarily creatures of the Silurian, 410 million to 438 million years ago, and were presumably extinct by the Late Triassic, around 250 million years ago. It is known that they inhabited only saltwater and that certain species aggregated in large numbers, but the function of the toothy apparatus is still a mystery. In 1973, William Melton and H. C. Scott were working the Bear Gulch Limestone of Montana when they found fossils of cigar-shaped creatures with a fin at one end, and believed they had found the long-sought conodont animal. The only problem was that the toothlike structures were not at the head end but in the gut, so they later realized that they had not found a conodont, but rather a creature that had *eaten* a conodont.

In early 1982, Derek Briggs and Euan Clarkson found fossils in the Carboniferous shrimp beds near Edinburgh that clearly showed the soft anatomy of the complete conodont animal. They referred them to Richard Aldridge of the University of Leicester, the leading world authority on conodonts, and in 1993, Aldridge, Briggs, et al. published "The Anatomy of Conodonts." In this review, 10 specimens were described as elongate, 21 to 55 mm (0.8 to 2.14 inches) in preserved length with a short head, a trunk with V-shaped myomeres (muscle bands), and a ray-supported dorsal fin. The head is characterized by two lobate structures, which are interpreted as hollow cartilages indicating the position of large eyes. One specimen preserves traces of possible otic (auditory) capsules and branchial structures. Ventral and immediately posterior to the eyes lies the feeding apparatus, with the ramiform elements at the anterior end. There is no evidence of tissue surrounding the apparatus, indicating incomplete preservation of ventral soft parts, at least at the anterior end of the specimens.

Because the inch-and-a-half-long, wormlike creatures with a headful of tiny teeth and a muscular flexible body had traces of a notochord (a primitive rod, the precursor of the spinal column), they might very well have been among the earliest of the vertebrates. Indeed, Aldridge, Briggs, et al. concluded that "the evidence of conodont anatomy, including the histology of the elements, indicates that conodonts are primitive vertebrates." Mark Purnell, a conodont specialist of the University of Leicester in England, built models of conodonts' hard parts and then photographed them to show what they would look like if they were squashed flat, like the fossils. He found that the places where the parts would have met were scratched and chipped, just as teeth would be, and concluded that "the evidence suggests that hard parts appeared in order to make the earliest vertebrates more efficient hunters and killers" (Abrams 1996). When the paleontologist Richard Fortey (1998) of London's Natural History Museum deciphered a mysterious fossil "plant," he discovered that it was actually the remains of a much larger conodont animal, this one about the size of a small fish, with muscle bars and fins, but also with globular bulbs at the head end that might have been eyes.

Even though the anatomy of the conodonts has revealed that they were probably vertebrates, their position in the hierarchy of vertebrate evolution is still unresolved, and, as with so many subjects paleontological, they have inspired much controversy.* In a 1998 article, Bergström et al. wrote, "The report of sensory organs for orientation (eyes or perhaps ears) in conodont animals is really important news. Much more dubious are other claimed chordates. One of them is the Burgess Shale (Middle Cambrian) *Pikaia gracilens.* Its V-shaped myomeres make it a chordate, but there is no evidence of either eyes or teeth. If this absence is real, *Pikaia* may be a relative of the modern amphioxus." John Maisey, an authority on early fishes, believes that the conodonts might be tied to early craniate evolution because the tail section appears to contain the short rods that resemble the fin rays of lancelets and fishes. If indeed the bulbs at the head were eyes, they were probably supported by cartilaginous rings, which are formed from the neural crest, a characteristic structure in vertebrates. Maisey wrote: "Conodonts may represent a group of simple craniates that diverged from the mainstream prior to the advent of more complex structures such as an

---

* When the Russian scientist A. P. Kasatkina found an inch-long (27.7 mm) chaetognath-like creature in a plankton sample collected in the Laptev Sea, north of Siberia, at a depth of approximately 2,000 meters (6,500 feet), she identified it as an "adult egg-bearing female euconodont" and named it *Panderiella viva,* a new genus and species. Her paper identified it as a living fossil, and her illustration showed a fish-shaped creature with a conglomeration of mouth parts and teeth attached at peculiar angles. Ichthyologists looking at the paper, however, recognized the specimen as a damaged gonostomatid (bristlemouth) fish, probably of the genus *Cyclothone.*

elaborate brain, an internal skeleton, and bony external plates and scales. But conodonts lacked olfactory organs and a labyrinthine organ of balance, and had no bones or scales in the skin. The main event was going on elsewhere, among creatures that peered at their environment through eyes encased by bones."

Indeed, there are those who think that conodonts might *not* be early vertebrates. In a 1989 discussion, Aldridge and Briggs wrote, "We identified similarities with two living groups: chaetognaths [arrow worms], and primitive chordates such as the lancelet, a sand-burrowing fishlike marine animal. But in the absence of definitive evidence we opted to retain a separate phylum for the animal, the Conodonta." Peter Forey and Philippe Janvier (1994) believe that the hard tissues (the "teeth") differ markedly in composition from the bone and enamel teeth of vertebrates, "so it is not at all clear that conodonts are structurally comparable to vertebrate teeth." Forey and Janvier speculated that "the conodont animal is the larval stage of an extinct agnathan." An agnathan (literally, "jawless one") is a jawless fish.

The origin of the vertebral column that defines the group is not so easily found. The larvae of tunicates possess a primitive notochord that they lose as adults, suggesting a common ancestor with the backboned animals. They are far removed from living vertebrates, however, because though they have a heart and a network of blood vessels and nerves, the distributive functions are performed by cells known as macrophages. There are more than two thousand classified species, found worldwide and at every depth. Most tunicates (the name is derived from the Latin *tunic,* referring to their flexible protective covering) attach themselves to the bottom or some other object, but the sea grapes (*Molgula*) are free-floating.

Tiny fossil fragments from what was then believed to represent the earliest known vertebrate were found in the Early Ordovician Valhalfonna Formation in Spitsbergen, Norway, by Bockelie and Fortey (1978). When they realized that these fragments did not resemble other heterostracan fishes such as *Astraspis* or *Eryptichius,* they named the new genus *Anatolepis* ("body scales"). Their announcement pushed back the date of the first heterostracan fishes (and therefore some of the first vertebrates) to around 470 million years ago, but only a couple of years later, John Repetski of the U.S. Geological Survey drove the date another 40 million years into the past. In the Late Cambrian (520 million to 505 million years ago) of Crook County, Wyoming, he found tiny fossils (the largest specimen is less than one tenth of an inch long) consisting of partial cylindrical shapes just like the Spitsbergen fossils, covered with what look like rhomboidal scales. It was another *Anatolepis,* and in his 1978 paper, Repetski says that he has subsequently recovered specimens from Texas, Oklahoma, New York, Utah, Washington, and Greenland. The minute *Anatolepis* fossils were identified as having been a part of an early jawless fish because they are composed of apatite, which is the substance of bone and is not found in any nonvertebrates, those who would dispute Repetski's identification claimed that similar "scales" have been found in *rocks,* and that this piece might have been from a lobsterlike arthropod. In 1996, however, Repetski and his colleagues showed that the apatite in *Anatolepis* contains dentine, a substance unique to vertebrates. (It is what your teeth are made of.) Repetski (1978) wrote: "The presence of these fish in the stratigraphic units listed above . . . all of which are of undoubted marine origin, is also strong evidence for a marine habitat for the earliest vertebrates."

In the Burgess Shale, which is 530 million years old, Charles Walcott found an inch-and-a-half-long fossil that he first described as a polychaete worm. He named it *Pikaia gracilens,* after nearby Mt. Pika and the fossil's delicate curvature (*gracilis* means "thin" or "slender"). Like the annelid worms, it appeared to have regular segments, but when Simon Conway Morris examined it closely 50 years later, he saw that it had a notochord. This made it the earliest known representative of the Chordata, the phylum that includes the vertebrates. The tail was expanded into a fin, suggesting that *Pikaia* swam just above the sea floor, probably with undulations of its body. "It is not a vertebrate," wrote Briggs, Erwin, and Collier (1994), "but it is similar to more primitive chordates like the living cephalochordate *Branchiostoma,* commonly known as the lancelet." In *Prehistoric Life* David Norman wrote, "*Pikaia* is the one solid link to the Cambrian, and even this species is rather frustratingly similar to modern *Branchiostoma.* How such relatively insignificant little fossils survived the extinctions that affected so many creatures in the Late Cambrian is uncertain, but survive they certainly did, and ultimately spawned groups that would dominate first the seas, then the land, and finally, the skies."

In 1995, the fossil *Yunnanozoon* from the Chengjiang Formation was described by Chen et al. Like *Pikaia,* it has a notochord and several typically vertebrate features, such as muscle blocks (called myomeres), a digestive tract, and a vascular system. But it has no true vertebrae, so we cannot classify it with the true vertebrates. Its position on the genealogy of vertebrates is unclear, and some researchers believe it should be placed among the relatives of early chordates known as hemichordates. The following year, Shu, Morris, and Zhang (1996) described *Cathaymyrus diadexus,* the "Chinese eel of good fortune," also found in the productive Changjiang shales, but unlike many of the other fossils, for which hundreds of specimens have been unearthed, there was only a single specimen of the 2-inch-long eel-shaped creature. *Cathaymyrus* is also believed to be a fossil relative of modern lancelets. The early appearance of these cephalochordates strengthens the case for their nomination as ancestors of true vertebrates.

And so to the most primitive of living chordates, the tiny *Amphioxus* (or *Branchiostoma*), also known as the lancelet. Less than 2 inches long, it is not unlike *Pikaia* in size and shape, but where the early chordate of the Burgess Shale had a pair of short tentacles at the head end and an obvious tail, *Amphioxus* has a buccal (mouth) opening at one end and only a small tail fin at the other. It resembles a tiny eel or a colorless anchovy filet. In some areas, such as Discovery Bay, Jamaica, there are reports of more than 5,000 per square meter of sand. In some parts of Asia, lancelets are important food items and are commercially harvested.

With about 25 species inhabiting shallow tropical and temperate oceans, the cephalochordates are a very small but important branch of the animal kingdom. They have all the typical vertebrate features, including a notochord. The musculature of the body is divided up into V-shaped myomeres, and there is a postanal tail. The gill slits—which number more than 100—are enclosed by folds of the body wall, the metapleural folds, to form a body cavity known as the atrium. Food particles in the water are trapped by mucus, after which water passes through the slits and out of the atrium through the atriopere, located toward the posterior end. The rest of the digestive system is fairly simple: a pouch, or hepatic caecum, secretes digestive enzymes, and actual digestion takes place in a specialized part of the intestine known as the iliocolonic ring. Cephalochordates also have a well-developed circulatory system and a simple excretory system. The sexes are separate, and both males and females have multiple paired gonads. Eggs are fertilized externally, and develop into free-swimming, fishlike larvae.

On the other hand, cephalochordates lack features found in most or all true vertebrates: the brain is very small and poorly developed, sense organs are also poorly developed, and there are no true vertebrae. Since they have no hard parts, their fossil record is extremely sparse. However, fossil cephalochordates have been found in very old rocks indeed, predating the origin of the vertebrates. The Middle Cambrian Burgess Shale Formation of British Columbia has yielded a few fossils of *Pikaia,* which appears to be a cephalochordate (although the fossils are still being studied). More recently, *Yunnanozoon lividum,* from the 525-million-year-old Early Cambrian of South China, was reported to be a cephalochordate, the earliest known (Chen et al., 1995). These inch-long fossils show that the chordate lineage appeared very early in the known history of the animal kingdom, and they strengthen the case for an origin of true vertebrates from a cephalochordate-like ancestor.

THE JAWLESS FISHES

Until very recently, the fossil record for the earliest vertebrates, the jawless fishes (class Agnatha), was good as far back as the Early Silurian period, 430 million years ago, but anything earlier was either nonexistent or questionable. The position of the conodonts is still equivocal, and the place of creatures like the Burgess Shale *Pikaia* are also unresolved. The *Pikaia*-like chordate named *Yunnanozoon* was described in 1996 by Shu, Morris, and Zhang from the fossil-rich shales of Chengjiang, in southern China, but it was not as dramatic as the discovery by Shu et al. (1999) of the most convincing Early Cambrian vertebrates so far, which set the origin of vertebrates back another 115 million years. The two new animals have been named *Haikouichthys* (from Haikou, a locality near Kunming, and *ichthyos,* the Greek word for "fish"), and *Myllokunmingia* (from *myllos,* another Greek word for "fish," and Kunming, the capital of Yunnan Province). Because the Chengjiang stones preserve not only the bones but also the soft anatomy, it was seen that the fossils had clearly preserved muscle blocks (myomeres), gill pouches, a heart, fin supports, and what may have been a cartilaginous skull. Both new animals are about 2 inches long, and both resemble modern agnathans, or jawless fishes; *Myllokunmingia* resembles a hagfish, and *Haikouichthys* bears a faint resemblance to a lamprey. In summerizing the implications of these finds, Shu and his colleagues wrote, "The discovery of these Lower Cambrian vertebrates has implications for the likely timing of chordate evolution.

The occurrence of *Myllokunmingia* and *Haikouichthys* in Chengjiang . . . shows that even more primitive hagfish-like vertebrates had almost certainly evolved by the beginning of the Lower Cambrian. Such a find may help to elucidate the transition between the cephalochordates and the first vertebrates." Or, as the *New York Times* (of November 4, 1999) put it, "Scientists have unearthed two half-billion-year-old fossils of fishlike creatures that could be the earliest known vertebrates."

And then there were more. In the same Chengjiang site, Chen, Huang, and Li (1999) found another "craniate-like chordate" that they named *Haikouella lanceolata.* Even though they had no skull, the scientists from the Nanjing Institute of Paleontology believe that *Haikouella*'s relatively large brain and the suggestion of eyes place it at the very beginning of the evolution of a backbone, and thus make it prominent in the history of vertebrates. The fine-grained rocks of Chengjiang preserved the soft tissue of more than 300 specimens, of which 30 were complete. Two inches long like *Myllokunmingia* and *Haikouichthys,* the fossils showed some of the hallmarks of chordates, including a notochord and an arched back that appears to contain muscle segments. As quoted in *Science* (Enserink 1999), Nicholas Holland of the University of California at San Diego said, "There's no question that these things are chordates." "This puts an end to the discussions of *Yunnanozoon,*" said Philippe Janvier of the Musée Nationale d'Histoire Naturelle in Paris.

Until the discovery of the Chengjiang fossils, the earliest unquestioned vertebrates were the ostracoderms ("shell-skins"), armored, jawless fishes whose remains have been dated from the Late Cambrian, about 480 million years ago. They were also boneless—that is, if one expects to find bones inside the body. The internal skeleton, including the notochord, was made of cartilage or gristle, and the only bones were found on the outside, in the form of bony head shields and/or small bony scales. Since cartilage, like flesh and viscera, decays rapidly, only these ossified parts have been left as fossil evidence of the existence of these small early fishes. With no jaws with which to seize their prey, the ostracoderms were restricted to grubbing along the bottom. Another ancient group is the cyclostomes, also jawless and represented today by the hagfishes and lampreys, considered among the most primitive of living vertebrates. Ancient oceans were populated by what sometimes look like failed attempts at fish design, but were of course, enduring and successful developments in the long history of marine vertebrate evolution.

Few except for invertebrate paleontologists would mourn the passing of the eurypterids, formidable scorpionlike predators that must have been the scourge of the Paleozoic seas. Arising in the Ordovician, about 500 million years ago, and disappearing by the Permian, 250 millions years ago, they were the largest arthropods that ever lived. They had blunt, narrow heads, tapering and segmented abdomens, and exoskeletons that were ornamented with knobs, scales and ridges. The tail, called the telson, was either rounded or long and spine-shaped, and it is because of this long tail and the appendage at the tip, poisonous in some species, that they have been called sea scorpions. They had a chitinous exoskeleton and four or five pairs of walking legs, a pair of sharp, pincerlike appendages called chelicerae, a pair of compound eyes, and a pair of simple eyes. (A scorpion has only four pairs of legs and a pair of powerful

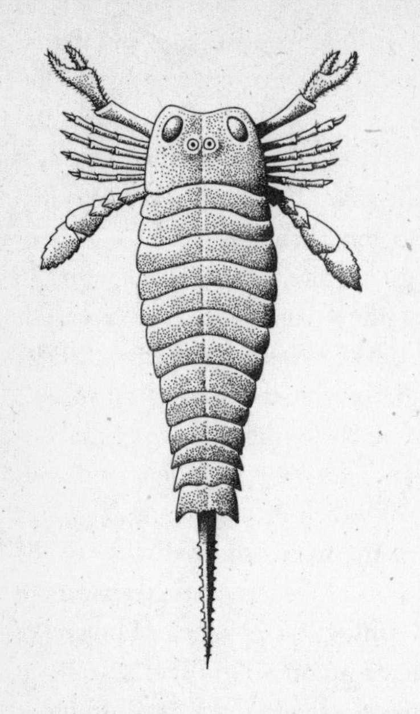

Arising in the Ordovician, about 500 million years ago, and disappearing by the Permian, 250 million years ago, the scorpionlike eurypterids were the largest arthropods that ever lived. Pterygotus, *which reached a length of 7 feet, had eight legs, a pair of swimming paddles, and a pair of grasping chelicerae, which it probably used to prey on the early jawless fishes.*

pincers.) Eurypterids, spiders, scorpions, and horseshoe crabs are known collectively as chelicerates, after the pincerlike "biting claws." Eurypterids are among the earliest animals in which the sexes can be differentiated: males had clasping organs on the underside; females had genital organs that fitted over these claspers. *Stylonurus* reached a length of 4 feet and had eight long legs like a spider; *Pterygotus,* which reached a length of 7 feet, had eight legs, a pair of swimming paddles, and a pair of grasping chelicerae, which it probably used to prey on the early jawless fishes. *Pterygotus* had a pair of large eyes embedded in the upper surface of its head end (the prosoma), which suggests that it hunted animals that swam above it. Eurypterid fossils are known from all continents, but because they are particularly plentiful in New York, *Eurypterus remipes* (maximum length: 8 inches) was named the state fossil in 1984.

Because the fossils of early jawless fish represent only tiny scraps of what is presumed to be the outer scaly integument, we have no idea what the animal looked like. The fishlike *Arandapsis,* however, was represented by nearly complete remains, and we therefore have a very good idea of what it looked like. It was small, not more than 6 inches in length, with a streamlined, flattened shape. The front of the body was encased in a bony shield, composed of upper and lower elements, and 14 or more paired plates covered the gills. The tiny eyes were located at the front of the head, like automobile headlights, and there were two small openings at the top of the head that may have been light-sensory organs. The jawless mouth was on the underside of the head, indicating that it fed on or near the bottom. Inside the mouth were small moveable plates. Lacking fins, *Arandapsis*'s swimming must have been erratic, rather like a tadpole's. The fossils, which have been dated from the Ordovician (470 million years ago), were found in 1959 near Alice Springs in the heart of the Australian continent; the name is derived from the Aboriginal Aranda tribe and the word *aspis,* Greek for "shield." A similar but slightly younger form was found in Bolivia in 1986 and was named *Sacabambaspis* for the town of Sacabambilla; at 450 millions years old, it is the oldest complete fossil fish known. North America is represented by *Astraspis* and *Eryptychius,* both found in Colorado by Charles Walcott in the nineteenth century. He found only bone fragments, but a complete specimen of *Astraspis* that was recently unearthed showed that the entire fish, including the tail, was covered in platelike bony scales.

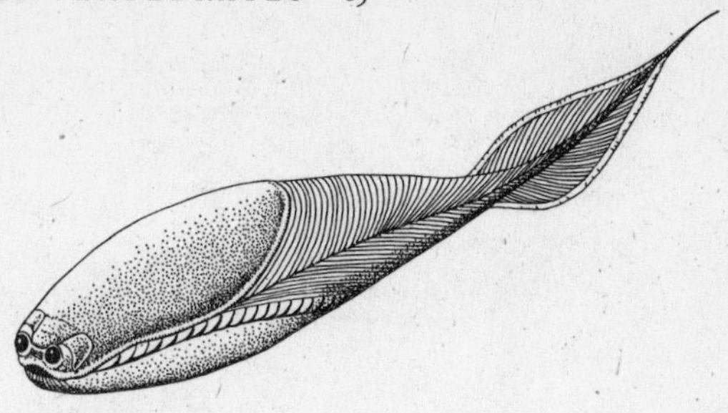

Sacabambaspis *was a jawless, armored fish that lived about 450 million years ago in seas that covered what is now Bolivia.*

Another 100 million years later, the jawless fishes radiated into more widely varying forms quite unlike the eel-like lampreys and hagfishes, the only living jawless fishes. The osteostracans ("bony shells") included the 8-inch-long *Hemicyclaspis* and also *Cephalaspis,* from Early Devonian England, with an odd-shaped head that Benton (1996) described as "shaped rather like the toe of a boot." On either side of the "boot," where the sole would meet the upper, were sensory organs that may have served as pressure sensors, or might even have enabled *Hemicyclaspis* to sense electrical fields, the way sharks do today. They had a single nostril like the cyclostomes (lampreys and hagfishes), but the head was covered with a bony shield, and they were bottom-dwellers, with the mouth on the underside. Their small, prominent eyes were set high on the head, and they had a long upper tail lobe and a rearward dorsal fin, but the most significant improvement over their predecessors was paired pectoral fins, which all modern fishes now have.

By the Siluriam Period (438 million to 410 million years ago), the heterostracans ("different shells") had arrived; they were as jawless as their predecessors, but their body shields were often segmented, and they had a single gill opening on each side of the head. They came in many shapes and sizes, and though most were small, some were almost 4 feet long. In the Early Devonian—which would come to be known as the Age of Fishes—the heterostracans included the pterapsids (*pteros* is Greek for "wing"), which had pointed spines extending back from the head shield instead of paired fins. The jawless *Anglaspis* were completely covered in plates and scales and had no fins, so it must have moved by wriggling its tail from side to side. Lacking jaws, it probably shoveled in detritus from the bottom. Another heterostracan was *Doryapsis,* which had an apparently movable rostrum resembling that of a sawfish, and horn formations that extended outward from the shield.

The thelodont ("nipple tooth") fishes were named for the shape of the scales, which were concave but rose to a point in the middle. These toothlike scales were composed of dentine and had a pulp cavity within, very much like the scales of sharks, which are known as "dermal denticles." The thelodonts also had multiple gill openings, but because they had a large stomach, absent in lampreys, they are considered more advanced than the anaspids, the forerunners of the lampreys and hagfishes. They were still jawless, but the large stomach suggests feeding on organisms suspended in the water, rather than bottom-grubbing. Found in Early Devonian deposits at scattered locations around the world, *Phlebolepis* was about 3 inches long,

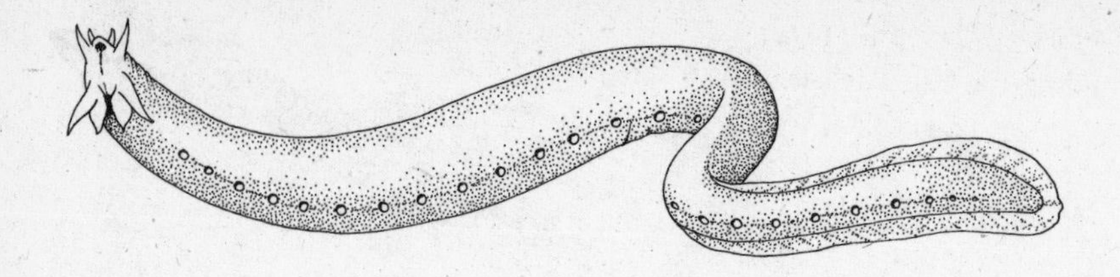

The hagfish, a primitive creature with no eyes and no jaws, makes a living by eating carrion, which it eats from the inside out. Hags are also capable of tying themselves into knots, and they can emit quarts of slime from special glands on their sides.

with a broad snout and an eye on either side of its wide mouth. It had dorsal and ventral fins, and a tail fin with a lower lobe that was longer than the upper. There were eight small gill openings, and no sign of head shields at all. Recently discovered thelodonts from Canada show a decidedly more fishlike form, with a laterally compressed body and a forked tail that consists of many backward-pointing spikes. Long (1995) has written, "The new finds from Canada support a suggestion that the thelodonts were far more advanced than previously thought, and may be a closer link to jawed fishes than other agnathans."

The anaspids were shieldless little fishes that did not exceed 6 inches in length. They had simple fins running the dorsal and ventral length of the body, and a definite tail with a lobe that pointed downward. Their bodies were covered with small, overlapping scales, and bony plates covered the head. The small, round mouth was located at the front, and they breathed through a row of holes along the side of the head; the species known as *Legendrelepis* had as many as 30 pairs of gills, the highest number known for any vertebrate. All anaspid fossils have been found on the ancient Euramerican continent, from locations like Canada, Spitsbergen, and Russia.

The living lampreys and hagfishes are considered primitive because we can trace their lineages as far back as the Carboniferous, 340 million years ago. Because they are living representatives of such an ancient line, they are of great interest to paleontologists and some ichthyologists, but because of their repugnant appearance and habits, they are reviled by almost everyone else. The hagfishes (families Myxinidae and Eptatredidae) are eel-shaped vertebrates with a cartilaginous skeleton, several barbels on the end of the snout, and two pairs of toothlike rasps on the top of a tonguelike projection. Eels are actually fishes, with a proper mouth, jaws, eyes, and fins. Hags have a single nostril, and degenerated eyes that are buried under the skin. There is a tail fin, but no other paired fins, and no jaws or bones. They have a typical vertebrate heart and three auxiliary hearts: one to pump blood to the liver, one in the head, and a tiny one in the region of the tail. About 40 species, ranging from 20 to 30 inches in length, are found in all the cold, deep waters of the world. Unlike the lampreys, which attack living fishes, hagfishes are scavengers, remaining buried until some sort of carrion drifts down, and then emerging to eat. From a series of glands along the side, hagfishes can emit gallons of nauseating, toxic slime, accounting for their com-

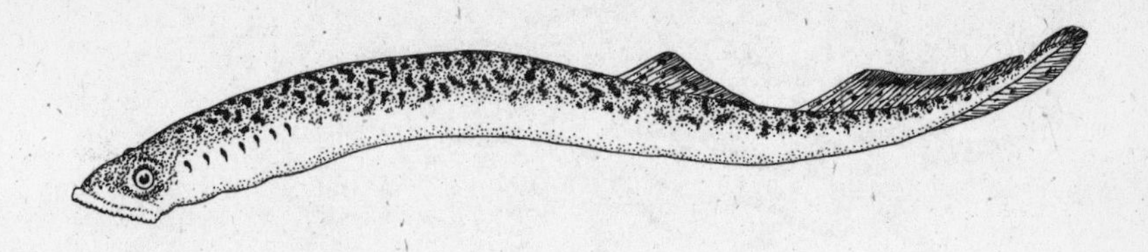

Lampreys attach themselves to living fishes with their suckerlike mouths and eat their way in. Along with the hagfishes, they are the only remaining representatives of the early jawless fishes (aguatha) that dominated the seas 400 million years ago.

mon name, "slime eels."* They can tie themselves into knots to provide leverage for pushing themselves into a dead fish, which they bore into and eat from the inside out. The so-called eel-skin leather, currently popular for expensive belts, wallets, and briefcases, is actually the skin of hagfishes that have been collected in New England waters and shipped to Korea for processing. After removing the skins and sending them to China to be made into leather goods, the Koreans eat the hagfish meat, consuming nearly 5 million pounds each year.

Lampreys (Petromyzontidae) are fishlike vertebrates that are placed with the hagfishes in the class Agnatha. They have a well-developed eyes (unlike hagfishes, which have none), a single nostril on top of the head, and seven gill openings on each side. They have no bones, no jaws, and no paired fins. Parasitic predators on fishes, they are equipped with a mouth that consists of a round, horny-toothed sucker-disk. Lamprey eggs hatch into burrowing larvae called ammocoetes that are completely unlike the adults they will become. Before the sucker-disk develops, the ammocoetes feed on suspended food particles that they capture on mucus-covered cilia in the mouth. Of the 22-odd species of lampreys, there are some that spend their entire lives in fresh water. In the early nineteenth century, when the Welland Canal was built to allow ships to sail to Lake Ontario into Lake Erie, sea lampreys (*Petromyzon marinus*) invaded all the Great Lakes and reduced the resident populations of trout to nearly zero. Attempts to restock the lakes have failed, and the lampreys have established themselves as permanent freshwater residents.

THE LUNGFISHES

In the known history of fishes, there are some deep ancestral types whose successors have survived into the present, yet still closely resemble their ancient predecessors.

* In the scholarly *Fishes of the Western North Atlantic*, the ichthyologists Henry Bigelow and William Schroeder (1948) wrote, "The Hag being of no value itself, is only a nuisance to the fishermen because of its habit of damaging better fish, and a loathsome one, owing to its ability to discharge slime from its mucous sacs all out of proportion to its size. One Hag, it is said, can fill a two-gallon bucket, and we think this is no exaggeration."

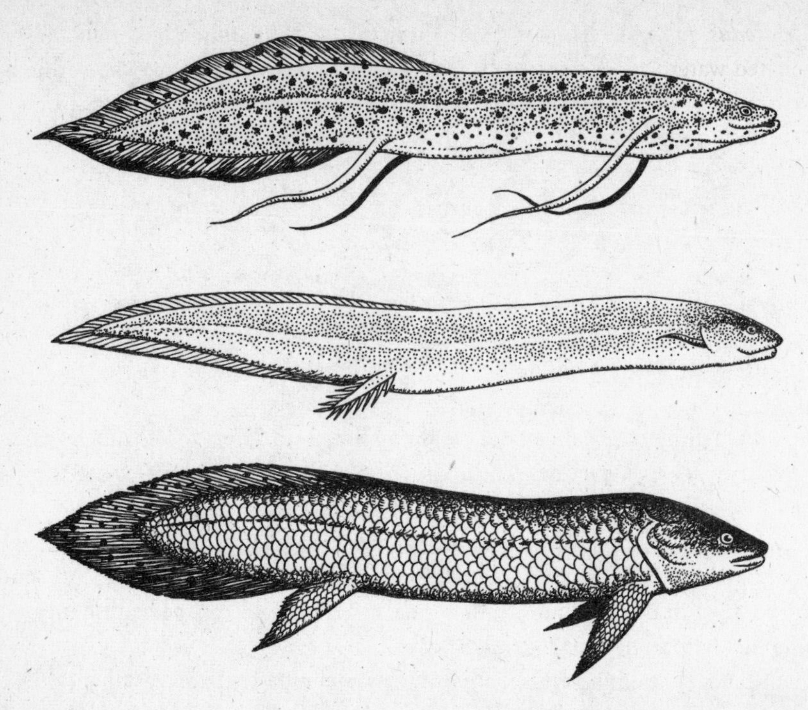

These are three species of lungfishes alive today: Protopterus, *the African lungfish (top);* Lepidosiren, *the South American (middle); and* Neoceratodus, *the Australian version (bottom).*

This virtually unbroken line of descent means the lungfish substantially antedates such long-extinct creatures as ichthyosaurs and mosasaurs.

The first lungfishes, dating back some 400 million years, lived in the sea, but by 340 million years ago, they became inhabitants of freshwater environments, and they have remained there ever since. What did the lungfishes do that enabled them to survive almost unchanged for 400 million years? For one thing, they learned to survive conditions so adverse that any other kind of fish would die. Some of the primitive lungfishes have been found encapsulated in fossilized mud, suggesting that they built burrows. They had two equal-sized dorsal fins and long paired pectoral and pelvic fins. Their mouths were equipped with powerful jaws containing hard, ridged plates that they used to grind or rasp their food. Early Devonian lungfishes had large, heavy scales that were covered with an enamel-like substance called cosmine, but this disappeared in later species to be replaced by thinner, rounded scales. Although lungfishes are now confined to the fresh waters of the Amazon, West and Central Africa, and the Mary and Burnett rivers of northern Queensland, whole and partial fossil specimens have been found in rocks from Greenland to Antarctica and Australia. The lungfishes represent an important step in the progression from water-dwelling to terrestrial vertebrates, but where they fit in is a matter of some uncertainty.

As the name suggests, lungfishes are actually fishes with lungs, and if they are kept moist, they can live out of water. The most primitive, the Australian lungfish

(*Neoceratodus forsteri*), has only one lung as well as functional gills. In well-oxygenated water it does not need to come to the surface, and since its riverine habitat does not normally dry up in the summer, it has no need to estivate (spend the summer in a dormant state) the way the South American and African lungfishes do. *Neoceratodus* retains many of the features of its Devonian ancestors, including large scales and lobed, paired fins. It is a carnivorous fish, feeding on frogs and small fishes, and can reach a length of 6 feet and a weight of 80 pounds. The fossil record of the Australian lungfish stretches back more than 100 million years, making it, in the words of John Long (1995), "the most enduring species of vertebrate on Earth."

When the Austrian naturalist Johann Natterer found the first living lungfish at the mouth of the Amazon in 1836, he could not decide whether it was a fish or an amphibian. It looked sort of fishlike, but it had only threadlike fins, and, most significant, it clearly had lungs, which no fish is supposed to have. The specimens were sent to the Natural History Museum in Vienna, where the curator, Leopold Fitzinger, announced that it was a reptile. But when the great British anatomist Richard Owen dissected a specimen from the White Nile in 1867, he concluded that it was a fish, and, unaware that the Amazonian specimen had been named *Lepidosiren,* he named the African version *Protopterus.* Upon examination, Owen saw that it had ribbed plates instead of teeth, exactly like some fossils of a fish that had lived in the Triassic, 220 million years earlier. Because the nostrils were not connected to the mouth, *Lepidosiren* could only smell with its nasal passages, and not breathe through them like mammals and reptiles, so it had to be a fish. In 1871, Albert Günther of the British Museum described the first Australian specimen, and named it *Neoceratodus.*

The African and South American species are placed in the same family, Lepidosirenidae, and it is believed that they were differentiated when the South American continent pulled away from Africa. (There are actually four similar African species, all in the genus *Protopterus.*) Both types have small, embedded scales and filamentous paired fins, and are eel-like in shape, whereas the Australian species looks more like a typical fish. They have two lungs, and must breathe air to survive—if held under water they drown. During the dry months, the African lungfish, *Protopterus,* digs a burrow in the mud and, after secreting a mucous covering, curls up inside with its tail over its eyes. The drying mud becomes rock-hard, and the fishes, which always orient head-up, breathe through a small, tubelike opening near the surface. Unlike hibernating animals, which use stored fat for nourishment, estivating lungfishes absorb muscle tissue. They can survive for as long as four years in this state of suspended animation, but in the wild they usually remain incarcerated for only a few months, until the rains come.

A recent study by John Power and colleagues of Flinders University in Adelaide (1999) has shown that the lungs of the Australian lungfish, the most primitive member of the lungfish family and one that relies on its gills more than the other two species contain cells that reduce the surface tension on the lung's internal surfaces, preventing the lungs' surface from sticking together when collapsed. Human beings, though only distantly related to lungfishes, have similar proteins in their lung tissue, and Power said in a 1999 *New Scientist* interview, "These proteins must have been conserved for 300 million years, suggesting that all air-breathing vertebrates use the same mechanism to keep their lungs open."

Ancestral lungfishes and other vertebrates developed the ability to breathe air independently, and so lungfishes are not considered ancestral to the amphibians. They are believed to have developed the ability to estivate in response to the drying-out of the ponds and streams in which they lived. With the coelacanths, lungfishes belong to a subclass known as the Sarcopterygii (meaning "flesh fin"), or the lobe-finned fishes.

THE COELACANTHS

When it was learned that the presumed extinct fish the coelacanth was alive instead of only a fossil, as everyone believed it to be, the world of vertebrate paleontology, not to mention ichthyology, was turned upside down. Previously it had been easy to separate the extinct fishes from the living species, because we knew about the extinct ones from the fossil evidence, whereas the living ones could be observed swimming around. With the discovery that a coelacanth was alive, however, it was as if the fresh carcass of a velociraptor had suddenly appeared in a butcher's shop.

On December 22, 1938, fishermen from the South African city of East London hauled in a 5-foot-long fish that was steely-blue in color, with large bony scales, paired fins that appeared to be on leglike stalks, and a peculiar little extra tail fin. It was examined on the deck of the fishing boat *Nerine* by Marjorie Courtenay-Latimer, a museum curator, who could not identify it, so she contacted J. L. B. Smith, a professor of chemistry at Rhodes University at Grahamstown and an amateur ichthyologist. Smith correctly verified it as a relative of a lobe-finned fish known as *Macropoma,* which had been extinct for about 70 million years. He named it *Latime-*

The coelacanth (pronounced "seel-a-canth") is the sole survivor of a line of bony fishes that were believed to have gone extinct 75 million years ago—until one was caught by fishermen off South Africa in 1938. Since that discovery, many more have been found off South Africa and the Comoro Islands, and in 1999, a completely unexpected population was found in Indonesian waters, some 1,300 miles from the original discovery.

ria chalumnae, after Miss Latimer and the Chalumna River, near which it was found. The second coelacanth, found in 1952, lacked the first dorsal fin and the epicaudal lobe (a small central organ) of the tail, so Smith decided it was a new species and named it *Malania anjouanae,* after D. F. Malan, the prime minister of South Africa, and Anjouan, one of the Comoro Islands. Later examination revealed it to be *Latimeria,* the fins missing from an old injury.

After the second coelacanth, many more were caught off the Comoro Islands, between mainland Mozambique and northern Madagascar. Curiously, almost every one was caught off the islands of Grand Comore and Anjouan, and none off the other islands of Mohéli and Mayotte. Local fishermen had often unintentionally caught them, usually while fishing for the oilfish (*Ruvettus pretiosus*), but since coelacanths were considered inedible, they were usually thrown back. More recently, between 150 and 200 coelacanths have been caught, and most of them have ended up in museum collections around the world. What was once believed to be a stable population of about 650 animals is now thought to number no more than 300, and this rare and zoologically significant creature is considered close to extinction . . . again.

The earliest fossil coelacanths appeared in Devonian rocks dating from 400 million to 350 million years ago, and although they flourished for the next 325 million years, they were believed to have become extinct around 75 million years ago, around the time of the disappearance of the terrestrial dinosaurs. The name coelacanth means "hollow spine," in reference to the fin rays supporting the tail, and was originally used in 1839 by Louis Agassiz to describe the fossil species. The early coelacanths were stocky in form, with relatively thin scales and a three-pronged tail. The dorsal and anal fins were fanlike, and the paired fins had the diagnostic stalks, or "lobes." The first coelacanths, like *Rhadoderma* of the North American Carboniferous, had a sharklike tail with a longer upper lobe, but most species later developed the extra-lobed tail that can be seen in *Latimeria.* The early coelacanths inhabited fresh water, but they later colonized brackish and saltwater habitats. *Latimeria* has prominent teeth in its upper and lower jaws, but early coelacanths were toothless, and their diet is unknown. Like many other fishes, coelacanths flourished during the Devonian Period, and fossils of 15 to 20 genera have been found, including the very early *Diplocercides, Dictyonosteus,* and *Miguashaia.* Some, like *Axelrodichthys,* were no more than 2 feet long, but there were some monster coelacanths, such as *Mawsonia,* which reached a length of 12 feet. There are minor points of difference between the living *Latimeria* and the extinct *Macropoma* (Forey 1984). The living coelacanth is neither a "missing link" nor a "living fossil," though both terms are often used to describe it. It is a member of a group originally recognized on the basis of fossils and only recently found to be still in existence.

The Bear Gulch Limestone of Montana, most famous for the dizzying variety of Paleozoic chondrichthyans uncovered there, principally by Richard Lund (see pp. 113–25), has also yielded a surprising number of coelacanths, about which Lund and Lund (1985) wrote, "The osteichthyan order Coelacanthiformes includes fish ranging in age from late Devonian to recent, which differ from each other in proportions or relatively small details of osteology. Essentially unchanged and essentially similar in all aspects are the basic body shape and fin disposition, or locomotor system, and the basic head and mouth shape, or feeding system. . . . The coelacanths, therefore, are the longest surviving group of virtually unchanged vertebrates known." Between 1969 and 1979, 177 coelacanth specimens were collected from the Bear Gulch site, which has the largest representation of coelacanths in the world. Fifteen percent of all the fish specimens collected there were coelacanths, including five new species. They differ in some details and proportions, and from the Lunds' analysis, we can learn where and how each species lived in the bay that became the Bear Gulch Limestone. The geology of the limestone indicates "a patchy environment consisting of floating algae, fringes of sponge thickets, and areas of open water within a narrow, shallow basin upon a very soft, fine substrate" (Lund and Lund 1985).

Many of the Bear Gulch fossils are exquisitely preserved, showing complete body outlines, fin shapes, and even coloration. Four of the five Bear Gulch coelacanths are built along the *Latimeria* plan: a long body, lobed pectoral and pelvic fins, and a broad, fanlike tail with a small extension. The fifth Bear Gulch coelacanth, *Allenypterus montanus,* is not rounded in cross-section like all other known coelacanths, but compressed, with a short, high head, and a tail that differs from that of all other coelacanths in that it just tapers to a point. *Allenypterus* was probably a slow swimmer that lived in sheltered weedy habitats, and its downward directed mouth suggests that it was a bottom feeder. When they were first discovered, the other four Bear Gulch species were given nicknames by Lund and his workers before they were officially named, and "Big Head" became *Polyosteorhynchus simplex*; "Bushy Tail" was *Lochmoceras aciculodontus*; "Short Stubby" was named *Hadronector donbairdi*; and "Long Body" came to be known officially as *Caridosuctor populosum.* ("Long Body" was by far the most abundant fish of any kind in the Bear Gulch Limestone Formation and was found in every habitat in equal abundance, making it a "generalist" in habitat preference.) "Tail outlines of all coelacanths," wrote Lund and Lund in 1985, "are high-acceleration-high-drag profiles indicative of minimal efficiency in sustained cruising, with the squarest and stiffest tails being the best suited for short bursts of speed. . . . The coelacanthiform swimming mode in general is best adapted for stalking or lunging at prey rather than active pursuit." The intercranial hinge of the skull, one of the characteristic features of coelacanths, present in *Latimeria* and all the Bear Gulch species, allows for a wider gape and enables the coelacanths to employ a type of suction feeding that has been postulated for *Latimeria* and its extinct relatives. Like the chambered nautilus, *Latimeria chalumnae* is the sole survivor of a group that became extinct millions of years ago and whose relatives are known only from fossils.

The discovery of a living coelacanth off South Africa in 1938 was one of the most surprising and rewarding events in recent zoological history, because it has given us an opportunity to observe and examine a living creature that was previously available only as a fossil. As Keith Thomson wrote (1991):

> We cannot take a perch or a cod as a model for understanding its early Devonian ancestors. But *Latimeria* looks very much like the Devonian *Diplo-*

> *cercides* or *Nesides*. Therefore, by studying *Latimeria* in detail, alongside the three lungfishes and in conjunction with the fossils and physical evidence that the rocks themselves provide about the ancient environments in which they lived, we might be able to reconstruct a lot about the biology of these long-distant Devonian forms. And not just the evolution of the skeleton, but the blood, the liver, how they breathed, how they reproduced, how they fed and swam; their whole biology.

All lobe-finned fishes had a vertebral column that consisted of bony rings around a stiff, hollow, fluid-filled notochord, a primitive backbone that is characteristic of the coelacanths. The heart is merely an S-shaped tube, simpler than that of any other living fish, and the brain is also tiny, occupying only 1.5 percent of the cranial cavity. The remainder of the braincase is filled with fats and oils. Coelacanths have paired nasal sacs, each of which contains papillae that increase the surface area, and a "rostral organ" that is located above the nasal sacs and is separate from them. Through three paired tubes this unusual organ is open to the surrounding water, and it is filled with a gelatinous substance. It is believed that the rostral organ is part of the fish's sensory system and may be used to detect weak electrical fields. Like that of sharks, the coelacanth's skeleton is made of cartilage, but its skull is bone, with an unusual hinge running across the top. Powerful jaw muscles indicate that this slow-swimming predator uses its short, sharp teeth for catching smaller fishes and squid. There are no true lungs, but the coelacanth has a fat-filled swim bladder that contributes to its buoyancy. From a pregnant female caught in 1962 it could be seen that the coelacanth is ovoviviparous, which means that its eggs—which are huge, approximately the size of an orange, the largest eggs of any living fish—hatch inside the body, and the young are born alive.

In addition to examining complete specimens, biologists have also entered the realm of the living coelacanth and filmed it in action, providing a heretofore unavailable view of the life of a Devonian fish. Hans Fricke, a physiologist from the Max Planck Institute in Germany, descended in a submersible off the Comoros, and on January 17, 1987, he became the first person to film a coelacanth swimming in its natural habitat. Although they are probably capable of short bursts of speed, coelacanths seem to spend most of their time hovering near the bottom, with their fins flared. According to Fricke, the extra fin at the tip of the tail is flicked back and forth "like a metronome," and acts as a sort of trim tab. The lobed fins encouraged biologists to believe that *Latimeria* might spend some time on the bottom, either "walking" or propping itself up on its fins, but Fricke's films showed only that it swam by moving its pectoral and pelvic fins alternately, rather like the walking motions of a tetrapod, such as a dog or a horse. But since alternating movement is a consequence of the way fishes normally move by bending their bodies, it does not show that *Latimeria* lent this movement to the tetrapods. Even though much of the popular literature places it on the list of mammalian ancestors, it is not a direct ancestor of the tetrapods, only a distant cousin. Adult coelacanths were roughly the size of an adult human being—nearly 6 feet long and weighing up to 175 pounds—which has been noted in the popular press, though it has absolutely nothing to do with the fish's recent or fossil history. J. L. B. Smith did little to discourage the idea of a walking coelacanth, and hence an ancestor of tetrapods, when he titled his 1956 book *Old Fourlegs*.

Comoran fishermen catch coelacanths on a hand line, usually in water that is around 2,000 feet deep. We have been able to ascertain that the fishes have been caught at a depth of 1,000 feet or less, and always at night. From his submersible, Hans Fricke has now filmed more than 100 adult individuals (he has never seen a juvenile), and they have been seen at depths ranging from 385 to 650 feet. They spend the day in lava caves 300 to 500 feet down and descend to around 2,000 feet to forage at night. They are seen to be passive drift feeders, opportunistically capturing fishes, squid, and octopuses, much in the manner of a large grouper. Fricke also observed a peculiar "head-down" behavior, where the fish spent several minutes at a time in a vertical position with its snout close to the bottom. The reason for this behavior is unknown but it might have something to do with the organ at the end of the snout, whose function has not been determined but which may be electrically sensitive.

Until 1992, all the coelacanths had been caught in the vicinity of the Comoros, with the exception of the first one, which was caught off East London, on the southeast coast of South Africa, 1,800 miles from these islands. But in 1992, a pregnant female was caught at a depth of about 150 feet off the coast of Mozambique, some 800 miles west of the Comoros, and in 1995, another one was caught off the southeast coast of Madagascar. The story changed dramatically in 1997. In July, while on their honeymoon, the zoologist Mark Erdmann and his wife, Arnaz Mehta, spotted a coelacanth in a fish market on the tiny island of Manado Tua off the northern tip of the Indonesian island of Sulawesi. At first the Erdmanns were unaware of the importance of the fish the Indonesians called *raja laut* ("king of the sea"); assuming that the coelacanth was already known from the western Pacific, they paid little attention to it. When they returned to the United States, however, they reported what they had seen, and learned of the significance of their sighting. With funding from the National Geographic Society, they came back to Sulawesi the following year, determined to gather the evidence they had passed by. For ten months they scoured the fish markets of northern Sulawesi, and hit pay dirt on July 30, when they were brought a specimen that was alive, but only barely. (Coincidentally, the Sulawesi coelacanths are caught by fishermen whose target is the same oilfish sought by the Comoran fishermen.) Before they released it, they swam with it and photographed it, hoping it would return to the depths, but it died. Some 6,000 miles from East Africa, it appears that there is another, completely unexpected, population of coelacanths. The Comoran coelacanths have all been dark blue, but the Indonesian ones were brown, suggesting a separate species or subspecies. Are there other undiscovered populations?

Erdmann donated the carcass of the coelacanth to the Indonesian authorities, and expected to be able to publish the definitive description. Erdmann's account of the finding of the coelacanth, which he suggested was *Latimeria chalumnae,* was published in *Nature* in 1998 (Erdmann, Caldwell, and Moosa 1998). Laurent Pouyaud of the French Scientific Institute and Indonesian coauthors published a second description in the April 1999 *Comptes Rendus de L'Académie des Sciences*, in which they claimed that "the Comorean and Indonesian coelacanths belong to distinct populations." They named the new species *Latimeria menadoensis*, after the village off which it was found.

The story took a surprising turn in July 2000, when a French team (Séret, Pouyaud, and Serre) submitted an article to *Nature* in which they claimed that they had actually found a coelacanth in Indonesia in 1995, three years before Erdmann saw one in the fish market of Manado Tua. Georges Serre said that the 1995 specimen had been caught in the Bay of Pangandaran, in southwest Java, about 1,200 miles from Manado, suggesting *another* undiscovered population of coelacanths. When the editors of *Nature* noticed that the fish in the photograph submitted by Serre was virtually identical to Erdmann's photograph of the Manado specimen, they concluded that Serre's photograph was nothing but a doctored version of Erdmann's, and refused to publish the article. Instead, a story entitled "Tangled Tale of a Lost, Stolen, and Disputed Coelacanth" was written by Heather McCabe and Janet Wright, and appeared in *Nature*, July 13, 2000. The last word came from Erdmann and Roy Caldwell, two of the authors of the original paper on the Sulawesi coelacanth, who documented the shameful hoax in a later issue of *Nature* (July 27, 2000), in a letter entitled, "How new technology put a coelacanth among the heirs of Piltdown Man." They wrote, "the Séret, Pouyaud and Serre image . . . is clearly an altered copy of a photograph taken by M. V. E. in 1998 off Manado Tua and printed in *Nature* soon afterwards," and they concluded, "the Indonesians and Comorans are rightfully proud of efforts in their two countries to preserve these rare and very special fish. What pride can we in the western scientific community take in this affair?"

Off the South African town of St. Lucia, in Zululand, in Natal, is Sodwana Bay, part of the St. Lucia Marine Protected Area. Two submarine canyons indent the continental shelf near Sodwana Bay to a depth of about 3,000 feet. On October 28, 2000, three recreational divers, Pieter Venter, Peter Timm, and Etienne le Roux, made a dive to 320 feet using a mixture of diving gasses. Beneath an overhang, Venter saw a fish about 2 meters long and realized that it was a coelacanth. He signaled to Timm, and both men saw two more. They had no cameras. "It was," said Venter, "like seeing a UFO without taking a photograph." Calling themselves "SA Coelacanth Expedition 2000," the group, with several additional members and cameras, returned to the spot in late November. On November 27, Venter, Gilbert Gunn, and the cameramen Christo Serfontein and Dennis Harding went down again to a depth of 350 feet using four different mixes of gas for a dive lasting 134 minutes. They had a bottom-time of 15 minutes. Twelve minutes into the dive they found three coelacanths. The largest was between 1.5 and 1.8 meters long, the other two 1.2 meters and 1 meter. The divers could not confirm whether these were the same three seen on the earlier dive or a different group, making a total of six. The fish swam head-down and appeared to be feeding from ledges (Venter et al. 2000).

The coelacanths of Sodwana Bay are far enough from East London and also from the Comoro Islands to suggest that they represent another heretofore unknown population. It is unlikely that the fish were strays from the Comoros, and so coelacanths may be more widespread than was previously thought. In the November 2000 issue of the *South African Journal of Science*, Phillip Heemstra of the J. L. B. Smith Institute of Ichthyology in Grahamstown stated that the sighting implies a viable population in Sodwana Bay.

One of the more curious aspects of coelacanth biogeographical lore was revealed in a 1966 article by Donald de Sylva in the magazine *Sea Frontiers*. Accompanying the article is a photograph of what was described as a "four-inch silver model of a coelacanth" that was found hanging in a church in Bilbao, Spain, and was bought by an Argentine chemist in 1964. It looks remarkably like a coelacanth, but of course, as far as anyone knows, there have never been any Spanish, or Atlantic, coelacanths. Another silver model of the same sort was found in a Paris antique shop, and this one was said to have come from Toledo, Spain. Keith Thomson (1991) wrote that "the simplest explanation is that an artisan merely made up a fanciful fish and by chance produced something looking a bit like a coelacanth," but the photograph and the drawing reproduced in Thomson's book certainly show a coelacanth. The model for the silver miniatures did not have to come from the place where it was found in 1964; one can find artifacts from all over the world in Paris antique shops. It is certainly possible that an African craftsman made one or both of the models. With the discovery of the Sulawesi coelacanths, another whole population must be considered as a possible inspiration for the models.

PISCES TRIUMPHANT

Fishes have been swimming on Earth for more than 450 million years. They predated the dinosaurs by hundreds of millions of years, and they were the first creatures to have an internal skeleton. In that sense, they are the ancestors of all vertebrates—amphibians, reptiles, birds, and mammals. Fishes are among the most successful animals ever. (The oldest hominid fossils that can be dated are from 2.3 million to 1.6 million years ago.) Nowadays, fishes are the most diverse of all vertebrates; more than 24,000 species inhabit the world's oceans, rivers, and lakes.

A major advance in the evolution of fishes was the development of jaws, which allowed them to prey on other species, and not depend on filter or bottom feeding. Living bony fishes (class Osteichthyes) may be divided into two subclasses, the Actinopterygii (ray-finned fishes, which dominate the world's fish fauna today), and the Sarcopterygii, or fleshy-finned fishes, which only number a few scattered groups. They are differentiated from cartilaginous fishes (class Chondrichthyes), such as sharks, skates, rays, and chimaeras, whose internal skeletons are composed of cartilage. Included in the Sarcopterygii are the Crossopterygii, the presumed ancestors of land vertebrates. The coelacanth, all of whose relatives disappeared along with the dinosaurs 70 million years ago, is a crossopterygian. The paired appendages (whether fins, legs, or something in between) of the early crossopterygians were attached to the shoulder or pelvis by a single bone, the humerus or femur, respectively, which allowed for greater freedom of movement, and also led to the designation of these creatures as "lobe-fins." Although there were many variations on the lobe-finned plan in the Paleozoic, only *Latimeria* and the lungfishes have survived into the present.

"Where did the bony fishes come from?" asks Per Erik Ahlberg, a paleontologist at the Natural History Museum, London, in a 1999 article in *Nature*, in which he discusses the discovery of *Psarolepis romeri*, an enigmatic species first described by Zhu, Yu, and Janvier in 1999. Found in a Lower Devonian stratum in Yunnan, China, *Psarolepis* ("speckled scales") is, at around 400 million years old, one of the earliest bony fishes known. *Psarolepis* appears to be a link between osteichthyan and nonosteichthyan groups, sharing some characteristics with the sarcopterygians (coelacanths, lungfishes, and tetrapods—the group that includes birds, reptiles, and mammals), and also the actinopterygians, the group that contains sturgeons, gars, and bony fishes. "Into this confusion," wrote Ahlberg, "drops Zhu and colleagues' discovery, the 400-million-year-old *Psarolepis* from the earliest Devonian and latest Silurian of Yunnan in South China." In their article, Zhu, Yu, and Janvier wrote that "this fossil fish combines features of sarcopterygians and actinopterygians and yet possesses large, paired fin spines previously found in two extinct gnathostome [jawed] groups (placoderms and acanthodians). This bony fish provides a morphological link between osteichthyans and non-osteichthyan groups . . . [and] offers new insights into the origin and evolution of osteichthyans." In other words, 400 million years ago, there was a kind of fish that shared the characteristics of the two major fish groups alive today, and may have been the ancestor of both.

Gnathostomes, the first jawed fishes, had head and shoulder girdles that were heavily armored with bony plates. Known first from the Early Devonian period, 400 million years ago, they were a diverse group that included 35 families and more than 270 genera. The class Placodermi ("plate-skins") includes the jawed fishes that were partially encased in a bony armor. Some of the earlier forms resembled the heterostracans in their armored body and trunk, with only the tail exposed, and most of them seemed adapted for living on the bottom. There were probably species that were not armored and that did not lend themselves to fossilization, and so far, we have no record of them.

The arthrodires ("jointed neck"), the dominant predators of Devonian seas, had a two-part armor covering: one segment covering the head and the other the thorax. These segments were connected by ball-and-socket joints, hence the name arthrodire (from *arthros*, "joint," and *dirus*, "fearsome"). The earliest known arthrodires, genus *Arctolepis*, were flattened with the eyes far forward, and in some specimens, a massively developed bony braincase. *Arctolepis* also had enormous spines that extended back from the trunk. Although these look like wings, they were probably only for protection.

The commonest fossil arthrodire is the two-foot-long *Coccosteus*, found throughout Europe, but particularly in the red sandstones of Scotland. *Coccosteus* was a fast-moving hunter with an upturned tail and a dorsal fin. It had no teeth, but its jaws had sharp tusklike projections that probably functioned like the beak of a bird of

The largest of the arthrodires was the formidable Dunkleosteus, *which may have reached a length of 30 feet. Although its foreparts were encased in heavy armor, its long, eel-like tail could propel it powerfully through the water. It was an apex predator, meaning it was able to capture and overwhelm anything that swam 400 million years ago in the oceans of the Late Devonian Period.*

prey. The external armor was made of bone, an arrangement that has no known analog in the bony fishes of today, which usually have most of their bones on the inside, not the outside. The largest was the formidable placoderm *Dunkleosteus* (also known, incorrectly, as *Dinichthys*), which may have reached a length of 30 feet. Although its foreparts were encased in heavy armor, its long, eel-like tail could propel it powerfully through the water. *Dunkleosteus* was an apex predator, able to capture and overwhelm anything that swam during the Late Devonian, including smaller placoderms, spiny sharks, and primitive ray-finned fishes. Active about 400 million years ago, the placoderms disappeared as the sharks were gaining dominance. Although it is difficult to imagine what sort of animal might pose a threat to *Dunkleosteus,* the more flexible cartilaginous sharks might have been more efficient predators, contributing to the downfall of the arthrodires; they had all gone extinct by the end of the Devonian.

In the far north of Western Australia are the remains of a giant barrier reef that teemed with life 370 million years ago. From the shales known as the Gogo Formation, which were formed in quiet inter-reef bays, have come exquisitely preserved, three-dimensional skeletons of the fishes that swam in these ancient seas. Their original bone preserved in limestone nodules within the shale, they represent the best-preserved early fishes in the world. Discovered by John Long in 1986 and described in 1995, *Mcnamaraspis kaprios* is a placoderm that in life would have been about 10 inches long. Like other placoderms it is characterized by a bony head shield that articulates in a ball-and-socket joint to a trunk shield. It had a sharklike body, with a single dorsal fin; broad, fleshy paired pectoral and pelvic fins; and an anal fin. *Mcnamaraspis* had cartilage in the snout, the first evidence for this structure in placoderm fishes. This is significant because the structure would have allowed water to flow between the nasal openings and the mouth, indicating that the fish possessed a well developed sense of smell. This, combined with the sharp, prominent teeth imply that

Mcnamaraspis was a voracious predator, probably feeding on the small, shrimplike crustaceans that abounded in the warm, tropical seas of prehistoric Western Australia.

Another group of Devonian placoderms was the antiarchs, which Romer described (1966) as "grotesque little creatures which looked like a cross between a turtle and a crustacean; a series of odd forms which are armored caricatures of modern skates and rays." Smaller than the arthrodires, the antiarchs also had their foreparts encased in heavy shell-like armor, but they also had jointed, bony flippers that looked like the claws of a crab. These appendages were probably useless for swimming, and might have been used to pull the animal along the bottom. The mouth was located on the bottom of the head, near the front. They are represented by *Bothriolepis,* a most peculiar animal indeed. In sectioning a particularly well-preserved specimen, R. H. Denison (1941) found a pair of sacs branching off the pharynx, which may have been auxiliary lungs. There was also a spiral gut filled with sediment, so *Bothriolepis* may have specialized in feeding along the shallow margins of streams, and may even have been able to drag itself overland with its jointed "arms." In the shallow-water deposits of the Bay of Chaleur on the southern shore of the Gaspé Peninsula in eastern Canada, thousands of specimens of *Bothriolepis* have been found. There are no placoderms known after the Early Carboniferous Period (350 million years ago), and no living fishes show any of the characteristics of these primitive creatures. Hugh Miller (1802–1856) was a Scottish geologist who described some of the first Devonian antiarch fossils in *The Old Red Sandstone* (1841). In his description of *Pterichthyodes,* he so misunderstood the joint between the head plate and the thoracic plate that he identified it as the mouth, placing it on top of the head and behind the eyes.

Around 365 million years ago, 5 million years before the end of the Devonian, a massive extinction event occurred that extinguished 35 of the 46 known families of fishes, including the last of the ostracoderms, all the placoderms, and at least 10

Among the Devonian placoderms was Bothriolepis, *which may have had auxiliary lungs.* Bothriolepis *may have specialized in feeding along the shallow margins of streams, and may even have been able to drag itself overland with its jointed "arms."*

families of lobe-fins. There are many theories about why this might have happened—falling sea levels, rising temperatures, reduced oxygen in the water, etc.—but no one really knows.

The oldest ray-finned fishes (order Actinopterygii) appeared in the later Silurian, about 410 million years ago. Found in the Middle Devonian of Scotland was *Cheirolepis,* a foot-long fish with a single dorsal fin, tiny scales, and fleshy fin lobes. Somewhat later is the Western Australian *Mimia,* characterized by a large mouth with teeth in the upper and lower jaws, and ganoid scales, which are thick bony elements covered with dentine and locked into regular rows, like the scales of gars today. Looming mysteriously in the distant past history of ray-finned fishes is *Leedsichthys,* a poorly preserved monster found in the Middle Jurassic of England. With a tail 15 feet high and an overall length that may have exceeded 40 feet, *Leedsichthys* was the largest bony fish that ever lived.*

Among the surprising fish fauna of the Late Cretaceous (100 million to 75 million years ago) was *Xiphactinus audax,* sometimes known as the bulldog fish for its prognathous lower jaw. It was an ichthyodectid, (the name means "biting fish," from the Greek *dektos,* "bite"), an extinct group of teleost fishes (those with bony skeletons) that flourished in the Cretaceous Period, and were characterized by powerful jaws and teeth. Shaped more or less like a tarpon, *Xiphactinus* reached a known length of 20 feet and may have gotten even larger. With premaxillary fangs protruding from the front of its mouth and a mouthful of nasty teeth, it was probably the dominant predator of the warm, shallow seas that covered much of interior North America 95 million to 65 million years ago. On display at the Sternberg Museum in Hays, Kansas, is a 12-foot fossil *Xiphactinus* with the complete skeleton of the smaller ichthyodectid *Gillicus* inside it. Other fossils of this species have also been found with undigested fish inside, so it would appear that the bulldog fish swallowed its prey whole.

Although a couple of jawless fishes are still swimming around today, along with some 700 species of cartilaginous sharks and rays, one surviving crossopterygian (the coelacanth), and three lungfishes, basically, the seas, rivers, and lakes of the world are today populated by the enormously successful ray-finned fishes. There are more than 24,000 species of bony fishes, compared to about 8,600 species of birds, 4,000 of mammals, 6,000 of reptiles, and almost 5,000 of amphibians; clearly, fishes outnumber all other living vertebrates combined. Of course, the abundance of fishes might have something to do with the amount of ocean; more than two thirds of the Earth's surface is under saltwater, and that is a lot of available habitat. Or, to look at it from

---

* Of today's teleosts, the longest is the long, skinny oarfish (*Regalecus glesne*), which has been measured at 26 feet, and the heaviest is the ocean sunfish (*Mola mola*), which has been recorded at over 4,000 pounds. Of the big game fishes, the largest black marlin was officially weighed at 1,560 pounds, and the heaviest tuna was well over 1,400. *Carcharodon megalodon,* a gigantic extinct lamnid (shark), may have reached a length of 50 feet, and certain living sharks, like the whale shark (*Rhincodon typus*) and the basking shark (*Cetorhinus maximus*), grow as large or larger than *Leedsichthys,* but they are cartilaginous fishes. According to the official record book of the International Game Fish Association, the largest fish ever caught on rod-and-reel was a great white shark (*Carcharodon carcharias*) that weighed 2,664 pounds.

Sometimes known as the bulldog fish for its prognathous lower jaw, Xiphactinus *flourished in the Cretaceous Period. Shaped more or less like a tarpon,* Xiphactinus *reached a known length of 20 feet—and may have gotten even larger. It was one of the dominant predators of the warm, shallow seas that covered much of the interior of North America 95 million to 65 million years ago.*

another point of view, 23,600 species of terrestrial vertebrates have to live on one third of the Earth's surface, of which many parts, like mountaintops and deserts, are virtually uninhabitable. With the exception of birds, which can fly above the surface, moles, which can burrow beneath it, and various arboreal mammals and reptiles, terrestrial vertebrates are generally restricted to an almost two-dimensional world. They can move east and west and north and south, but there is not much up and down available to them. For fishes, however, up and down is the definition of their habitat; their oceans are deeper than the highest mountains are high, providing an abundance of available living space. Of the 71 percent of the Earth's surface that is covered with ocean, 90 percent is more than 2 miles deep. The predominant habitat on Earth is the deep ocean.

And the fishes have had plenty of time to colonize this habitat. Since the Devonian "age of fishes," many species have arrived and departed. Nevertheless, today's oceans are populated by a richness of piscene fauna that is staggering in its diversity and abundance. The descendants of the early actinopterygians now include gars, tarpon, eels, swallowers, sardines, herrings, anchovies, carps, minnows, catfishes, pikes, smelts, salmons, hatchetfishes, viperfishes, dragonfishes, lizardfishes, lanternfishes, trouts, codfishes, toadfishes, anglerfishes, frogfishes, batfishes, clingfishes, flying fishes, grunions, oarfishes, squirrelfishes, flashlight fishes, whalefishes, pipefishes, seahorses, trumpetfishes, scorpionfishes, pigfishes, lumpsuckers, snooks, groupers, moonfishes, snappers, tripletails, grunts, porgies, threadfins, drums, croakers, goatfishes, butterflyfishes, angelfishes, roosterfishes, jacks, remoras, hawkfishes, mullets, threadfins, damselfishes, anemonefishes, wrasses, parrotfishes, wolffishes, icefishes, stargazers, sanddivers, blennies, dragonets, gobies, rabbitfishes, surgeonfishes, barracudas, mackerel, tunas, swordfishes, lefteye flounders, righteye flounders, soles, spikefishes, triggerfishes, leatherjackets, cowfishes, porcupinefishes, puffers, ocean sunfishes, and thousands and thousands more.

THE EVOLUTION OF BIOLUMINESCENCE

It is not difficult to imagine the adaptations necessary to colonize the deep ocean, the largest biosphere on the planet. It can be predicted that creatures that live their entire lives in darkness would become relatively small; they would probably develop large eyes and teeth to enable them to see and capture their prey more efficiently; and some species might even evolve an unusual method of breeding, where the males permanently affix themselves to the females so as to avoid having to wander aimlessly around the dark abyss looking for a mate. It seems logical—even necessary—that organisms of the abyss would develop a way of seeing, or hunting, or communicating in the dark, but the solution is so far beyond the boundaries of traditional evolutionary theory that it can hardly be imagined: they grew lights.

The seventeenth-century scientist-philosopher Robert Boyle (1627–91) is generally considered the discoverer of bioluminescence. In 1667 he reported "stinking fish" and "a luminous neck of veal" that gave off light. Although he could not have known the cause of the glow, it was luminous bacteria that produced it in decaying animal matter. Boyle also found that air was required for the light emitted by "glow worms" (larval fireflies), but since his experiments took place more than a hundred years before the discovery of oxygen by Lavoisier and Priestley, he obviously could not recognize oxidation. Benjamin Franklin suggested that the common phenomenon of phosphorescence in the sea might be caused by microscopic organisms, but he was unable to explain the mechanism whereby animals could light themselves up.*

Some deep-sea fishes have glittering rows of light-producing organs, called photophores, along their sides; others are equipped with filaments that glow; some have light organs in their eyes and others have them on their tongues. One species has a glowing rectal gland. On one occasion, a deep-sea fish was observed to be feeding by shining a 2-foot beam of light into a swarming school of shrimplike crustaceans. Bioluminescence has intrigued biologists for years, but there is still no agreement on its function or the explanation of how so many different creatures—not only fishes, but also many squids, certain sharks, starfishes, sea-cucumbers and crustaceans—have developed these structurally similar mechanisms. Although the structures differ from phylum to phylum, and even from species to species, many deep-water denizens have light organs, composed of cells that light up and serve various purposes in the lives of animals that would otherwise have to function in total darkness.

There are three ways marine animals can generate "living light." The commonest is *intracellular,* where certain cells are organized into special structures such as lanterns and photophores. In the cells of these organs, energy is released in the form of chemoluminescence, or "cold light," when the oxidization of a substance known

* Although the terms "phosphorescence" and "luminescence" are sometimes used interchangeably to describe chemically generated heatless light, "phosphorescence" is more accurately applied to the combustion of the chemical element phosphorus, which glows in the dark, and takes fire spontaneously when exposed to air. It is not found free in nature.

as luciferin is stimulated by the catalytic activity of an enzyme known as luciferase. (Because hardly any heat is generated by the luciferin-luciferase reaction, this "cold light" of chemoluminescence has become a subject of great interest to physicists concerned with the preservation of energy.) The organs that produce the light may be a simple luminous element surrounded by a layer of black pigment cells, or cups with a reflecting layer, or even some complex structures with lenses and color filters. In some instances, the light is produced extracellularly and can result in the discharge of a luminous cloud into the water.

Some fishes harbor luminescent bacteria that provide a light source that glows from the fish's body, but the light is not actually produced by the fish. This association is deemed symbiotic because the fish provides the luminous bacteria a sheltered environment and a supply of nutrients and oxygen, while the bacteria serve as a continuous source of light, which the fish uses for a variety of purposes. Two strains of luminous bacteria are currently recognized, *Vibrio* and *Beneckea. Vibrio* requires a "host" before it can luminesce, whereas *Beneckea* is free-swimming. Deep-sea anglers have a "lure" dangling in front of their mouths, and the bacteria are usually maintained in this lure, called the esca. The "sea-devils" (*Ceratias* and *Cryptopsaras*) have additional bacterial bulbs in the form of caruncles, which are modified dorsal fin rays. One of the most intriguing problems in evolutionary biology is that luminous bacteria must be transferred from generation to generation, or, as Peter Herring wrote in a 1977 study of bioluminescence, "The light organs of successive generations must be reinfected with the appropriate strain." The luminous bacteria cannot exist outside the host, but they must somehow pass to the offspring externally, but how this might work has not been explained. In 1993, Herring wondered "where the bacteria come from to reinfect successive generations of host. We know they can leak from the host into the surrounding sea water, but do they survive there? Adult and juvenile anglerfishes live at different depths; how does a larval female acquire the right bacterium, even if it is leaked by the adult, if the adult lives several hundred metres deeper?"

When I asked him in March 2000 whether anyone had resolved the question of the acquisition of luminescent bacteria, Herring answered, "Your puzzlement is similar to mine. The answer to part of your query is that some species of fish and squid have easily cultured recognizable species of luminous symbiont, identical to those cultured from seawater. Maggie McFall-Ngai has done some really neat work explaining how a squid system works and collects its bacteria, *Vibrio fischeri,* from the surrounding seawater. Anglerfish and anomalopids are different in that symbionts from the light organ cannot (yet?) be cultured, and are genetically distinct from known species. Anglerfish don't acquire their bacteria in the light organ for at least several weeks after hatching, so what triggers their acquisition and where from is still a complete mystery."

The flashlight fish known as *Photoblepharon palpebratus* from the Indian Ocean has a unique structure immediately below each eye, consisting of a light organ that it can "turn off" by raising an opaque lower lid that covers the light organ but not the eye. "Like a number of other bioluminescent fishes," wrote McCosker and Lagios

(1975), "*Photoblepharon* does not manufacture its own light, but relies on symbiotic bacteria to produce it. Enormous numbers of these bacteria are packed into special nutitive compartments, akin to living incubators, within the fish's light organ." McCosker and Lagios attempted to culture the bioluminescent bacteria in the lab, but they failed. Kenneth Nealson of the Scripps Institution of Oceanography in La Jolla explained to them that the light-generating microbe of *Photoblepharon* is so specialized that it cannot survive outside its host. "Unlike more specialized bacteria," Dr. Nealson related, "the *Photoblepharon* bacterium can only partially metabolize its host's sugar, glucose. . . . It is postulated that the host cells which form the living incubator compartments in the lid of *Photoblepharon* supply the metabolic machinery which the bacterium lacks and thus permit its symbiotic survival."

In a discussion of the luminescent bacteria in the deep-sea angler *Oneirodes acanthias,* William O'Day (1974) wrote, "I have cultured bacteria from the esca [lures] of *O. acanthias* [but] I have not seen them luminesce." Other attempts to cultivate the bioluminescent bacteria have shown similar results: the bacteria may be cultured, but the host's glucose is required for bioluminescence. Work done by Margo Haygood and Daniel Distel (1993) has revealed that the bacteria taken from the light organs of certain deep-sea fishes are not known luminous bacteria (e.g., genus *Photobacterium*), but they are new groups related to the *Vibrio* species. Haygood and Distel wrote: "These results show that the anamalopid [flashlight fishes] and ceratioid [deep-sea anglerfishes] symbionts appear to have arisen from the same common ancestor that was the source of the facultative *Photobacterium* light organ symbionts and other free-living luminous marine *Vibrio* species, and the physiological differences that make them difficult to culture are probably due to evolution occurring after the establishment of symbiosis, rather than to a dramatically different phylogenetic organ." The two species of flashlight fish, *Anomalops* and *Photoblepharon,* were collected in the same area (although not at the same depth), but their bioluminescent symbionts are completely different, each species synthesizing a different symbiont for its own requirements.

In captivity, flashlight fishes sometimes lose their bioluminescent capabilities, but they do not become reinfected with the necessary bacteria by other individuals in the tank (Herring 1993). How then does the fish find and capture the symbionts required for its bioluminescence? Fishes have retained their secrets, but an answer has been found in the lifestyle of the little Hawaiian squid, *Euprymna scolopes.* In 1983, when C. T. Singley discussed *Euprymna* (1983), he wrote that it was "a relatively unknown species . . . [about which] there is much information to be gained by further study." This iridescent, shallow-water squid has now become the poster boy for bioluminescent symbiosis, or, as Montgomery and McFall-Ngai wrote in 1994, "The association of the Hawaiian sepiolid squid *Euprymna scolopes* with its luminous bacterial symbiont *Vibrio fischeri* provides a model system to study the initiation and development within symbiosis-specific host tissue."

Many other squids have luminescent organs, known as photophores, but these are not bacterially inspired. Endemic to the Hawaiian Islands, the inch-and-a-half-long *Euprymna scolopes,* commonly known as the Hawaiian bobtail squid, buries itself in the sand during the day but comes out at night. Adult females lay 50 to 450 eggs, and around 20 days later, the juveniles hatch from their egg cases with light organs that are devoid of bacteria, "and if these juveniles are not exposed to *V. fischeri*, a common constituent of the free-living bacterioplankton, the light organ remains sterile. However, under normal conditions, *Vibrio fischeri* is abundant (approx. 200 cells/ml) in seawater where adults are found, most likely because of a diel [during the day] behavior characteristic of the adults in which they release 90% of their bacterial culture into the water column each day at dawn" (McFall-Ngai 1998). The bacteria, therefore, are essential for the normal development of the symbiotic organ, which involves changes in the epithelial tissue where the bacterial culture will take up residence.

The little Hawaiian bobtail squid is born without bioluminescent bacteria, but shortly after hatching it acquires the bacteria necessary for its light organ to function.

A few hours after hatching, the squid must acquire the bacteria that will populate its light organ, but the turbulence of the marine environment and the vanishingly low concentration of microbial cells present a serious challenge. How can the squid find and incorporate just the right bacterium, and in numbers large enough to create the bioluminescent organ? In a study published in August 2000, Nyholm, Stabb, Ruby, and McFall-Ngai found that the newly hatched squidlets exude from ciliated appendages a viscous material that binds only the bacterium *V. fischeri*, and as a result of unknown chemical signals, causes them to aggregate on the mucuslike matrix. Currents created by the ciliated appendages (which will be recycled after accomplishing their mission) amass the bacteria, which become embedded in the mucuslike material suspended just above the light organ pores. After a period of several hours, the captured cells migrate along the mucous strands and colonize the epithelia in the internal spaces of the light organ. The presence of *V. fischeri* is required for the production of the mucous matrix, so it appears that both the bacterium and the squid play an active role in the process.

As in the bacterial bioluminescent organs of some fishes, the symbiotic development of the light organ of *Euprymna scolopes* is essential for the organ's development. When the bacteria are withheld in the laboratory, the light organ does not develop. Once the bacteria have colonized the light organ, it changes shape and the ciliated cells die off. This cell death, according to Margaret McFall-Ngai and her student, Jamie Foster, of the University of Hawaii, is not caused by the bacterial infection, but rather, the bacteria seem to signal the cells to undergo apoptosis, a sort of cellular suicide. The microbiologist Edward Ruby, also of the University of Hawaii, who worked with McFall-Ngai on the original studies of the bioluminescent squid and the bacterium, now studies the bacterial side of the squid-*Vibrio* association. The little squids can be raised successfully in captivity, so students all over the world can examine and analyze this amazing example of symbiosis. According to Roger Hanlon of the Marine Biological Laboratory at Woods Hole, *V. fischeri** produces a thousand times more light in the squid than it does when cultured in the laboratory. The light organ of the squid is located on its underside, and it is believed to provide the squid with counterillumination that protects it from being seen from predators below it. The symbiotic "benefit" to the bacterium is not known.

In their study, Ruby and Nealson identified *Vibrio fischeri* in the Japanese pinecone fish; it was the first time this bacterium has been found in association with a luminous fish. But they were as baffled as anyone else as to how the fish might acquire the bacteria. In the language of science, they wrote, "The way in which the luminous organ obtains its bacterial innoculum is not known; nor is the degree of isolation of the organ culture from exchange and/or contamination with the external environment known. Clearly there must be host mechanisms involving selection for symbiont characters, but as yet we have no knowledge as to the specific features and mechanisms which operate."

Many species of deep-sea fishes have bioluminescent organs, but depending on their location, size, brilliance and structure, they probably serve different purposes for different species. In a 1967 article entitled "The Significance of Ventral Bioluminescence in Fishes," D. E. McAllister listed the function of photophores in interactions with members of the same species, and with members of different species. In intraspecific interactions, he wrote, lights might be useful to attract potential mates, to serve as recognition signs for members of the same species, to indicate the sex of the possessor, to use in courtship displays, and to allow individuals to distribute themselves in space that otherwise provides no fixed reference points. In interspecific interactions the luminous organs might serve to lure prey within range, illuminate the items of prey, startle or divert predators, reveal predators to larger predators, or mislead predators through mimicry.

Bioluminescent squid have been known for some time, but the first glow-in-the-dark octopus was not discovered until 1999. (Prior to that, a female deep-sea octopus of the genus *Eledonella* was found to have a ring of bioluminescent tissue around its mouth, which Robison and Young [1981] suggested might "act to attract a mate.") During a research cruise in the Gulf of Maine, a bright orange octopus that had button-shaped light organs instead of suckers on its arms was collected. When the foot-long specimen, which had been collected alive, was placed in a tank, it was seen that the suckers did not adhere to anything but glowed with a blue-green light. Sönke Johnsen, Elisabeth Balser, and Edie Widder (1999) published a description of *Stauroteuthis syrtensis* and proposed that "these modified suckers may have two functions: communication and attraction of prey." Examination of the stomach contents showed that the octopus had been eating tiny copepods, probably attracted by its bioluminescent suckers, then trapped in the web between the arms and brought

---

* A closely related bacterium, *Vibrio cholera,* causes the human intestinal disease cholera, by releasing a toxin similar to that released by *V. fischeri.* Another *Vibrio* species causes paralytic shellfish poisoning.

to the mouth. The researchers suggested that *Stauroteuthis syrtensis* evolved from a shallow-water species, and as it colonized deeper waters, its suckers, with nothing to stick to, developed other functions.

The most startling use of bioluminescence is that employed by the large squid *Taningia danae.* It has eight appendages instead of the ten that most squid species possess, and on the ends of two of these are lemon-sized and lemon-colored photophores—the largest light organs of any known animal—which can be flashed at will. Each of the photophores is equipped with a black, eyelid-like membrane that can be opened or closed. The function of these stroboscopic flashers is unknown, but they may be used to startle their prey or to confuse a predator. On the suckers of its arms *Taningia* has retractable claws, like those of a cat. Little is known about this species, which can reach a length of 7 feet and a weight of 135 pounds, but it has been collected, mostly from the stomachs of sperm whales, from all the subpolar, temperate, and tropical oceans of the world, making it one of the most widely distributed of all squids.

As Hastings wrote (1998), "Bioluminescence may be thought of as a bag of tricks: the light can be used in different ways and for different functions. Most of the perceived functions of bioluminescence may be classified under three main rubrics: defense, offense, and communication." The lighted lures of the ceratioid anglers serve to attract prey items toward the mouth of the predator. The "mousetrap" arrangement of *Thaumatichthys,* where the fish has a light organ in its mouth, is the most sophisticated variation on this theme. The lanternfish *Neoscopelus* has photophores on its tongue, allowing it to attract prey to the proper location. Light organs around the eyes of some species of lanternfishes probably function as "headlights" to illuminate prey items immediately before they are consumed, but the illumination, even from a photophore the size of a pea, cannot show the fish where to go, but only what to do when it gets there. One of the most unusual demonstrations of bioluminescence has been observed in *Searsia,* a black, bathypelagic fish that can discharge a luminous secretion into the water from a subcutaneous gland just behind its head. (Members of the family are commonly known as "tube-shoulders.") *Searsia* appears to be the only fish that bioluminesces outside its body, but the squid *Heteroteuthis dispar* is also known to emit luminescent ink clouds, and is the only cephalopod to do so. In addition to photophores, the stomiatoids *Stomias, Idiacanthus,* and *Chauliodus* can envelop themselves in a sheath of luminous, gelatinous tissue that glows bright pink when stimulated. As William O'Day (1973) wrote, this kind of luminescent silhouetting does the opposite of camouflaging the fish, and "may aid in mating, spacing themselves out as they hunt, maintaining conspecific aggregations, warning potential predators of their own formidable size, or perhaps allowing them to escape from predators by temporarily blinding them." He concludes by saying, "These functions, however, remain speculative."

In his 1983 overview of the general functions of bioluminescence, R. E. Young is careful to identify the possible dangers of lighting up under water. He writes, "Attracting a mate with luminescence has a number of inherent drawbacks. An animal hoping to lure a mate may attract a predator instead, or an animal searching for a mate may be attracted to a predator instead." Furthermore, wrote Young, "even in the clearest oceanic water, fine detritus (marine snow) is abundant and light scattering by such molecules may well alert potential victims before they are exposed by the beam." Young concludes, "I view life in this dark environment as a peculiar battle in which stealth and luminescence are major weapons. It is a struggle unlike any found elsewhere on this planet." M. D. Burkenroad (1943) suggested an imaginative and somewhat convoluted scenario in which the flashing might serve as a sort of "burglar alarm," and attract not only the first predator, but another predator to prey upon the first, thus protecting the original prey.

Not all bioluminescence functions in the darkness of the depths; sometimes it is employed when there is downwelling light from above the surface. Most luminous organs are located on the ventral surfaces of the fish, and no simple explanation readily springs to mind as to why this location is so heavily favored. It has been proposed that ventral bioluminescence serves as a sort of countershading (although "counterlighting" might be a more appropriate term), in which the luminescence counteracts the daily shadow that the fish appears to be against surface illumination. The characteristically dark coloration of the dorsal surface of most fish enable them to match the background when lit from above. Many species of squid are equipped with an organ known as the "extra-ocular receptor," which enables them to read the downwelling light and adjust their luminescence accordingly, and there is a possibility that some fishes have a comparable apparatus. J. V. Lawry suggested just this (1974) with regard to *Tarletonbeania* (also known as the blue lanternfish), which "may see downwelling light and bioluminescence from a supra-orbital photophore and adjust the output of its ventral photophores to match environmental illumination."* In order for such a system to work, the fish has to be able to switch its photophores off at night, since in darkness, instead of camouflaging the fish, the luminous organs would draw attention to it. In the experiments with *Tarletonbeania,* Lawry observed that captured specimens kept in total darkness did not luminesce, but when they were illuminated from above, the photophores emitted a bluish light that was extinguished whenever the light source was.

The blue lanternfish has a series of lights along its flanks that may help it to locate a potential mate or may serve as counterillumination to hide it from predators.

Even more fascinating, male and female *Tarletonbeania* were once thought to be different species, because only one, now known to be the male, has lumi-

* The supraorbital photophore in fishes appears to correspond to the extraocular receptor in squids, as well as the pineal eye in some species of deep-sea sharks. This is a good example of convergence, in which completely unrelated creatures have developed similar solutions to a particular problem, in this case, reading the ambient light level in order to adjust one's own luminescence to it.

nous caudal organs, and the males were only known from the stomachs of predators. Those now known to be females could be captured in nets, but never the males. "The apparent explanation," wrote J. Woodland Hastings of Harvard, who has been studying bioluminescence for almost 50 years, in 1998, "is that when a predator attacks, the males dart off in all directions with their dorsal lights flashing, like a police car, leading the predators on a wild goose chase (and sometimes getting caught), but leaving the females, who remain in place, safe from the predator in the cover of darkness, yet easy to catch in a net."

Hastings was among the first to notice that ventral lighting might help to camouflage the silhouette of a fish during the day. In 1971 he wrote an article called "Light to Hide By," in which he described the bioluminescent pony fish, *Leiognathus equulus,* caught in New Guinea waters in 1969, which emits a "diffuse light over much of the ventral surface . . . due to symbiotic bioluminescent bacteria." As with *Photoblepheron,* the bacteria are always "on," and the pony fish has an internal shutter that directs the light into the swim bladder, which is equipped with silvery, reflecting guanine crystals that permit the light to be diffusely emitted over a broad portion of the fish's lower body. George C. Williams of SUNY at Stony Brook was so impressed by Hastings's work that he titled his 1997 book *The Pony Fish's Glow (and Other Clues to Plan and Purpose in Nature)*. About pony fish he wrote,

> The pony fish is never in bright light. It stays in water dark enough to make it unlikely to be seen by its enemies, but if seen it will be from below. If there is any light at all coming from above, a predator below would see a pony fish as a silhouette moving across that overhead light—unless, of course, the potential prey can extinguish its silhouette by making its belly glow in a way that matches the light coming from above. . . . J. W. Hastings did this work in the 1960s, and it is now generally assumed that this is a common adaptation in many species of open-ocean fishes, even those that have discrete external photophores.

N. B. Marshall (1954), a well-known British ichthyologist, has suggested that piscene photophores are ventral so that they will not reflect sunlight when illuminated from above, but again, this would apply only during the day. Numerous authors have suggested that the ventral photophores disrupt the silhouette of the fish, but this occurs only when the predator approaches from below, and does not explain why dorsal light organs would not serve the same purpose. In his discussion of the function of bioluminescence, W. D. Clarke (1963) suggested that since most fishes are nearsighted, they cannot see the photophores as separate lights, but from a distance, they appear as an indistinctly-glowing monochromatic cloud. In his 1967 survey, McAllister wrote, "As the specific functions attributed to bioluminescence do not seem to require a predominately ventral position, one must look elsewhere for explanations of this pattern."

There are a few aspects of bioluminescence that we now understand, but its origins are still one of biology's great mysteries. It is so much a part of the deepwater existence that almost every kind of deep-sea animal has representatives that light up. There are bioluminescent starfishes, sea cucumbers, squid, octopuses, jellyfishes, and comb jellies. The naturalist-adventurer William Beebe estimated that of the fishes he caught down to 13,000 feet in Bermuda waters, more than 95 percent were bioluminescent, and Fitch and Lavenberg (1968) cited two studies in which bioluminosity was present in 96 and 98 percent of the species collected. Hastings wrote in 1998, "At midocean depths (200–1200 m) . . . bioluminescence in fish may occur in over 95% of the individuals and 75% of the species. The midwater luminous fish *Cyclothone* [a bristlemouth] is considered to be the most numerous vertebrate on the planet."

How on Earth—or in the depths of the ocean—could so many fishes acquire lights, especially as it now appears that it must have occurred independently on as many as 40 different occasions? We may not know the answer, but it is obvious that in Darwinian terms there was a strong selection pressure for optical signals, perhaps concurrent with the development of complex eyes in deep-sea fishes. In his 1954 *Aspects of Deep-Sea Biology*, N. B. Marshall tried for an explanation:

> Through evolutionary processes, mesopelagic fishes, euphausiids, prawns and cephalopods have "learned how to acquire" invisible cloaks of bioluminescence. Have they "learned" through their eyes? Certainly there are no midwater users of luminescent camouflage with poorly developed eyes. Perhaps the users "learned" through their eyes two outstanding features of submarine light; that as the depth increases light becomes more monochromatic (towards blue) and that the main axis of the light field moves towards the vertical. Vertically migrating animals might be the best "learners," but some users of ventral camouflage are nonmigrators. And what were the steps in the evolution of the necessary photophores? Perhaps simple ventral lights were first evolved to enable animals to see one another, especially during sexual congress. Later, as such lights provided some ventral camouflage, the way might have been open for evolutionary steps towards more effective camouflage. Eventually, prey disappeared from hunters' eyes, despite their keen sensitivity to blue luminescent light and twilight. Even so, certain fishes and squid, which have yellow lenses in their eyes, may "see-through" luminescent camouflage. Deep-sea red camouflage has also been exposed by certain predatory fishes.

Edie Widder of the Harbor Branch Oceanographic Foundation, an authority on bioluminescence, wrote in 1999 that "disparate chemical systems and diverse phylogenetic distribution patterns are most easily explained by the late emergence of bioluminescence, after the evolution of photoreceptors. Since bioluminescence presumably has its greatest utility in dimly-lit environments, it follows that the ability to produce light must have first evolved where organisms had adopted nocturnal or crepuscular lifestyles and therefore had highly sensitive photoreceptors."

Marshall and Widder seem to be saying that bioluminescence arose because it would prove to be useful, especially, as Marshall says, to animals that already had photoreceptors. As she is wont to do, Lynn Margulis approaches the problem from a completely different angle, the standpoint of molecular evolutionary history. She comes much closer to providing an explanation. In *Symbiosis in Cell Evolution* (1998), she wrote:

> Bioluminescence, a scientific curiosity for more than half a century, may be related to oxygen detoxification. Bacteria, dinomastigotes, fireflies, fungi, coelenterates, pyrosomes (tunicates), and fish radiate cold light. The luminescent system of the bacterium *Photobacterium fischeri* requires oxygen and is sensitive to an oxygen pressure of 0.0005 mm Hg. When peroxide it formed, it reacts immediately with luceferin aldehyde to form the corresponding organic acid and water. No toxic radicals are produced, and light is emitted. Since the tasks of detoxification and, eventually, utilization of oxygen were taken over by other oxygen-handling metabolic agents . . . luminescence was lost on the ancestors of most extant organisms.

On land there are only a few bioluminescent life forms, and none at all among the higher terrestrial vertebrates. There are luminescent bacteria, toadstools that glow on rotting logs, centipedes and millipedes that glow in the dark, and of course, fireflies (which are not flies at all, but beetles). Bioluminescence in fresh water is extremely rare, so it is in the ocean—particularly in its deeper reaches—that living light is a common adjunct to existence.

At the Catholic University of Louvain (Belgium), the biologist Jean-François Rees and his colleagues believe they have solved another part of the puzzle of light production. In their analysis of the luciferin-luciferase interaction, Rees et al. (1998) said that the old theory that luciferase reacts with oxygen in the presence of luciferin "just doesn't make sense." When Rees examined coelenterazine, the most abundant kind of luciferin, he recognized that in a terrestrial context it actually mopped up the dangerous free radicals, such as superoxide and hydrogen peroxide, that can destroy cell membranes on contact. But in the depths of the sea, where most bioluminescence occurs, these free radicals are scarce because the UV light that generates them cannot penetrate. Joanna Marchant wrote (2000), "Rees thinks that as animals gradually moved deeper, the lack of light provided increasing pressure for the evolution of bioluminescence at just the time that a diminishing danger from free radicals freed up the ideal molecule for making light." Marchant quotes Peter Herring as saying, "At first this was thought to be pretty off the wall as an idea. Now people are beginning to take him [Rees] seriously. . . . It is very difficult to prove what was going on millions of years ago, but Rees has done a great job of coming up with circumstantial evidence."

In their 1998 study, Wilson and Hastings wrote,

> An old idea, that bioluminescence did not first evolve for the production of light, but as a mechanism for detoxifying an atmosphere becoming danger-

> ously aerobic, keeps resurfacing in various forms. The finding that coelenterazine, widespread in luminous and non-luminous marine organisms, is a potent antioxidant suggests that protection against reactive oxygen species could have been its primary role. . . . Enzymes that would rapidly channel the energy liberated by the oxidation of coelenterazine into harmless photons might then have been selected.

Even though bioluminescence has some drawbacks, for example, betraying one's presence, it is obvious that overall, it serves the species, if not the individual. If it didn't, it would surely have disappeared. And survival of the species no matter how achieved or how erratic—is *critical.*

SHARKS, ETC.

Everybody "knows" that sharks have been swimming in the world's oceans for 300 million years, and that these cartilaginous creatures, less sophisticated than the bony fishes, have barely changed during that time. Indeed, their primitive nature proves that they are perfectly designed predators. While everything else had to change, often in a futile attempt to escape the voracious sharks, the sharks swam silently, steadily onward through time and tide (if they stopped swimming they would sink)—perfect killing machines, rulers of the seas from the deep Devonian to the present. Everybody may know that, but not a word of it is correct. Sharks have indeed been around for a long time, but the early ones didn't look much like the current species. Some of the early versions were so bizarre that they look like failed attempts to figure out where the fins go, or the spines, or even the teeth. You think that teeth have always been arranged in uppers and lowers that fit together more or less snugly and can be used to bite off pieces of grass or twigs or legs? Wait until you meet *Stethacanthus,* with teeth on top of its head, or *Helicoprion,* with its buzz-saw tooth-whorls.

With their cartilaginous skeletons, sharks developed more or less simultaneously with the bony fishes (teleosts), and are in no way more "primitive." In fact, bony fishes are probably older than sharks. Sharks are vertebrates with a boneless support structure, skin made of toothlike protrusions known as dermal denticles, and multiple gill slits—five, six, and one species with seven pairs. The cartilaginous skeletons of sharks do not lend themselves to fossilization, but their teeth do, and since sharks replace their teeth regularly—and most species have multiple rows—the record of sharks through history is mostly written in teeth. Shark teeth are the most common of all vertebrate fossils.

Sharks (or at least their teeth) appear in the fossil record at least as early as the Lower Silurian Period (about 425 million years ago), represented by tiny isolated fossilized dermal denticles in marine rocks from what is now Siberia that were assigned to the shark genus *Elegestolepis.* Shark denticles of similiar age are also known from what is now Mongolia; they belong to sharks of the genera *Mongolepis* and *Polymerolepis.* The oldest sharklike denticles date back to the late Ordovician Period, about 455 million years ago, and came from Colorado, but these denticles differ from those of the earliest known sharks in several important respects, and not all paleontologists agree that they came from true sharks. Since most sharks, ancient and modern, have these toothlike protrusions on the skin, it is not possible to identify the species, but it is clear that sharks were around in pre-Devonian times. The best-known of the early sharks is *Cladoselache* from the Late Devonian to the Mississippian, a slim, torpedo-shaped fish about 4 feet long, with a heterocercal tail (upper lobe longer than the lower, as in most modern species), and narrow-based fins. The teeth of *Cladoselache* were multicusped on a broad base (called "cladodont," hence the name of the species). Sometimes there is one cusp flanking the central spire, sometimes there are two cusps on either side, and sometimes there are three.

Pleuracanths are early sharks. They may have been derived from the cladoselachids, but they had a long dorsal fin that ran almost the length of the back, and some species had a long mobile spine that extended outward from the rear of the brain case. A typical species was the Permian *Orthacanthus*, which reached a length of 13 feet and had a long, serrated neck spine. They were all freshwater sharks with most unusual teeth: the central cusp was usually *smaller* than the lateral cusps, exactly the opposite of the arrangement in cladodonts and modern sharks, in which the smaller cusps flank the larger central blade. From the Late Devonian to the Late Triassic, pleuracanths were found from North America to Australia and Antarctica.

The Bear Gulch Limestone Formation in Montana, dating from the period known as the Mississippian, 320 million years ago, is a sequence of bedded limestone layers up to 90 feet thick and approximately 8 miles in east–west extent that was deposited under shallow-water, tropical marine conditions. Bear Gulch Bay was an arm of the Mississippian Seaway, which covered a sizable portion of what is now North America, but over the ensuing 320 million years, the continental landmasses have undergone some rather dramatic relocations. The latitude of the bay was approximately 10 degrees north of the Equator at the time of deposition, and therefore the climate of the surrounding area must have been tropical and may even have been arid. Fish fossils were first discovered in Bear Gulch, Montana, in 1968, and since that time, more than 4,500 specimens, representing 113 species, have been excavated from this deposit, first by the late William G. Melton Jr. and subsequently by Richard Lund of Adelphi University in New York, and his field parties. The site also contains well-preserved arthropods, sponges, starfish, and soft-bodied organisms, as well as shelly fossils such as brachiopods, bryozoans, and mollusks.

Bear Gulch is in Montana, but "Montana" is only a recent idea. If one were able to look at the Earth as it existed 320 million years ago, no part of it would be recognizable. A single supercontinent known as Pangaea was being formed by the collision of Laurentia (North America and Europe) with the old southern supercontinent of Gondwanaland, which included India, Australia, and Antarctica. The landmass that became North America had, like all the other continental landmasses, spent the previous 280 million years wandering the globe, subjected to the movement of the tectonic plates that made up the skin of the Earth and were themselves moved by seismic forces deep within the planet. Only by the Early Cretaceous, 130 million years ago, did "North America" arrived at its current location, but at that time, rising sea levels inundated the central and western portions of the continent, creating the Western Interior Seaway. This separated the Rockies in the west from the plains and mountain ranges to the east, until it gradually dried up by the close of the Cretaceous, 65 million years ago. The Bear Gulch Formation was reinundated during the Pennsylvanian and again in the Later Jurassic, an inundation that continued into the Cretaceous.

The site at Bear Gulch, Montana, is extraordinary because of the diversity of organisms that can be seen here. The fishes of this prehistoric seaway include over 60 species of sharks and chimaeras, several coelacanths, and a host of ray-finned bony fishes. Most of these species had never been seen before by science and none had ever been seen intact before until their discovery in Montana. The high diversity of fish and the wide range of body forms is evidence of a complex ecosystem most similar to modern bay or estuarine communities.

Whatever the explanation for the extraordinary preservation, the best-preserved animals and plants were killed and buried in one event. Richard Lund explained to me the extraordinary nature of this site:

> The fossils of the Bear Gulch Limestone are not ordinary fossils in an ordinary rock. The rock is a very fine-grained limestone that preserved very fine impressions of very small organisms from an ancient bay. The fossils include soft-bodied invertebrates and plants, shrimp that looked good enough to barbecue and eat, very small fish that just show up as pairs of eyeballs and a smile where the jaw bones had ossified, and some fish with color patterns, livers, and most of their veins preserved as pigments. That small and soft-bodied animals were clearly preserved meant that they must have been killed and buried by the same circumstances, or they would have rotted, or been scavenged. The Bear Gulch Limestone is also not ordinary because virtually the entire bay is present in an outcrop, rather than just a patch of rock totally out of context. So the fossils are extraordinarily preserved, in the context of the bay in which they lived and died, and are a clear window into a time and a part of the history of life on Earth we know very little about.

Stethacanthus, the shark with a toothed anvil on its back, appeared in the Late Devonian of North America, and has been puzzling paleoichthyologists since it was discovered in the Ohio Shales. Where other species have a first dorsal fin, the males of these 3-foot-long sharks had a flattened structure that was covered with a brush-like arrangement of dermal denticles. The purpose of such an arrangement is an utter mystery, and various theories have been proposed to explain it. It might have been some sort of a visual signal to females. (When an entire shark is preserved, male sharks are easy to differentiate from females. Males of all species have paired "claspers" emanating from the pelvic fins, which are used as intromittent organs to transfer sperm to the females.) In addition, the dorsal surface of the head was also equipped with a covering of spines, facing backward, like a brush haircut of tiny

Stethacanthus, *a shark with a toothed anvil on its back, appeared in the Late Devonian of North America, and has been puzzling paleoichthyologists since it was discovered. Where other species have a first dorsal fin, the males of this 3-foot-long species of shark had a flattened structure that was covered with a brushlike arrangement of dermal denticles.*

teeth. In his 1981 summary of the Paleozoic elasmobranchs ("strap-gills"; all sharks, skates, and rays). Rainer Zangerl suggested that this structure might be of use as a threat, "the top of the head and the spine assembly simulating tooth-studded jaws of a very large mouth." In 1984, obviously fascinated with this idea, he devoted an entire paper to the "possible function of the spine-brush complex of *Stethacanthus.*" If the shark arched its back and raised its head so that the flattened surface of the "anvil" was brought closer to the head, the teeth on the anvil would be pointing down, while those on the head would be pointing up, not unlike the teeth in the mouth of a much larger fish.

The caption for the illustration in Zangerl's paper reads, "Spine-'brush'-head-denticle pavement complex of *Stethancanthus* in threat posture, as it might have appeared against a dark-mud background to a potential predator approaching head-on from above." A "dark-mud background" as well as a dark skin color that would highlight the "teeth" as shown in the illustration seem to require the contingent occurrence of too many convenient variables, and besides, even if it existed, this mechanism might be useful only to males, because there was pronounced sexual dimorphism in the stenacanthids. Furthermore, for the "two denticle patches" to look like anything other than two serial denticle patches, the shark would have had to arch its back and raise its head in a manner that is unknown in any living species, and, given the inflexibility of the spinal column in other sharks, probably impossible. In *Sharks of the World* (1985), the paleontologist Rodney Steel makes a most unsharklike suggestion:

> Could *Stethacanthus* have used its "brush" and its head denticles to clamp a hold on larger fish to hitch a ride and perhaps share leftovers from its bene-

factor's meals? Certainly this metre-long Paleozoic elasmobranch was an indifferent swimmer, with relatively weak fins, quite apart from the drag the "brush" must have offered to the water, so hitch-hiking may have been its favoured means of getting around.

The species known as *Symmorium* resembles *Stethacanthus* but it lacks the "anvil," so it might be a female *Stethacanthus*—or it might be a different species. At more than 9 feet in length, *Stethacanthus productus* was one of the larger predatory sharks of the Mississippian. It is known from one complete skeleton and from jaws and isolated teeth of other individuals. The cladodont teeth were large and sharp, with bony bases housed in strong, well-muscled jaws. Because so few specimens have been found, it has been suggested that *S. productus* might have been only an occasional visitor to Bear Gulch Bay, but it is also known from Australia (Vickers-Rich and Rich, 1999), indicating that these weird creatures were not necessarily uncommon, and may have been widely distributed around the world.

Other stethacanthids were equipped with first-dorsal fins that were almost as peculiar as the "brush" of *Stethacanthus*. A species known as *Damocles serratus* received its name because its dorsal spine curved forward until its serrated leading edge was underneath and hanging over its head, like the sword that was suspended over the head of Damocles to remind him of the uncertainty of life. *Falcatus falcatus* ("falcate" means curved like a sickle) also had a spine curving forward over its head, but like that of *Damocles,* its function is unknown. Richard Lund believes that the head spines of the stethacanthids are a manifestation of sexual dimorphism, and their purpose was to attract females. A fossil found by Lund in the Bear Gulch Limestone of Montana included a male and a female of this species, with the female riding the back of the male, holding the curved dorsal spine in her mouth. Lund maintains a Web site on Bear Gulch. Of *Damocles* he wrote (1998) that it is

> the most abundant shark in the Bear Gulch. Ubiquitous, but schools of this shark have been found in the central basin, in open water. Size range: to

The primitive shark known as Damocles serratus *received its name because its dorsal spine curved forward until its serrated leading edge was underneath and hanging over its head, like the sword that hung over the head of Damocles in ancient Greece, suspended by a single thread to remind him of the uncertainty of life.*

> about 10–12 inches. . . . A small, generalized predator, known to have fed on shrimp. A large soft rostrum suggests strongly that they had an ampullary sensory system for electric detection of prey like that of modern sharks. Large eyes suggest a strongly visual predator as well. Females lack scales. Pectoral fin with a trailing whip for maneuverability as in all stethacanthids, otherwise thick and relatively stiff. Countershaded darker above and lighter below; probably similar to the sediment color (tan)—with head and spine probably colorful in males, for advertisement. Sexual dimorphism: dramatic! The sexual dimorphism and the schooling allows us to deduce a great amount of behavior about these fish. Note also that they had very large eyes and sclerotic bones in the covering of the eyes, unusual for a shark.

Other early sharks found at Bear Gulch included the ctenacanths (pronounced "*teen*-a-canths," and meaning "comb-spines"). They were so named because the spines that projected from the skin in front of the two dorsal fins were ornamented with comblike rows of denticles, called tubercles. They had strong jaws like modern sharks, but their rate of tooth replacement was probably slow. The common fossil shark *Hybodus* was a ctenacanth, a 3-foot-long creature that looked much like some sharks today, except that this species and many of its hybodont relatives had spines on either side of the head, resembling horns. (The first of these fossil "horns" were identified as teeth, but subsequent complete specimens revealed their proper location.) Like the spines of *Damocles* and *Falcatus,* these horns are believed to have been related to mating, although their actual purpose can only be guessed at.

In his 1990 summary, "Chondrichthyan Life History Styles as Revealed by the 320 Million Years Old Mississippian of Montana," Lund wrote:

> Chondrichthyans comprise 55 of the 94 vertebrate species found to date in the Mississippian marine Bear Gulch Limestone 6 km by 19 km tropical bay. The chondrichthyan-osteichthyan faunal proportions differ strongly from those of modern marine fish faunas. Secondary sexual dimorphism, size partitioning, life period segregation, and sexual segregation, reinforce the considerable morphological specializations among the chondrichthyans that were involved in the maintenance of this diversity. The large number of rare species suggests that many species are represented only by occasional recruitment, reflecting differential life history styles as well as habitat limitations for these species. Repeated fluctuations in environmental parameters through time, producing fine-scale spatial and temporal non-equilibrium, were probably ultimately responsible for the maintenance of this broad-scale high diversity community.

The Bear Gulch fauna is but a small window into the Mississippian Period, when so much was going on, and it is possible that similar mixed assemblages might be found elsewhere. So far, however, we are extraordinarily lucky that this spot, so remote in space and time, was uncovered.

Rainer Zangerl describes (1981) the Eugenodontida as among the most unusual animals ever found, and one of them remains a total mystery to this day. It is *Helicoprion* (sometimes misspelled *Heliocoprion*), which has been found in Russia, Idaho, Japan, Greenland, Australia, and elsewhere. (*Helico* means "spiral" or "curl," and *prion* means "saw," so *Helicoprion* actually means "circular saw".) It was originally described by the Russian paleontologist A. P. Karpinsky (1899) in a German geology publication. Only a year after Karpinsky's description appeared, the American paleontologist Charles Eastman wrote, "Helicoprion is a fitting title for the peculiar 'spiral saws,' coiled in three and a half whorls and armed with upwards of 150 sharp teeth, which have recently been discovered in the Permo-Carboniferous (Artinsk Series) of the government of Perm."

Like everyone else, including Karpinsky, Eastman could not figure out how the tooth whorls might have been used, nor, for that matter where on (or in) the shark they went. He wrote, "Of the two leading theories as to the position of [the] 'spines,' he [Karpinsky] first ascribes them to the jaws of a shark or skate, and the other to the median line of the back, some distance in advance of the dorsal fin. Vladimir Obruchev, in his 1952 analysis of the tooth-whorl, decided that if it "was placed in the lower jaw, it would only prevent the fish from feeding," and he therefore placed it in the upper jaw, "where it could serve as an effective protection . . . acting as a shock absorber and protecting the head of the animal."

The mysterious fossilized "tooth-whorl" of Helicoprion. *Although there have been some good guesses, no one really knows how this 280-million-year-old shark used its teeth, or for that matter, where on the shark they were located.*

Svend Bendix-Almgreen described (1966) *Helicoprion* as having flat crushing teeth in its upper jaw with the circular-saw blade made of teeth somehow inserted into its lower jaw. Where the two halves of the jaws come together in the front is known as the symphysis, so the coiled buzz-saw is technically known as the "symphyseal tooth spiral." Bendix-Almgreen wrote:

> With regard to the role and function of the symphyseal tooth-spiral many different opinions have been published, in which it has been suggested as a defense weapon, as a cutting device, or as a crushing organ. If it was principally a crushing organ, really strong wearing marks would be expected to occur . . . The proposal that the tooth spiral acted as a defense weapon was mostly based on its assumed location in the upper jaw symphysis, now an invalid assumption. The most probable function and use of the tooth-spiral on the evidence of wearing marks seems to have been as a cutting, and probably

> to some degree, a tearing device, in combination with the rows of small upper jaw teeth.

A veritable cottage industry is devoted to the solution of the *Helicoprion* tooth-whorl mystery. In the April 1, 1973, issue of the *Journal of Insignificant Research* (note the issue date) Michael Williams and Kathy Elbaum published a paper called "Bendix-Almgreen's Recent Investigations on *Helicoprion* and the Biomechanical Significance of Karpinsky's Reconstruction," in which they demonstrated that the whorl was the toothed equivalent of a New Year's Eve noisemaker, and that *Helicoprion* could extend it by the use of hydraulic force. (That they tested it on strawberry Jell-O might provide an idea of the tone of their paper.) In *Planet Ocean: Dancing to the Fossil Record* (1994), the author, Brad Matsen, and the artist, Ray Troll, devoted an entire page to the "vexing fossil shark." Ray Troll has an "obsession" with *Helicoprion*'s tooth whorl, and has now made a model.

> There was no upper half whorl of sharp teeth for the whorl to cut against, only rows of small crushing teeth. In all of the whorls and skulls Bendix-Almgreen examined, no other upper teeth were found besides the crushing teeth. The lower jaw support jutted out a bit at the very end of the whorl as well. After going back and rereading the paper he wrote as well as Obruchev's 1953 papers, I could begin to "see" the outline of the skull: long and extremely narrow. I realized this thing had a long nose on it much like a modern day Goblin shark. Basking sharks also have a somewhat elongated rostrum as well (personal communication, 2000).

Later analysts have done a better job of describing the fossil, but have come no closer to placing the tooth-whorl or describing its function—if it had one. It is not at all obvious, by the way, in which direction the teeth grew. They are usually unaffected by wear, so we cannot tell which are newer. Romer (1966) seemed confident that he knew the answer, writing, "In *Helicoprion* the teeth, in contrast to the usual shark conditions, did not drop out after they had reached the edge of the jaw, but curved downward and inward to form the core of a complex growing spiral." But Ray Troll, who has been trying to figure out *Helicoprion* for years, wrote me, "The new teeth are the *big* teeth coming in from the back of the jaw. This is a bit mind-blowing and counter-intuitive. . . . When I finally realized how they worked it was an epiphany for me. Rainer [Zangerl] and J. D. [Stewart of the Los Angeles County Museum of Natural History] had been telling me this for years. The oldest teeth are the tiniest teeth. When the shark was petite it had small teeth. As it grew it produced bigger teeth . . . so what you're seeing is really a fossilized growth ring."

In *The Rise of Fishes* (1995) John Long includes an illustration of *Helicoprion,* which shows the shark's *lower* jaw curling downward into a tooth-studded spiral, but, he says "there is much speculation about how these whorls were used in life." In this illustration—which was the basis for a painting in the September 2000 issue of *Na-*

Artist's reconstruction of Helicoprion. *Here the tooth-whorl has been placed in the lower jaw because that seems to be the most likely location, but exactly how the shark used this "buzz-saw" is not evident.*

tional Geographic—it looks as if the shark might be able to uncoil its jaw and lash at prey items with a sort of a toothed whip. Long then writes, "It has even been suggested that they may have mimicked ammonites, coiled shellfish that were abundant at the time, and thus could have attracted prey to the shark, if used in a particular fashion. It seems more likely that these sharks used the jagged tooth-whorls when charging into a school of fish or ammonites and thrashing about to snag prey on the projecting array of teeth." He compares this activity to that of the saw sharks, which have a flattened, tooth-studded rostrum. In fact, however, *their* feeding technique is not well understood either. It is believed that saw sharks (Pristiophoridae) swim into a school of fish and slash at them with the rostrum, but Compagno (1984) has written, "The lateral rostral teeth, flat snout and head . . . are evident modifications that allow these sharks to use their rostra as offensive weapons to kill their prey and stir up bottom sediments to rouse prey organisms, but unlike the batoid saw-fishes (Pristidae) this behavior has not been observed."

Because of the mystery that surrounds it and the tooth-whorl's perfect logarithmic spiral in those fossils where it is complete, *Helicoprion* stands as one of the most intriguing and enigmatic creatures of all time. The bizarre fossils of *Helicoprion* tooth-spirals have been placed by scientists all over the body of the shark and have also been identified as snakes, snails, ammonites, or even entire hypothetical animals. Though their positive affiliation with Paleozoic sharks narrows the field somewhat, their secrets remain locked in the stone.

Other fossils of similarly inexplicable teeth have been found: *Parahelicoprion, Sacroprion, Campyloprion,* and *Toxoprion* are all lower-jaw tooth-whorls (Zangerl 1981), and their function is no more obvious than is *Helicoprion*'s. In the case of *Parahelicoprion*, the evidence of its existence is a segment of a tooth row found in a Lower Permian deposit in Russia. But what a segment! It consists of a 10-inch chunk of rock with a single row of six damaged, banana-sized teeth that undamaged would have been sharply pointed and with knifelike cutting edges fore and aft. Another, somewhat larger specimen has been found in Bolivia. Measured vertically, these teeth are considerably larger than those of *Carcharodon megalodon,* the much-publicized giant relative of the great white shark and would be in the same range as the serrated teeth of *Tyrannosaurus rex,* usually considered the most formidable dental equipment in history. Giving scientists nothing more to go on but this single tooth row, *Parahelicoprion* remains a complete enigma, but unless it was an animal with a gigantic head or outlandishly oversized teeth, it had to have been a monster, at least 100 feet long and maybe more.

Another genus with problematic dentistry is *Edestus,* which had jaws that sported two single rows of teeth arranged—maybe—like a pair of toothed scissors. Where the teeth of *Helicoprion* overlapped like a splayed fan of playing cards, those of *Edestus* were slotted together in a line and did not overlap. There are more than a dozen nominal species (*Edestus vorax, E. heinrichi, E. minor,* etc.), from deposits in Germany, Russia, and America. Zangerl (1981) wrote that "*Edestus giganteus* indicat[ed] an animal of immense size" but didn't say how big it might have been. Lund suggests that with modern "normal shark" proportions, it might have gotten as large as *Carcharodon megalodon,* in the 50-foot range (personal communication).

Then there was *Ornithoprion,* which translates as "bird saw." Its beaklike jaws explain the "bird" part; the "saw" consisted of a miniature tooth-spiral located midway in the lower jaw. Three skulls were found in the Logan Quarry shales of Indiana, and described by Zangerl in 1966. Some of the fossils were embedded in the dense shale and were too fragile to be removed, so the only way to examine them was to X-ray the stone. Thus, as Zangerl wrote, "The specimens themselves have thus not been seen except as shadow pictures on film." Others were found when the shale was split, but even then, the fossils invariably separated into the part and counterpart, and were therefore very difficult to study.* The X-rays revealed a skull with a greatly extended lower jaw and a short, beaklike upper, with enlarged symphyseal teeth and tiny rod-shaped pavement teeth. Zangerl compared the skull of *Ornithoprion* to that of the living half-beak fish, *Hemiramphus,* which feeds near the surface, but then wrote that the teeth of *Ornithoprion* made surface feeding unlikely. "The functional significance of the grossly elongated mandibular rostrum . . . remains a mystery, but it is possible that the animal used this device to stir up (and perhaps flip up) potential food items from the bottom, and then proceed to grasp them." The skull of *Ornithoprion* was less than 6 inches long, and the entire shark was about 3 feet long.

As described by Zangerl and Case in 1976, *Cobelodus aculeatus* was a "widely distributed and common shark of the Pennsylvanian central basin complex of North America." ("Pennsylvanian" refers to the geological period also known as Late Carboniferous, from 320 million to 290 million years ago.) Originally described in 1891 by Edward Cope from a partial specimen found in Illinois, *Cobelodus* is now represented by more than 100 specimens, and is therefore the best-known Paleozoic elas-

---

* Fossils found in sedimentary shales, like those at Holzmaden in Germany or Bear Gulch in Montana, are often formed in layers that are carefully separated to reveal the fossil between them. The process produces a fossil that was trapped between two layers, and the positive element, with the bones protruding from the matrix, is known as the part, or plate, and the negative image is the counterpart, or counterplate.

mobranch. Five to 6 feet in length, these sharks had multicusped (cladodont) teeth in the lower jaw and needle-shaped, cuspless (monodont) teeth in the upper. As reconstructed from the fossilized cartilaginous skeletons, *Cobelodus* was a big-eyed, blunt-snouted shark, with a single dorsal fin set way back, and a deeply forked, almost perfectly homocercal tail (having two equal-sized lobes). There was no first dorsal and no anal fin. In general appearance, this shark was similar to *Simmoriam,* the species that was once held to be a female *Stethacanthus,* and since it has been shown that only the males of these early sharks, such as *Stethacanthus, Damocles,* and *Falcatus,* sported spikes, spines, and other decorations, there is a strong likelihood that *Cobelodus* was not a separate species at all, but the female of an identified species.

Although not found in the Bear Gulch Limestones, the iniopterygians ("neck wings," from *inion,* "nape," and *pteron,* "wing") are another group of small, very odd fishes that resemble early chimaeroids. They were first described by Zangerl and Richardson in 1963, but the fossil material was scrappy and badly mutilated. Not for another 10 years were many more specimens, in a much better state of preservation, discovered in the carbonaceous sheety black shales of Nebraska and Iowa. Like the Bear Gulch fossils, the seven known species of iniopterygians date from the Pennsylvanian period, roughly 320 million years ago. As reconstructed by Zangerl and Case (1973), these foot-long creatures were large-headed, large-eyed cartilaginous fishes with tiny teeth, and relatively enormous claspers in the males. The teeth are arranged in "families" of separate denticles, or in various whorls. In both sexes the pectoral fins were so large that they made the iniopterygians look like round-bodied airplanes, but with the wings located in the position that airplane designers call "high wing." Janvier (1998) calls this configuration "the almost dorsal position of the pectoral fins." A male of one of the species (*Iniopteryx tedwhitei*) had rasplike spines on the anterior rays of the pectoral fins, which was probably an example of sexual dimorphism and may have provided the male with a means of grasping the females. It now appears that most, if not all, mature iniopterygian males had some display of anterior pectoral denticles. Some of them had prominent single lower symphysial and paired upper symphysial tooth-whorls, miniatures of those of *Edestus* (personal communication from Lund).

The iniopterygians ("neck wings": from inion, *meaning "nape," and* pteron, *"wing") were a group of small, very odd fishes that resembled early chimaeroids. These foot-long creatures were large-headed, large-eyed cartilaginous fishes, with tiny teeth, and relatively enormous claspers in the males. They date from the Pennsylvanian Period, roughly 320 million years ago.*

In Greek mythology, the Chimaera was a fire-breathing she-monster with the head of a lion, the body of a goat, and the tail of a serpent. When not being used by ichthyologists, the word means any kind of grotesque creation made of disparate parts, or an imaginary, fearsome monster. The living chimaeras are far from fearsome monsters, but they do appear to be assembled from different kits. Chimaeras are cartilaginous fishes generally less than 3 feet long that are found in the world's temperate to cold waters, from the shallows to the depths. They are characterized by platelike teeth, often fused together in a beaklike arrangement, which they use to feed on small fishes and invertebrates, especially mollusks. Where most sharks have sharp teeth arranged in multiple rows, the chimaeras have slow-growing tooth plates, which has led to their being known as bradyodonts (*brady* means "slow-growing," and *dontus* means "tooth"). Except for the cartilaginous skeleton and the presence of claspers, there is little to affiliate the chimaeras with sharks, and although most popular books say that they are "related" to sharks, they are quite different and are placed in their own order, the Holocephali, which means "whole head" and refers to the fact that the upper jaw is fused to the cranium.

Chimaeras have large, emerald-green eyes and two dorsal fins, the first of which has a serrated, venomous spine in front of it. Their skin is smooth and scaleless. They have multiple gill openings, like sharks, but theirs are covered by a plate, the operculum, like those of teleosts (all bony fishes and all placoderms). Like some sharks, skates, and rays, they lay large eggs encased in horny capsules. The males also have claspers, and there is an additional clasper in front of the eyes whose function is not clearly understood. It may serve no other purpose than to advertise the male's sexual availability. Chimaeras move by slowly flapping their broad pectoral fins, which results in a bobbing, ungraceful swimming style. There are some 28 species of chimaeras living today, ranging in length from 2 to 6 feet and separated into three groups differentiated by their snouts, which may be rounded, pointed, or plow-shaped. Most species of the genera *Chimaera* and *Hydrolagus* are round- or short-nosed and have a long, tapering tail, which probably accounts for the common name ratfish. The long-nosed versions (*Rhinochimaera*) have a nose like that of a jet fighter, and the plow-nosed (*Callorhynchus*) have an appendage that hangs down and points back toward the mouth. As inoffensive, deepwater species, they bother nobody, but perhaps because of their enigmatic nature and less than beautiful appearance, they have acquired an array of pejorative common names and nicknames, including, in addition to ratfish, ghost sharks, spook-fishes, and elephant fishes, the last inspired by the most unusual trunklike proboscis of the plownose chimaera. In *The Encyclopedia of Fishes,* Paxton and Eschmeyer (1995), for reasons completely unexplained, call chimaeras "ghouls of the sea."

If today's species are a confusing lot, the chimaeras of the past are so confounding that many scientists are not sure how to classify them, nor for that matter, how to define a chimaera in the first place. After coelacanths, the most numerous fishes in the Bear Gulch limestones are chimaeras, and one of the most numerous of these is the jaw-breaking *Harpagofututor volsellorhinus,* of which 89 specimens have been found. Richard Lund named and described the species in 1982 from *harpagos,* Greek for "grappling hooks"; *fututor,* "copulator"; and *volsellorhinus,* which translates as "pincer-snout." It is therefore a copulator with pincer-shaped grappling hooks on its snout. The names refer to the bizarre secondary sexual apparatus of the males, which

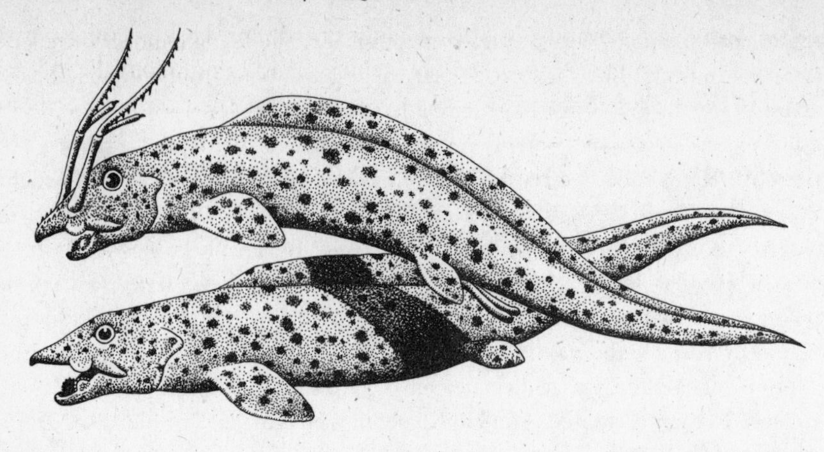

The jaw-breaking Harpagofututor volsellorhinus *was found by Richard Lund in Montana's Bear Gulch Formation. The name comes from the Greek for "grappling hooks," "copulator," and "pincer-snout"; thus it is a copulator with pincer-shaped grappling hooks on its snout.*

consists of paired, jointed spikes with forked ends projecting from the top of the snout, a region technically known as the ethmoid. It is not clear how the male used these toothed pincers; Lund (1982) wrote, "The conclusion is inescapable that the ethmoid claspers had to play a critical role in mating." Female chimaeras, both extinct and extant, have no claspers on the head and certainly none in the pelvic region. In 1997, Lund described an incredibly well preserved specimen of *Harpagofututor volsellorhinus* from Bear Gulch in which the soft-tissue pigments of the internal organs were visible, providing "otherwise unavailable evidence for the internal reproductive features of these sexually mature Paleozoic chondrichthyans." It was also clear that the animal died of asphyxiation. *Harpagofututor* was an anguilliform (eel-shaped) creature with no prominent fins save for the grappling hooks of the males.

Another spectacular fossil from Bear Gulch is *Debeerius ellefseni,* named and described in 2000 by Lund and by Eileen Grogan of St. Joseph's University in Philadelphia. (*Debeerius* is for Sir Gavin DeBeer, a pioneer in the study of skull morphology, and *ellefseni* for Gissur Ellefsen, Richard Lund's father.) Like the holocephalans, it

Described and illustrated by Dick Lund and Eileen Grogan in 2000, Debeerius ellefseni *is a chimaeralike creature from the Bear Gulch Formation in Montana. The fossils were so well preserved that the actual color pattern on the skin could be seen.*

had an operculum covering its gills, a spiny first dorsal, large pectoral fins, and a "moderately low-angle herterocercal tail" (Grogan and Lund 2000). Eighteen specimens of *Debeerius* were collected from the Bear Gulch Limestone beds, and of these, some were so beautifully preserved that the authors were allowed an unprecedented view of the skeleton, particularly the jaws, which, because they are autodiastylic (suspended from two points), may be a forerunner of the jaw arrangement in later sharks, chimaeras, and coelacanths. *Debeerius* is special in many ways (it has been placed in its own family, genus, and species), and one fossil is especially remarkable because it has been preserved with its color pattern intact. It consists of a series of dark saddles on a light ground, not unlike the pattern of the living leopard shark, *Triakis semifasciatus.*

The genus name *Echinochimaera* means "spiny chimaera." Janvier writes that it is "strikingly chimaera-like in overall aspect," and like so many of the Bear Gulch species, it is an exceptional creature indeed. It is (or was) the only chimaeroid with skin scales, all the other fossil and extant forms being smooth-skinned. Also, the males were decorated with an array of fin and head spines that differentiated them from the males of any other species. The smallest specimen of *Echinochimaera meltoni* was less than half an inch in length, but the mature males described by Lund in 1977, were 6 inches long. They had a dorsal fin that resembled the sail of a windsurfer, cocked forward and supported by a powerful spine with a raspy leading edge, and a row of large, tooth-shaped denticles that ran from the second dorsal to the base of the tail. Characteristic of this species were the supraorbital (above the eye) spines, which in the males were branched like the antlers of deer; in the females they were much less complex. Although *Echinochimaera* became extinct 300 million years ago, many of its salient characteristics are visible in the living chimaeras.

In the phylogeny of the six-gill (*Heptranchus* spp.) and seven-gill (*Notorhynchus* spp.) sharks, there are fossil species that are so anatomically similar to the living species that the latter have been awarded the title living fossils. The morphology of these living species—a single dorsal fin located far posteriorly along with the anal and pelvic fins, and more than five gill slits—has led to their regular characterization as primitive. According to Maisey and Wolfram in *Living Fossils* (1984) the frilled shark, *Chlamydoselachus,* is another candidate for that appellation, but its "biggest problem is its lack of a fossil record." The early hexanchoid sharks were classified by Cuvier as "Notidanus," and although this name is no longer in use, it was chosen as a convenient title by Maisey and Wolfram's article in *Living Fossils* (1984). As with most fossil sharks, the primary evidence for their existence and affinities is their teeth, but there have been some fairly complete hexanchoid fossils found in the Late Cretaceous shales of Solnhofen and Eichstadt. There are enough differences in the fossil species—*Hybodus,* for example—and the living hexanchoids to demonstrate to Maisey and Wolfram that "there is no evidence to support a relationship between modern hexanchoids and Mesozoic or Late Paleozoic hybodonts. If hexanchoids are living fossils, it is only because they have a record extending back to the Early Jurassic, not because of their putative relationship with some other fossil sharks." Still, a species whose direct lineage can be traced back for 200 million years surely qualifies—

along with the crocodiles, horseshoe crabs, and coelacanths—for the registry of ancient and honorable species.

Many people think that most sharks look like the great white shark in *Jaws*: a large, man-eating, powerful creature with an evil black eye, a grinning mouthful of razor-sharp teeth, and a dorsal fin knifing through the water. In fact, of the 350-odd species of sharks alive today, very few of them fit this description. To be sure, some, such as the bull shark, tiger shark, and mako, are powerful swimmers with an occasional interest in human flesh, but most sharks are less than 3 feet in length and prey on small fishes and squid. There are sharks alive today with pavement-like teeth for crushing shellfish, and others that are exclusively plankton-eaters.

Fossil teeth of the Cretoxyrhinidae, another related family of lamnid sharks, have been collected and analyzed for years. Because of their similarity to mako teeth, the teeth of the cretoxyrhinids are believed to have come from similar sharks, extinct since the close of the Cretaceous Period. The teeth of the six known species suggest fast, powerful predators, similar to the mako or great white in shape and feeding habits, with streamlined bodies, large tail fins, and excellent senses of sight, hearing, and smell. Welton and Farish (1993) wrote that "sharks of the Family Cretoxyrhinidae clearly were the largest and most voracious of all fish predators in Cretaceous seas." Fossils have been collected around the world, and some of the best come from North America. Examination of a single vertebral centrum from a shark found in the Kiowa Shale of Kansas (Shimada 1997) indicated a shark that was 32 feet long. In 1992, while excavating a mosasaur skeleton from the Kansas shales, Mike Everhart and his wife, Pam, found unmistakable evidence of a shark attack, consisting of tooth fragments in some of the mosasuar's bones and some bones bitten clear through. The attacking shark was identified as an 18-foot *Cretoxyrhina mantelli,* which Mike Everhart nicknamed the "Ginsu shark," because it "sliced and diced" its prey.

The great white shark, *Carcharodon carcharias,* is surely the most infamous of all sharks, the star of Peter Benchley's bestselling novel, *Jaws*, four Hollywood movies, and innumerable television specials. The great white has also been the eponymous subject of several books (including one of mine, *Great White Shark* [1991], written with John McCosker). Even though it was fun to think so, *Carcharodon megalodon* is not an ancestor of the great white, but rather, its phylogenetic great-uncle. Megalodon was one of the most fearsome predators ever to have lived on Earth, and its disappearance remains one of the great mysteries of evolutionary history. (For convenience, this giant fish is often called simply megalodon, which translates as "great tooth.")

The ancestry of *C. carcharias* is convoluted, largely because of the various names that have been given to this species over time and those given to its putative ancestors. Applegate and Espinosa-Arrubarrena (1996) identify *Cretolamna* as the "possible ancestor" of *C. carcharias.* Known from the Late Cretaceous of England and Morocco some 65 million years ago, *Cretolamna appendiculata* had teeth characterized by lateral denticles, which become progressively smaller through the intermediate species (*C. orientalis, C. auriculata, C. angustidens*), and disappear altogether in the two branches that diverge from *C. subauriculatus,* the putative ancestor of *C. carcharias* and *C. megalodon.* In their phylogenetic chart, the white shark and megalodon branch off about 20 million years ago, and megalodon disappears altogether 2 million years ago.

Whether or not it was an ancestor of the great white, or even a close relative, is still a subject of debate among elasmobranch taxonomists. Before these debates began, the similarities in the triangular, serrated teeth seemed more than sufficient for scientists and nonscientists to lump them together in the genus *Carcharodon,* and indeed, many still do. Led by Henri Capetta of Montpelier, France, there are those who would place megalodon in the genus *Carcharocles,* claiming that this name has precedence because it was erected in 1923 by Jordan and Hannibal. There is another school that favors the genus *Procarcharodon,* and another that supports *Megaselachus,* but the taxonomic infighting will not be resolved here.

Megalodon's teeth, our primary souvenirs of its existence, are often larger than a grown man's hand, but in general form they closely resemble those of its smaller relative, *Carcharodon carcharias.* The differences are nonetheless sufficient to justify separating the extinct and recent sharks as distinct species: besides their obviously greater size, megalodon teeth have more and relatively smaller serrations and, above the root, a scar or "chevron" that is lacking on the teeth of adult *C. carcharias.* That they are conspicuously different in color as well is simply a consequence of fossilization: the teeth of all living sharks are white, whereas the teeth of extinct or fossil species range from light brown to black.

After dinosaur bones and trilobites, the teeth of the huge extinct shark megalodon are probably the most popular of all fossils. These teeth—always fossilized, despite some claims to the contrary—are heavy and triangular, with serrations along the blade. The largest known megalodon tooth is 6.5 inches long. These teeth are shaped very much like those of today's great white shark, so megalodon comes to us with its reputation already in place. Abundant fossil evidence shows that megalodon lived during the Middle and Late Tertiary, as far back as 50 million years ago, and probably wandered the same seas as the modern white shark, plying pretty much the same trade—but on a substantially larger scale.

In 1909, a gigantic re-created megalodon jaw was placed on exhibit in the American Museum of Natural History, and the curator, Bashford Dean, wrote, "At the entrance of the Hall of Fossil Fishes, there is now exhibited a restoration of the jaws of megalodon, which lived along the coast of South Carolina in Tertiary time. There can be no doubt that this was the largest and most formidable fish living or extinct of which we have any record. The jaws of a fully grown specimen measured about nine feet across, and must have had a gape of five or six feet."* The jaws of the

* A photograph of this jaw, usually with one or more gents in lab coats sitting in it, appears in almost every book about sharks published up to 1996. The re-creation was, however, a considerable stretch, and the formidable jaw has been removed from exhibit and a downsized version installed in the museum's Hall of Vertebrate Origins. The exaggeration was caused by the curators' erroneous assumption that all the teeth in a shark's jaw are the same size; they therefore placed six-inch teeth at the rear of the jaw and reconstructed a jaw large enough to accommodate such teeth. In fact, the rear teeth would have been much smaller (Randall 1973).

shark as re-created by the museum were large enough for their owner to have swallowed a horse.

The examination of a fossil baleen whale in the Smithsonian's collection showed that the whale had been killed or at least viciously attacked by a gigantic shark between 12 million and 15 million years ago. Bretton Kent, a biomorphologist at the University of Maryland, has compared the teeth of megalodon to those of the great white. He found that the sharp, comparatively thin (in cross-section) teeth of *C. carcharias* are ideal for slicing through the flesh of pinnipeds and small cetaceans. The massive teeth of megalodon are also serrated, and were therefore good cutting tools, but in addition they are thicker and have much deeper roots, which suggests a different attack strategy than that employed by the great white. With his student Monica Newell, Kent examined the location of the bite marks on the bones of the Smithsonian's baleen whale and other whale bones with bite marks, and found that the giant sharks hit in the region of the shoulder and flipper, trying, Kent and Newell speculated, to crush the bones and damage the heart, lungs, and other delicate organs (Riordan 1999). There was obviously nothing subtle about the attack of this 50-foot-long shark, one of the most ferocious predators of all time.

We do not actually know what megalodon looked like, since all we have to go on are its teeth, but we assume from good paleontological practice that it looked a lot like the living white shark, although it was considerably larger and bulkier. In Gottfried et al. (1996), a reconstruction drawing shows a fish with a massive head and jaws and a body that is much more robust than the puny little white shark's. A very large shark swam in the sea as recently as 2 million years ago, and fed on large prey, probably whales, and is now—mercifully for swimmers, anyway—extinct. But which seas did megalodon inhabit? Robert Purdy (1996) segregates white sharks into "small-toothed species," which includes the current great white, and the "giant-toothed," which include *Carcharodon orientalis, C. auriculata, C. angustidens, C. subauriculatus,* and the giant of the giant-toothed sharks, *Carcharodon megalodon.* From the fossil evidence Purdy deduces that the various species of white sharks "first appeared in the Paleocene seas of southern Russia, Morocco, Angola, and the United States." The fossil teeth of giant-toothed sharks do not necessarily coincide with the bites on fossils of cetacean prey, because these sharks evolved in the Paleocene, before the appearance of whales. Indeed, the eventual appearance of larger and larger whales may have been a factor in the exponential increase in the size of the giant-toothed sharks. The areas most productive in fossils of the giant-toothed sharks are the southeastern coast of the United States from Florida to Maryland and, to a much lesser extent, southern England and the shore of what is now Belgium, although they also appear elsewhere.

Megalodon probably reached a length of around 50 feet, but everywhere there are those who want to increase its length. In a 1964 discussion of fossil teeth found at Uloa, South Africa, David Davies, the director of the Oceanographic Research Institute in Durban, wrote: "Estimates of size made from the fossil teeth of this wide-ranging shark obtained in various parts of the world indicate that it may have reached 60 to 80 ft. in total length." People seem more than a little reluctant to concede the disappearance of some "prehistoric" animals (think of the "plesiosaur" of Loch Ness) and of these, the one that surfaces most regularly is the giant shark. Not in the silly novels that play on the success of the *Jaws* phenomenon ("If they made all that money with a twenty-five-foot-long shark, think of what we could do with a hundred-footer"), but rather in works that might be taken seriously. The best-known of these is an account by David Stead (1877–1957), an Australian ichthyologist, of a 1918 incident:

> In the year 1918 I recorded the sensation that had been caused among the "outside" crayfishmen at Port Stephens [New South Wales], when, for several days, they refused to go to sea to their regular fishing grounds in the vicinity of Broughton Island. The men had been at work on the fishing grounds—which lie in deep water—when an immense shark of almost unbelievable proportions put in an appearance, lifting pot after pot containing many crayfishes, and taking, as the men said, "pots, mooring lines and all." These crayfish pots, it should be mentioned, were about 3 feet 6 inches in diameter and frequently contained from two to three dozen good-sized crayfish each weighing several pounds. The men were all unanimous that this shark was something the like of which they had never dreamed of. In company with the local Fisheries Inspector I questioned many of the men very closely and they all agreed as to the gigantic stature of the beast. But the lengths they gave were, on the whole, absurd. I mention them, however, as an indication of the state of mind which this unusual giant had thrown them into. And bear in mind that these were men who were used to the sea and all sorts of weather, and all sorts of sharks as well. One of the crew said the shark was "three hundred feet long at least!" Others said it was as long as the wharf on which we stood—about 115 feet! They affirmed that the water "boiled" over a large space when the fish swam past. They were all familiar with whales, which they had often seen passing at sea, but this was a vast shark. They had seen its terrible head, which was "at least as long as the roof of the wharf shed at Nelson's Bay." Impossible, of course! But these were prosaic and rather stolid men, not given to "fish stories" nor even to talking at all about their catches. Further, they knew that the person they were talking to (myself) had heard all the fish stories years before! One of the things that impressed me was that they all agreed as to the ghostly whitish color of the vast fish.*

Despite the thousands of megalodon teeth that have been dredged from the ocean floor or found embedded in the chalky cliffs of California, Maryland, Florida,

* Stead's recounting of the 1918 story is found in the 1963 publication *Sharks and Rays of Australian Seas*, edited after his death by Gilbert Whitley. Stead also predicted 80-foot sharks in a little book published in Australia in 1933 called *Giants and Pygmies of the Deep*. In the section devoted to the white shark, he wrote: "It reaches at least 40 feet, as far as observed specimens have been recorded, but teeth of a similar kind have been seen by me, which must have come from a specimen no less than 80 feet in length. Such a sea devil as this could accommodate one hundred humans at one meal."

North Carolina, Belgium, and Morocco, *not a single white one has ever been found.* All the teeth that have been unearthed or dredged up are brown or black, indicating that they are fossils, not from living sharks. The fact that they are not white does not mean they are not bone. A fossil bone is still bone, but one that sometimes contains a hard and heavy infilling of other minerals as well. In the same way that the "bones" of the dinosaurs in various museums are not exactly bones, but bones in which minerals have gradually replaced the inorganic material (apatite) in the bones over the eons. Should someone, then, dredge up a *white* megalodon tooth, we would know that the giant shark became extinct quite recently—or is flourishing somewhere in the vastness of the oceans and has simply lost a tooth.

Sharks of all species have multiple rows of teeth; teeth currently doing the biting are replaced regularly in a process that has been likened to the action of a moving escalator. Behind the functional front rows of teeth, there are other rows, waiting to move forward as those in the front rank fall out or are otherwise dislodged. A shark will therefore have many more teeth in its mouth during its lifetime than would any other vertebrate, the total number from a given animal across a lifetime numbering perhaps in the thousands. Thus, we can be certain that hundreds of thousands or even millions more teeth remain in the seafloors or in the ground, most of them never to be seen.

We are probably fortunate that megalodon is extinct. Consider what our attitude toward a day at the beach would be if there were hungry sharks large enough to swallow two or three of us at once. One of the more interesting questions of shark evolution concerns the disappearance of this 50-foot-long monster that is known to have fed on whales. The whales continued, but megalodon, one of the most fearsome hunters that ever lived, became extinct. Why? Did the whales develop more speed and maneuverability? Did the oceans cool off? Did killer whales replace megalodon as the oceans' apex predator?

LAND HO!

The fossil record often consists of unique specimens that are said to be transitional to other unique specimens, but they are often separated by vast tracts of geography or, more often, by vast tracts of time. The concept of transition would seem to require some sort of congruity or succession in the same general area over however long a time period is required for one form to evolve into another. In other words, transitional forms between fishes and amphibians have been found in Pennsylvania, Scotland, or Latvia, suggesting either that they migrated during or after the transition, that the land itself migrated, or that more digging is necessary.

"The vertebrate invasion of land," wrote Daeschler and Shubin (1995), "is one of the major events in the history of animal life. For almost 60 years after the discovery of *Ichthyostega,* paleontological approaches to tetrapod origins contrasted low diversity Late Devonian tetrapods with diverse and antomically advanced Carboniferous assemblages. New fossil discoveries . . . demonstrate that Devonian tetrapods were morphologically varied and globally distributed by the end of the Fammenian [367 million to 362 million years ago]. These fossils demonstrate that the origin of tetrapods was not a simple progression from aquatic fish to terrestrial amphibian—it involved a series of evolutionary experiments with structure, function, and ecology." The oft-repeated claim that "there are no transitional forms" is refuted—at least partially—in the case of the transition from fish to amphibian, even though the critical process whereby water-breathing gills were transformed into air-breathing lungs remains unexplained. Amphibians (the word "amphibia" means "both lives") are now defined as cold-blooded vertebrates that undergo a metamorphosis from an aquatic, water-breathing limbless larva (the tadpole stage) to a terrestrial or partly terrestrial air-breathing, four-legged adult. So the amphibian appears to be the transition between fishes and reptiles, but tens of millions of years passed between these forms, and there is no obvious connection, other than a fossil fish with legs, or an amphibian with a long, finned tail.

What did the early aquatic vertebrates look like? *Panderichthys,* a lobe-finned fish found in Latvia, had bones in its "arms," which resembled those of the tetrapods, but they finished in fins, not hands or feet. This creature had lungs and nostrils as well as gills, and might have been able to walk on dry land for a short while. In the Late Devonian strata (355 million years ago) of Blossburg, Pennsylvania, a fragmentary fossil (*Sauripteris*) was found with a humerus connected to the radius and ulna, like that of the later tetrapods, but the fin finished in eight fingers, rather like the early tetrapods *Acanthostega* and *Ichthyostega* (see below). This eight-digit plan did not perdure, and the five-finger arrangement seems to have dominated the future of reptiles and mammals.* According to Maisey (1996), "Having left the water, tetrapods took about 150 million years to return as fishlike ichthyosaurs, and almost 300 million years to return as whales and dolphins. Tetrapods have repeatedly failed in their challenge to fishes for mastery of the oceans."

The rhipidistian ("fan-finned") fishes, such as *Eusthenopteron* and *Osteolepis,* were much like the coelacanths in that many of them had a triple-lobed tail and paired fins. But they are believed to be much more closely allied to the land vertebrates than either the coelacanths or the lungfishes. They became extinct about 290 million years ago. The skeletal components of their fins were very like those of the early amphibians, and both groups had teeth whose enamel covering is folded in toward the center of the tooth, giving them the additional name labyrinthodonts. At a length of 4 feet, *Eusthenopteron* was one of the larger rhipidistians; *Osteolepis* was less than a foot long. These fishes had a lacrimal (tear) duct, perhaps related to a near-surface lifestyle in which the eyes had to be kept moist if they were out of the water for any length of time. The skull of *Eusthenopteron* was hinged to bend back, so the animal could open its mouth wider and attack larger prey. The skull bones of the

* Stephen Jay Gould wrote an essay on seven- and eight-fingered tetrapods in which he suggested that polydactly was a stepping stone on the way to the five-finger plan that so dominates vertebrate morphology. He was so taken with the idea that he entitled an essay on the subject "Eight Little Piggies"—as well as the 1993 book in which the essay appears.

rhipidistians closely resembled those of the earliest tetrapods, like *Ichthyostega* and *Acanthostega,* and even the bones in the forelegs were homologous; the pyramidal arrangement of the bones in the paired fins is strikingly similar to the arrangement of the limb bones in land animals. Although it was certainly a fish, *Eusthenopteron* is often shown crawling out of the water, preparing to give rise to the amphibians—and eventually to us.

The earliest known tetrapod so far is *Elginerpeton,* found from the Late Devonian (c. 400 million years ago) of Scat Craig, Scotland, by Per Ahlberg in 1991. The skull and jaw fragments and shoulder and leg bones that have been found suggest that it was adapted for an aquatic existence rather than a terrestrial one. According to Ahlberg, this 5-foot-long creature "may have evolved limbs in order to move more effectively in shallow, weed-choked waters." But Carl Zimmer wrote, "With only fragments of its limbs, it's impossible to know if there were toes yet, but *Elginerpeton* shows many signs of being intermediate between lobe-fins like *Panderichthys* and later tetrapods." In August 2000, Ahlberg, of London's Natural History Museum, announced the discovery of two fossil fragments of fishlike lower jaws with seven rows of teeth that he had "unearthed" in museum collections in Latvia and Estonia. The teeth differentiated them from any known fishes or tetrapods, and Ahlberg believes they are from a tetrapod that lived at the same time as *Panderichthys,* but its fused jaws indicate that it wasn't very fishlike, and therefore may have branched off from fishes even earlier than had previously been supposed.

Although it appears counterintuitive, the fish-to-amphibian transition was not a transition from water to land, but rather from fins to feet, a change that took place in the water. The very first amphibians seem to have developed legs and feet to scud around on the bottom in the water, not to walk on land. This aquatic-feet stage meant the fins didn't have to change very quickly; the weight-bearing limb musculature didn't have to be very well developed, and the axial musculature didn't have to change at all. Eventually, of course, amphibians did move onto the land. This required attaching the pelvis more firmly to the spine, and separating the shoulder from the skull. Lungs were not a problem, since lungs are an ancient fish trait and were present in the Mid- to Late Devonian *Eusthenopteron,* an early lobe-finned fish roughly intermediate between early crossopterygian fishes and the earliest amphibians. Its skull was very amphibian-like, and it had a strong backbone. Its fins were like early amphibian feet in the overall layout of the major bones, muscle attachments, and bone processes, but there are no perceptible "toes," just a set of identical fin rays. *Panderichthys* was a fish with feet, a lobe-finned fish that actually looked like a tetrapod in overall proportions, with its flattened body, dorsally placed orbits, frontal bones in the skull, a straight tail, and remarkably footlike fins.

Frogs, toads, salamanders, and caecilians* represent the class Amphibia today,

---

* The caecilians, a little-known amphibian group, are legless, burrowing creatures that resemble earthworms. They have vertebrate characteristics like jaws and teeth, and range in size from 7 inches to 4 feet. Their bodies are ringed, but the skin contains scales, a primitive amphibian trait. Their eyes are either so reduced as to be useless or completely non-existent. They are found in tropical swamps around the world, but are rarely seen because of their burrowing habits. The fossil record of caecilians is very poor.

but in ancient times, a much greater variety prevailed. They appear to have been descended from crossopterygian fishes, with the fins evolved to form legs. Fishes breathe water by means of gills, and amphibians in their adult stages breathe air, so this required a major modification. Living amphibians supplement their oxygen requirement by breathing through moist skins, but the earlier versions had scales or armor covering their bodies, so they could not respire through this heavy covering. Moreover, there has been no satisfactory answer to the question of why the amphibians left the water in the first place, given the much greater temperature range on land and dangers of desiccation. One suggestion is that air breathers could leave a drying-up pond or stream and travel overland to one that might contain water, but this hardly seems sufficient to explain the profound change from water- to air-breathing. Another idea is that they "escaped" from the sea to avoid predators and found a rich supply of worms, snails, and other invertebrates in the littoral zone, but this doesn't make much sense either.

The Devonian Period is sometimes known as the Age of Fishes, but most of the fishes that dominated the Devonian seas have disappeared. These included the heavily armored placoderms; the antiarchs, grotesque creatures that looked like a cross between a turtle and a crustacean; armored skates and rays; and creatures shaped something like sharks, but with anatomical features that significantly differentiated them from modern sharks. By the Late Devonian, some of the ancestral tetrapods had evidently emerged from the water. Early analyses suggested that they evolved as fishes escaping seasonal droughts, but more recent studies have suggested that they were actually fishes that already had legs. Romer (1966) imagines a rather unlikely scenario:

> it is easy to see how a land fauna might have eventually been built up. Instead of seeking water immediately, the amphibians might linger on the banks and devour stranded fish. Some types might gradually take to eating insects (primitive ones resembling cockroaches and dragon flies . . .), and finally, plants. The larger carnivores might take to eating their smaller amphibian relatives. Thus a true terrestrial fauna might eventually be established.

The first animals to walk on land were probably primitive amphibians known as labyrinthodonts, named for the complex folding of the dentine layers of the teeth. The first known beachhead for terrestrial vertebrates is in Australia. In the Devonian sandstones of the Grampians in Western Victoria, a fossil trackway more than 400 million years old shows where a labrynthodont amphibian trudged along in mud that would eventually turn to sandstone (Vickers-Rich and Rich 1999). We do not know what prompted these creatures to abandon an aquatic for a terrestrial environment, but the reason could have been an occasional escape from predators, or perhaps the colonization of a new and unoccupied ecological niche where competition was nonexistent. In an article (McLeod 2000) about the emergence of tetrapods from the water, Per Ahlberg is quoted as saying, "The key step during the mid-Devonian was the appearance of wood, which allowed land plants to grow much larger than previously." This led to the evolution of trees, which, if they fell into the water, created areas of flooded forest, a new kind of environment that previously didn't exist. Into this new environment came the panderichthids, lobe-finned fishes that were shaped like tetrapods but were equipped with both lungs and gills, which meant that they could leave the water. The conversion from extracting oxygen from the water to getting it directly from the air was one of the substantial changes that had to occur before water dwellers could function on land. Eyes had to be modified to see in a new medium, and new devices for support had to evolve for animals that previously had been rendered weightless by the buoyancy of water. All of these modifications except air-breathing would have to be reversed when the future marine descendants of these early terrestrials—the marine reptiles, sirenians, pinnipeds, penguins, and cetaceans—reversed the process and gave up a land existence for life partially or totally in the water.

The labyrinthodonts were descended from the lobe-finned fishes and are represented in the fossil record by *Ichthyostega,* a low-slung, stout-bodied, semiaquatic animal about 3.5 feet in length. As befits its tetrapod designation, *Ichthyostega* had four limbs splayed out from its sides, but, like many of the early fish-amphibian transitions, it had more than the traditional five toes on each limb: eight toes on each forefoot and seven on each hind foot. It probably swam better than it walked, and as a holdover to its ancestry, it had a fin running the length of its tail, which probably helped propel it through the water.

Concerning the animal's locomotion, in her 1997 study of Devonian tetrapods, Jenny Clack of Cambridge University wrote:

> In *Ichthyostega,* the forelimbs seem to have been weight-bearing, but the forearm was able to extend fully. Only one articulated specimen shows a femur in association with a body and forequarters. The shoulders were massive compared with the much smaller and paddle-like hind limb, in contrast with the more conventional proportions which it is usually given. The proportions of an elephant seal appear the closest analogue among living animals. Perhaps *Ichthyostega* hauled itself up on shelving beaches, moving its forelimbs in parallel and dragging its hind limbs.

It can be differentiated from the fishes by the bones of its skull, where the cheek region and the roof of the skull are fused, unlike the looser arrangement in fishes. *Ichthyostega* had a short neck, unlike fishes, which have no neck at all. Its long, broad ribs overlapped each other, forming a barrel-shaped ribcage that protected its heart, lungs and digestive organs. The ribcage was so solid that it probably could not expand, so the animal probably breathed by lowering the floor of its mouth to draw air in, and then lowering it again to pump air down the windpipe into the lungs.

First discovered at Celsius Bjerg in Greenland in 1933 by the Swedish paleontologists Gunnar Save-Soderbergh and Erik Jarvik, *Acanthostega gunnari* was described by Jarvik in 1952. The fossils consisted only of the roof of the skull, which had two

Acanthostega gunnari *had bony gill arches and long limbs with no articulated wrists or ankles. This early tetrapod had a vertically flattened tail—well suited for swimming—a fish's spine, paddlelike limbs, and a full set of gills, which means that it breathed water like a fish. Count the fingers and toes.*

rearward-facing prongs on the rear margin (*Acanthostega* means "spine-plate"). In the 1970s, some Cambridge University paleontologists collected more material from the same location, but it was ignored until 1983, when Jenny Clack began to study it and subsequently went to Greenland and collected more material herself, from the slopes of Stensiö Bjerg. With her graduate student Per Ahlberg and the paleoichthyologist Michael Coates, Clack began the systematic description of *Acanthostega.* The exceptionally well preserved material consisted of a complete skull and articulated jaw, one complete and two partial skeletons, which showed that it had bony gill arches (suggesting internal gills), and long limbs with no articulated wrists or ankles. Neither the leg bones nor the spine were designed to support the weight of this 3-foot-long creature, and the ribs, unlike the heavy bones of *Ichthyostega,* were thin and weak. Acanthostega had paddlelike limbs with eight digits on the front extremities and seven on the hind. It was clearly a tetrapod, but a poor excuse indeed for a land animal. *Acanthostega* had a vertically flattened tail, well suited for swimming, a fish's spine, paddlelike limbs, and a full set of gills, which means that it breathed water like a fish.

Another contemporary early tetrapod was *Tulerpeton,* named by Soviet paleontologist O. A. Lebvedev for the Tula region of Russia. It had six digits on its front feet and seven on its hind, and its shoulder bones were somewhat more robust than those of *Acanthostega,* suggesting that it was more suited for walking on land. Michael Coates has said (McLeod 2000), "When I started to work on *Acanthostega,* I didn't know if I'd find any digits at all—let alone eight. Then, following up a lead in obscure Russian papers on *Tulerpeton,* I started on the *Ichthyostega* hind limb and found the seven-digit pattern." The pectoral girdle lacked branchial lamina (plates), indicating that *Tulerpeton* did not have gills. The muscle attachments on the femur hint at an adaptation for swimming, but because the only known fossil was missing the tail, vertebrae and ribs, we still don't know for sure how *Tulerpeton* moved, and whether it did so on land or in the water.

The conventional view holds that the early tetrapods, such as *Acanthostega, Ichthyostega,* and *Tulerpeton,* were partially aquatic creatures that had developed "hands" in the course of their emergence from the water, but Jenny Clack believes that they never left the water at all, and that their multidigit limbs evolved for use in the water, only later to be used on land by descendant species. She wrote (1997), "All of these characters suggest that not only was *Acanthostega* aquatic, but that it was primitively so, and not derived from a more terrestrial forebear." There is debate as to whether some of these early tetrapods lived in fresh water or brackish lagoons, but they certainly were not fully marine.

It would be convenient to be able to identify a progression, as Romer did, from animals that were purely aquatic to those that were semiaquatic (amphibian) and finally to those that were completely terrestrial. Daeschler and Shubin (1995) believe that this is not possible. They wrote,

> Popular conceptions of the origin of tetrapods envision a progressive evolution of characteristics suited for life in land. Paleontological and functional approaches to basal tetrapods have often involved the search for features indicative of a terrestrial existence. So pervasive is this approach that the entire question of the "origin of tetrapods" is often assumed to be synonymous with the "invasion of land by vertebrates." Recent phylogenetic and paleontological discoveries suggest that these two issues need to be decoupled.

Rather than demonstrating a progression from water to land, the differences between, say, *Acanthostega* and *Ichthyostega* merely show different possible limb functions: "There is no unequivocal evidence that limbs necessarily developed to aid in strictly terrestrial locomotion," says McLeod, adding, "For the past 70 years, a single research group in Stockholm has held most of these fossils and much of the material has not been scrutinized. 'The current reconstructions are wrong in almost every respect,' says Clack. She and Alhlberg will first focus on the brain case. 'We already know how weird it was and we've got some idea about what it's going to tell us. . . . Potentially we'll have to rethink the whole thing again with the new *Ichthyostega* study coming up" (McLeod 2000).

Paleontology depends to a great extent on serendipity; even in a well-known fossiliferous region, a good fossil is usually a happy surprise. Also, connections can only be made by means of the careful examination of a new specimen and comparisons with other known specimens. More than any other science, paleontology depends upon publication. Since you cannot expect to see these long-extinct animals walking or swimming around, the only way you are going to know about forms that you haven't collected yourself is to read the published descriptions or examine the collections in museums.

The little *Casineria* was found by Nicholas Kidd, an amateur collector in Cheese Bay, Scotland, in 1992, but five years would pass before a paleontologist examined it and published its description. *Casineria* lived 20 million years later than the primitive tetrapods *Acanthostega* and *Ichthyostega.* It bore certain similarities to them, but it was considerably smaller and had five digits on each of its limbs instead of seven or eight. Whereas *Acanthostega* and *Ichthyostega* would have been awkward and ungainly on land—if indeed they ever came out of the water—*Casineria* was clearly designed to spend more time out of the water than in. For one thing, it was only about 6 inches long, not nearly as big or heavy as the others, and its fingers were designed to flex separately, an adaptation for walking on irregular terrain. The bones of the hands and feet of *Acanthostega* and *Ichthyostega* could only move together, more conducive to swimming or walking on level terrain, such as beaches or mudflats. *Casineria*'s vertebrae locked together in such a way as to provide a strong support for its body when it was lifted off the ground, unlike that of the earlier tetrapods, whose looser backbones were more like those of fishes. It has therefore been suggested that *Casineria kiddi* "pushes back the known occurrence of terrestrial vertebrates closer to the origin of tetrapods" (Paton et al. 1999), which means that it is the earliest known animal in the amniote progression that would eventually lead to birds and mammals. It may indeed be the first known reptile.

In 1998, Jenny Clack described a newly discovered Carboniferous tetrapod from the black shale of the East Kirkton Limestone of Scotland. It was 8 inches long from nose to pelvis (the tail was missing). Another skull found nearby was somewhat larger. Because this tetrapod shares characteristics with both reptiles and amphibians, it represents that moment in time when the reptiles (and hence birds, and mammals) diverged from the frogs and salamanders. The name she bestowed upon it was *Eucritta melanolimnetes,* the "critter from the Black Lagoon."

The amphibians led the charge that resulted in the conquest of the land by four-legged animals, but while their successors—the reptiles, birds, and mammals—have proliferated, the amphibians no longer rule the Earth. The first frogs (*Triadobatrachus*) appeared in the Early Triassic, about 200 million years ago, and the first urodeles (newts and salamanders) are found in somewhat later Jurassic rocks. Today, some amphibians lay their eggs on land, or even in trees, but most lay them in the water. In some cases the young undergo a metamorphosis in the egg and emerge as miniature adults, but most amphibians pass through a legless, larval (tadpole) stage in the transition from a water-breathing juvenile to an air-breathing adult.

Adult amphibians differ from reptiles in having moist skins and either no scales at all or tiny scales embedded in the skin. The newts and salamanders most closely resemble their ancestors in that they are long-tailed and four-legged. Some, like the hellbender (*Cryptobranchus alleganiensis*) and the Japanese giant salamander (*Megalobatrachus japonicus*) reach sizes large enough—3 and 5 feet, respectively—to give us an idea of what the prehistoric amphibians might have looked like. Frogs and toads (anurans), by far the predominant amphibians, are usually smallish; the body of the largest one, the goliath frog, *Gigantorana goliath,* is less than a foot in length. Frogs and toads all have muscular hind legs for jumping, and some have developed a completely arboreal habitat, some incubate their eggs on their backs, and others are brightly colored to advertise their poisonous skin glands.

It is almost miraculous that marine creatures were modified to emerge from the sea and take up residence on land. What then are we to make of their descendants who, having become fully adapted to a terrestrial lifestyle, readapted to a marine existence? To accomplish this, various unrelated vertebrates were profoundly retrofitted, some of them losing the legs their ancestors had worked so hard to acquire, others forsaking modifications such as fur or external ears that made life on land possible. We all know that evolution is nondirectional and does not aim for anything in particular. But this volte-face was executed by many groups—not only the marine mammals, but also penguins, sea snakes, turtles, and assorted extinct marine reptiles—not once, but many times.

RETURN TO THE SEA

MARINE REPTILES

During the great Age of Reptiles, group after group returned in part or as a group to the sea. Among the most ancient of reptile groups, the turtles produced the familiar sea-dwelling forms. So did the crocodiles (in the teleosaurs), and so did the lizards (in mosasaurs). Most completely marine of the extinct groups were the ichthyosaurs and the plesiosaurs, the former with a startlingly fishlike body form, and the latter like gigantic turtles with vastly elongated necks. Among existing reptiles, the snakes have in their turn produced a true sea snake, but as this is confined to the tropics, it is little known outside of zoological circles. Far more remarkable even than the sight of a whale is the occasional encounter in the Pacific or Indian Ocean with an aggregation of sea snakes.

—Schmidt and Inger,
Living Reptiles of the World

While walking on a Yorkshire beach in 1758, Captain Chapman and his friend Mr. Wooler came upon some fossil bones embedded in a cliff. They identified the bones as belonging to an "allegator," and sent their written descriptions to the Royal Society of London, where they were read into the society's journal *Philosophical Transactions.* (The fossil, now classified as the primitive crocodilian *Mystriosaurus bollensis,* is in the Natural History Museum in London.) The first mosasaur fossil was found in 1780 in a limestone mine in Maastricht, in what is now the Netherlands. It was not described until 1808, when Baron Georges Cuvier decided that it belonged somewhere between iguanas and varanus lizards. In 1821, Britain's first vertebrate paleontologist, the Anglican cleric Reverend William Daniel Conybeare (1787–1857) of Bristol, examined some fossils found at Lyme Regis, Dorset, in southwestern England, and with Henry de la Beche published the first descriptions of long-extinct reptiles that they called plesiosaurs ("near lizards"). Conybeare described the first one from a skull found in Somerset. Later, Mary Anning or another member of the Anning family found a complete skeleton at Lyme Regis in 1823, and this discovery sparked a nationwide interest in the extinct marine reptiles.

Mary Anning was born in 1799 in Lyme Regis, the daughter of a cabinetmaker and amateur fossil collector who died when she was 11, leaving the family £120 in debt and without a source of income. Mary walked the cliffsides looking for fossils that she might sell. (The tongue-twister "She sells seashells by the seashore" is about her.) In 1817, Thomas Birch, a well-to-do collector, befriended the family and even arranged to sell his personal collection at auction for the Annings' benefit. Joseph Anning, Mary's brother, actually found the first ichthyosaur in 1811, and Mary found the remainder of the skeleton nearly 12 months later. She also discovered the first complete skeleton of the reptile Conybeare and de la Beche had named *Plesiosaurus dolichodeirus,* which fossil was ultimately bought by the duke of Buckingham. Both Mary and Joseph Anning made many important fossil finds, but it was Mary Anning who was famous throughout Europe for her knowledge of the various fossilized animals she was uncovering. In addition to the ichthyosaurs and plesiosaurs, Mary Anning also found Britain's first pterodactyl. In 1838, for services rendered, Mary Anning was given an annual stipend from the British Association for the Advancement of Science, and she was made the first honorary member of the Dorset County Museum in 1846.

Reptiles evolved on land, but after their dispersal to various terrestrial habitats, some returned to the sea. In geological time, this turnaround occurred soon after the conquest of the land. There are records of lizards known as mesosaurs that had evidently achieved an aquatic existence as early as the Early Permian, almost 300 million years ago. *Mesosaurus* ("middle lizard"), the only known genus, was a 3-foot-long, fully aquatic lizard that propelled itself under water with its long tail and webbed hind feet. Some thought that their elongated jaws were equipped with fine, sharp teeth that were too delicate for snagging fish, so they might have formed a sieve to strain plankton from the water, but modern analyses render this an unlikely explanation. The mesosuars died out about 250 million years ago, and evidently left no direct descendants. Mesosaur fossils have been found in southern Africa and eastern South America, strong evidence for continental drift.

While the thecodonts (now known as synapsids) were lumbering toward their destiny as dinosaurs, the "fish-lizards" and plesiosaurs were beginning their domination of the ocean, and the pterosaurs, another group of reptiles, took to the air. Several groups of marine reptiles didn't make it through the Permian Extinction, but the first ichthyosaur skeletons appear in early Triassic formations, and while their ancestors remain unidentified, they were beginning to fill the same jobs in the water that the dinosaurs did on land: the apex predator positions. The ichthyosaurs lived from the Middle Triassic (about 200 million years ago) until about 85 million years ago, 20 million years before the mass extinctions of Late Cretaceous. They developed a basic body plan that consisted of a dolphinlike shape with a tooth-filled beak; a vertically oriented, lunate tail fin like that of a shark; and in most of the later species, a dorsal fin. The earliest ichthyosaurs were about a yard long, but they later grew to impressive lengths, like the 45-foot-long *Shonisaurus,* which got as big as a humpback whale, and a still unnamed ichthyosaur from British Columbia that was even larger. The ichthyosaurs of the Jurassic Period are the most numerous, suggesting a flowering of these animals some 140 million years ago. The dominant genus was *Ichthyosaurus,* best preserved in southern German fossils but also known from England, Greenland, and western Canada.

Jurassic seas were also occupied by plesiosaurs, marine reptiles with long necks, and their relatives, the pliosaurs, marine reptiles with short necks and powerful jaws and teeth. Giant marine crocodiles, with long, eel-like tails, competed for prey in Triassic seas, but unlike ichthyosaurs, which were pelagic and never left the water (although they had to surface to breathe, like the dolphins they resembled), the crocs occupied a near-shore environment and probably came ashore to rest and lay their eggs. The ichthyosaurs and plesiosaurs had already been gone for 25 million years by the Cretaceous-Tertiary Extinction. The mosasaurs too had gone extinct by the end of the Cretaceous, and at least one authority (Lingham-Soliar, 1994) believes their disappearance can be directly tied to a catastrophic event.

There are many living reptiles that are almost completely aquatic. Sea turtles live most of their lives in the water; females come out only to lay their eggs. (With the exception of their first overland journey to get from the place where they hatch to the water, male sea turtles live their entire lives at sea.) Crocodiles spend much of their lives in the water, but like the turtles, they have to lay their eggs on land, because reptile embryos must respirate through the egg's permeable shell. Crocodiles also come out to lie in the sun. Many modern lizards, such as the monitors, are particularly good swimmers, and there is even one commonly known as the water monitor.

Only one living lizard leads a semiaquatic existence. The marine iguana (*Amblyrhynchus cristatus*) is found only in the Galápagos Islands. (The Galápagos land iguana, *Conolophus,* is a completely different reptile that spends all of its life on land, dining on plants, particularly cacti.) Marine iguanas swim by propelling themselves with their tails, and they can dive to depths of 40 feet and remain submerged for nearly an hour at a time. On land, which is mostly lava rock, they bask in the sun in large groups. When Darwin visited the Galápagos in the *Beagle,* he described the iguanas thus: "The rocks of the coast are abounded with great black lizards between three and four feet long; it is a hideous-looking creature, of a dirty black color, stupid and sluggish in its movements." They vary in size from island to island; the largest ones, up to 5 feet in length (including tail) and weighing up to 26 pounds are found on Isabela Island. On Genovesa, the males do not weigh more than 2 pounds. They have blunt snouts to enable them to graze underwater on the red and green algae that make up almost all of their diet. To remove the salt from their systems, marine iguanas have a very efficient salt-excreting gland just above the nostrils, and they sneeze frequently, sometimes giving themselves a whitewashed appearance. Although primarily sooty black in color, some of these lizards show patches of red or green. Martin Wikelski of the University of Illinois at Urbana-Champaign, who has been studying *Amblyrhynchus cristatus* for more than 20 years, has explained this phenomenon: "This coloration is not, as many might suspect, a signal to other lizards. . . . Females don't choose males according to color, and males don't react to the color patterns of other males. Rather, pigments from the plants they eat build up in the unshed skin, so body color reflects the type of algae they are feeding on" (Wikelski 1999).

The males are on average larger than the females, and they posture and fight for the attention of their prospective mates. When impregnated, the female digs a nest in the sand, where she lays one to six leathery white eggs. After an incubation period of about three months, the eggs hatch and the newborn lizards emerge from the burrows. When they appear, they are preyed upon by almost every predatory animal in the Galápagos: snakes, lava gulls, hawks, owls, and on some of the islands, rats, and feral cats and dogs.

A recent survey of marine iguanas by Wikelski and Corinna Thom of the University of Würzburg has shown that they are able to make a remarkable adjustment to the scarcity of food that results from the El Niño events that affect the Galápagos every three to seven years: they shrink. Martin Wikelski has been keeping size records of the iguanas since 1982; when he noticed a significant shrinkage in adults after a two-year food shortage, he first attributed it to an error in measurement. "At first we didn't believe our own data," he said in an interview that appeared in *Science News* in January 2000 (Milius 2000a), but then he realized that a drop of up to 20 percent of total length in some females was way beyond any measurement error. "Reduction in body length," they wrote in *Nature* in 2000, "has been observed previously, and growth rates set to zero by definition, but this is the first report of the shrinkage of adult vertebrates." They concluded:

> Adult marine iguanas can switch repeatedly between growth and shrinkage during their lifetime, depending upon environmental conditions. In humans, bone regrowth after osteoporosis is largely impossible, causing major health problems (for example, in postmenopausal women). It is not clear whether other vertebrates can shrink and regrow. We propose that growth rates should no longer be set to zero, as shrinkage can confer a biologically important advantage.

Even though it is tempting to view the marine iguana as an aquatic animal in the making, it seems more reasonable to see it as a terrestrial creature that feeds in the water, rather like a hippopotamus. Like the hippo, which also spends a considerable amount of time out of the water, the iguana has not developed any of the specialized modifications known to benefit aquatic tetrapods, such as bone ballast, the density of bone that enables animals to submerge. The sirenians have particularly heavy bones, as does the sea otter (*Enhydra*). Some aquatic animals may have developed bone ballast during the course of their transition from a terrestrial to an aquatic lifestyle, but this feature may be secondarily lost in fast-swimming predators, such as ichthyosaurs and cetaceans, which tend to have both reduced lungs and lighter skeletons.

The Galápagos served as a natural laboratory for Darwin—his observations of the finches and tortoises on the various islands contributed to the formulation of his theory of the origin of species—and these islands also offer an unparalled opportunity to observe animals that have returned to the water, or are in the process of doing so. On these islands that Melville called "the enchanted isles" we find the marine iguana, and also penguins, those most aquatic of birds, who, having traded their wings for flippers, do their only flying under water. Here too are two species of eared seals, the Galápagos sea lion and the Galápagos fur seal, and the flightless cormorant. Aptly named *Nannopterum,* "little wings," this large cormorant's wings are so stunted that it cannot fly at all but can only waddle awkwardly on its short, stout legs. Like all cormorants, it is a fish eater; it hunts its prey under water by swimming with powerful thrusts of its broad, webbed feet. In the past, the loss of flight did not pose a threat to Galápagos cormorants because there were no land predators, but the introduction of dogs, cats, rats, and pigs to the islands has placed these ungainly birds at risk. Found only on the islands of Fernandina and Isabela, the cormorants now number only about 1,000 birds and the species is considered endangered.

On January 16, 2001, the Ecuadorean tanker *Jessica*, carrying 243,000 gallons of oil, ran aground off San Cristobal Island, spilling thousands of gallons of fuel oil into the sea around the islands and threatening the rare Galápagos penguins and the already-endangered flightless cormorants.

Marine iguanas, although limited in range, are alive today. The nonavian dinosaurs, the aquatic ichthyosaurs, plesiosaurs, and mosasaurs are all extinct. The ichthyosaurs and mosasaurs that lived in the western Interior Seaway, which bisected North America some 87 million years ago, went extinct before the seaway dried up. Otherwise, the extinctions remain unexplained. Several groups of living marine reptiles can trace their ancestry back hundreds of millions years. It is believed that sea turtles, sea snakes (the only fully aquatic reptiles), and marine crocodiles are all descended from terrestrial forebears, and in their partial return to an aquatic existence, they had to lose some of the terrestrial modifications, and reacquire the ability to spend more time in the water. Fossils have told us much about primitive snakes, early crocodilians, and early turtles.

ENCASED IN ARMOR PLATE

No order of reptiles of the past or present is more sharply and unequivocally distinguished from all others than the Chelonia or Testudinata. No order has had a more uniformly continuous or uneventful history. None now in existence has had a longer known history, and in none is the origin more obscure.

—Samuel Williston, 1914

Modern sea turtles are presumed to be descended from land-based reptiles that returned to the sea, but hardly any transitional forms have been identified. There are fossil land turtles, which have columnar legs and feet with claws, and fossil sea turtles, which have flippers, but there is nothing in the fossil record that looks like a semiaquatic turtle.* The suggestion has been put forward that sea turtles never left the water at all and developed directly from amphibians or early reptiles in the ocean, but this seems less likely than the idea that they had terrestrial origins. Carroll (1988) wrote that "no trace of earlier or more primitive turtles has been described, although turtle shells are easily fossilized and even small pieces are easily recognized. Apparently the earlier stages in the evolution of the shell occurred very rapidly or took place in an environment or part of the world where preservation and subsequent discovery were unlikely."

Turtles are not dinosaurs because they are not descended from the first dinosaurs, and they have a shell, which no dinosaur ever had. They represent a separate group of vertebrates, and one of the oldest of all continuous vertebrate lineages. Of the living vertebrates, only sharks and bony fishes have a lineage that is older than that of the turtles. There were sea turtles in the Middle Triassic, about 230 million years ago, and there are turtles today. According to Elizabeth Nicholls (1997), "Turtles are one of the great success stories of marine reptiles. They are the only living reptiles fully adapted to the marine environment, returning to shore only to lay their eggs." In his 1997 discussion of the evolution of sea turtles, Peter Pritchard wrote,

> The successful penetration of aquatic niches by both early and modern turtles was probably made possible by a remarkable example of preadaptation. Other living aquatic reptiles, including sea snakes . . . marine iguanas, crocodilians, etc., as well as the extinct ichthyosaurs, swim (or swam) by means of body and tail undulations not dissimilar to those of typical fishes. Turtles, on the other hand, lost the capacity for this form of propulsion when they developed the shortened, rigid body form and corselet that has characterized the group since the Triassic. This body form offered armored resistance to attack by predators, but the tradeoff was reduced speed and agility, obliging those terrestrial chelonian species surviving in a world with increasingly sophisticated predators to adopt specialized life styles.

Sea turtles used to be much more varied and diverse than they are today, but some of the earliest turtles, known from the Late Triassic, 200 million years ago, had a shell and an anapsid skull: a solid block of bone, with no openings for jaw muscles. The first known turtle was *Proganochelys quenstedi,* which had a fully developed shell and a turtlelike skull and beak, but it also had several primitive features not found in turtles today, including tiny teeth on its palate, a clavicle, and a simple ear. The early turtles were unable to withdraw their head or legs into their shell, but by the middle of the Jurassic some had developed this ability. By then, turtles had split into the two main groups of turtles found today, the side-necked turtles (pleurodires) and the

* Many living turtles, such as the familiar box turtles, terrapins, and snapping turtles, lead an amphibious existence, laying their eggs on land, and feeding on land and in the water. The land "tortoises," including the gigantic Galápagos tortoises, do not come to the water except to drink.

arch-necked turtles (cryptodires). They probably had terrestrial ancestors, but the sequence of descent is not evident.

The leatherback is the largest living turtle, weighing over a ton. Instead of a shell, it has smooth, leathery skin that covers the bony ridges on its back and belly. Leatherbacks feed almost exclusively on nutrient-poor jellyfishes, and pursue their prey at depths down to 3,000 feet.

The steps required to get from a Permian anapsid reptile to a fully armored creature with flippers for feet are also unclear. The turtles are probably descended from reptiles that inhabited swampy areas, where their bodies widened while their feet became webbed and they lost the articulation of their digits altogether. (Terrestrial tortoises have clawed toes, not unlike those of crocodylians.) The shell, which consists of an upper element, the carapace, and a lower, the plastron, is unique to turtles and tortoises. The path from Permian proto-turtle to recent turtle or tortoise was not without detours, and there were numerous variations on the armored reptile theme that did not perdure. *Proganochelys* is no longer with us, nor are the horned turtles of the genera *Niolamia* from Argentina, and *Meiolania* from Lord Howe Island, an isolated speck of land 300 miles east of Sydney, Australia. *Meiolania* had, in addition to "horns"—actually flanges of bone behind the eyes—a spiked tail that was also fully armored. Then there was *Stupendemys geographicus,* a side-necked turtle that lived only 10,000 years ago in South America, whose shell reached a length of 8 feet. In 1976, Roger Wood described a specimen from the Late Tertiary Urumaco Formation of northern Venezuela, which he labeled "the world's largest turtle," but in the paper he says that "*Dermochelys* [the leatherback] is reputedly the largest of all turtles, living or fossil." It also is not clear whether *Stupendemys geographicus* was fully aquatic.

As Samuel Williston, a paleontologist who worked with the Kansas fossils, wrote in the 1914 book *Water Reptiles of the Past and Present,* "Were there no turtles living we should look upon the fossil forms as the strangest of all vertebrate animals—animals which had developed the strange habit of concealing themselves inside their ribs, for that is literally what the turtles do." If it weren't for the 250 species of turtles living today, trundling around on land and swimming in the sea or in lakes and ponds, these animals encased in mobile homes could easily be viewed as bizarre evolutionary experiments that were ordained for failure.

One idea that did "fail"—but only in the sense that the exemplars are extinct—is that of putting the shell *inside* the skin. *Placodus,* which appeared and disappeared in the Triassic, was characterized by a massive skull and the presence of gastralia, an extra set of belly ribs that formed a ventral support system under the animal's underside and, with the ribs, enclosed it in a complete shield of internal bony armor. Placodonts looked something like a cross between a walrus and a turtle, and they came in armored and unarmored varieties. They fed on shellfish, harvesting them with their shovel-like front teeth and crushing them between the heavy, platelike teeth on the roof of their mouths and their lower jaws ("placodont" means "plate teeth.") *Placodus* may have reached a length of 10 feet, including the tail, which may have been used for propulsion.

Similar in form but with an even more specialized skull was *Placochelys* ("armored turtle"), which, despite its name, was not a turtle at all. The placochelyids were reptiles with adaptations convergent with those of the turtles, such as a flattened body covered with tough, knobby plates and a beak that allowed them to pluck food from the rocks. But where the turtles are toothless, the placochelyids had powerful jaw muscles and a mouthful of broad, crushing teeth that allowed it to crush the shells of its prey. Another armored placodont was *Henodus* from the Late Triassic, which looked like a square-headed turtle that tried to grow wings. Its back was covered with a mosaic of polygonal horny scutes (plates) but these extended into flanges so broad that the animal was as wide as it was long. With its uniquely reduced dentition, its square-snouted skull where the region in front of the eyes is tremendously shortened, and extensive carapace, *Henodus* appears to be most closely related to the placochelyids.

In 1870, E. D. Cope, a Philadelphia paleontologist, unearthed in the Niobrara Chalk of Kansas the fossil remains of a gigantic sea turtle that he named *Prostega gigas.* It had a huge head, unlike *Archelon ischyros,* also from Kansas, which, although much larger, had a proportionately smaller head. Since then, numerous fossil skeletons of these turtles have been found in Kansas and South Dakota. In 1914 Samuel Williston described what he believed to be the modus vivendi of these two Volkswagen-sized turtles:

> As regards the habits of these ancient sea turtles, we may offer some tolerably certain conjectures. In the opinion of the writer, the less reduced plastron indicates a bottom-feeding habit, a view that is strengthened by the more rounded form of the shell, like that of a river turtle. All in all, it would seem that *Prostega* and *Archelon* lived habitually on the soft bottoms of the shallower seas, feeding upon the hordes of large shell-fish, for which their powerful parrot-like beak was admirably adapted. That the species of *Prostega* did not commonly frequent the deeper oceans is indicated by the general absence of their remains in the deeper water deposits.

At a length of 12 feet and a weight in excess of 3 tons, *Archelon* was considerably larger than the largest living turtle, the leatherback (*Dermochelys coriacea*), which reaches a length of 8.5 feet, and can weigh 2,000 pounds. Like the leatherback, *Archelon* had a leathery covering over an inner framework of transverse struts—*"Dermochelys,"* "skin turtle," refers to this covering—not a hard, multiplated carapace like other sea turtles. Its skull was of the anapsid type, with no openings on the top or sides, but there are two large openings at the back. (In some modern turtles these openings have moved onto the top, near the rear of the skull.) Because we have a direct descendant to observe, we can make some educated guesses about the lifestyle of the ancient *Archelon:* it probably "flew" through the water with powerful strokes of its broad, paddle-shaped flippers, and fed on jellyfishes with its weak jaws and hooked, toothless beak. Because the Western Interior Seaway was probably never deeper than 1,000 feet, however, *Archelon* could not have competed with *Dermochelys* for the depth record for turtles; the leatherback is capable of reaching depths of a half mile or more.

Although J. R. Hendrickson (1980) didn't much like the use of the word "strategies" to describe the ecological development of sea turtles, he used it in the title of his 1980 discussion because the term has come into general usage, as in "reproductive strategies," "foraging strategies," "sex-ratio strategies," all used in reference to processes about which we agree that entities cannot make rational choices. The general "strategy" selected by the ancestral sprawl-legged chelonian was the "armored tank" mode, presumably for defense against large predators. About a dozen vertebrae were eliminated from the trunk, the ribs were freed from the trunk and spread laterally, and the axial skeleton was fused into a bony shell box. Cloacal expansion was limited by this bony box, so the chelonians could not give birth to living young and had to return to the land to lay their eggs. The "marine commitment" involved stiffening the foreflippers that enabled the turtles to "fly" through the water, but this meant that "all possibility of any reasonably efficient mobility on land had to be sacrificed." The "strategy" of digging unprotected nests on isolated beaches probably led to a most unreptilian "numbers game," whereby large numbers of hatchlings have to scramble toward the relative safety of the ocean, only a small number of which survive. Because of the necessity for traveling long distances to reach nesting beaches, large size is another requisite; "Apparently," wrote Hendrickson, "there can be no small-sized marine turtles as we know them."

According to a study by Renous, Bels, and Davenport (2000), presented at the 1996 Poitiers symposium on secondary adaptations to life in water, "Occupation of an open aquatic environment [by turtles] has required morphological, physiological and behavioral changes driven by natural selection in response to new constraints." Their study suggests that the vast oceanic environment offered new ecological opportunities to the armored ancestors of modern marine turtles, and that early turtles developed the foreflipper method of swimming in the ocean, which enabled them to cover great distances and also to crawl out of the water for nesting on land. *Dermochelys* pulls itself forward with its foreflippers, raises its head and shoulders, and then falls forward. This, say the authors, "is the result of interaction with the ground, in response to gravity." Other living sea turtles retain alternate movements of their forelegs on land (obviously the vertebrate norm), but they swim this way as well. Furthermore, "In juveniles of [the loggerhead turtle] *Caretta caretta,* the forelimbs and mouth are employed to hold and tear food," but *Dermochelys* does not use its limbs in eating its usual food, which consists almost exclusively of jellyfish. It is in the adaptation to the marine environment, with its requirements for long-distance swimming, that the turtles made their most striking modifications: they developed forelimbs that could act as synchronous hydrofoils. Unfortunately, this sacrificed their usefulness in manipulating food.

In a 1982 article on the behavioral ecology of sea turtles, Archie Carr of the University of Florida wrote that all living female sea turtles use their hind feet in exactly the same manner to dig a nest. Since no other land turtles or tortoises indulge in such "rigidly stereotyped behavior," Carr concluded that all six recognized living sea turtle species are descended from a common ancestor. The leatherback differs from the others (hawksbill, green, loggerhead, flatback, Kemp's ridley), in its size and leathery shell. Nevertheless, Zangerl (1980) believes they are indeed descended from a common ancestor, but that two lines diverged around the Late Cretaceous, about 100 million years ago. In the absence of fossilized evidence, it is assumed that the living sea turtles do not differ substantially from their ancestors and, like the coelacanth and the horseshoe crab, may be "living fossils."

Several species of large marine turtles are found in the tropical and subtropical oceans of the world. They all have flippers for forelimbs, lightweight shells, and heads that cannot be drawn into their shells. They spend most of their lives in the water, but the females come ashore to dig a nest in the sand and lay their eggs. All species are declining in numbers, because the eggs are frequently gathered for human consumption, and because turtles' meat is also eaten. The shell of the hawksbill turtle is the source of tortoiseshell, still popularly used for decorative objects, even though plastic has replaced most uses. The United States and 115 other countries have banned the import or export of sea turtle products.

The largest of the sea turtles, and also the heaviest living reptile in the world, is the leatherback (*Dermochelys coriacea*). It is the deepest-diving sea turtle, capable of reaching depths of 3,000 feet; of all the vertebrates, only the elephant seal and sperm whale can dive deeper. Where all other sea turtles have a shell or carapace, the leatherback has smooth, leathery skin that covers seven prominent bony ridges on the back and five on the underside. Leatherbacks store heat in their muscle tissue, a process known as "thermal inertia." They can thereby maintain a swimming body temperature that can be as high as 32°F higher than the surrounding water, an accomplishment unique among living reptiles and rendered even more surprising by the leatherback's diet, which seems to consist almost exclusively of jellyfish, not known for their nutritive value. The leatherback's mouth and throat are equipped with backward-pointing spines that enable it to eat the otherwise slippery jellyfish. (Unfortunately, the turtles sometimes mistake floating plastic bags for their usual prey, and choke to death.) Females gather on the beaches of Indonesia, New Guinea, Central America, the Guianas, and the Pacific coast of Mexico to lay their eggs. Marked turtles have been found as far as 3,000 miles from their nesting beaches.

The loggerhead turtle (*Caretta caretta*) has an oversized head, hence its unusual name. A loggerhead is a ball of iron on the end of a long handle used in the melting of tar or pitch, or, by extension, an unusually large head. (Also called a loggerhead was a wooden bitt or post in a whaleboat around which the harpoon line was wrapped as it ran out after a whale had been struck.) The loggerhead turtle reaches a length of 38 inches and can weigh 200 to 400 pounds. It feeds on crabs, mollusks, and shrimps, whose shells it cracks with its powerful jaws. The loggerhead can also be distinguished from the other sea turtles by the presence of two claws on each of its foreflippers. Many female loggerheads come ashore on Florida beaches to lay their eggs during the late spring and summer. Until recently no one had any idea where the hatchlings spent the years after hatching. Archie Carr referred to it as the "lost year," but recent studies of tagged hatchlings (Dellinger 1998) have shown that the "year" is more like 10 years, during which the young loggerheads drift in ocean currents to areas of seaweed, such as the Sargasso Sea, and eventually fetch up across the Atlantic on islands like the Azores and Madeira. For the rest of their lives, these turtles wander the open oceans. Radio tagging has answered few of the puzzling questions about their lives.

Named for its narrow, sharply downcurved beak, the hawksbill (*Eretmochelys imbricata*) is one of the smaller sea turtles, reaching an average length of about 30 inches and a weight of about 120 pounds. It is found throughout the world's tropical and subtropical waters, and its favorite habitat is rocks and ledges in relatively shallow water. It feeds primarily on sponges, but it also consumes mollusks and sea urchins. Unlike many other sea turtles, the hawksbill does not nest in groups; instead, the females come ashore singly to dig their nests and lay their eggs. The beautiful marbled shell of the young hawksbill has long been sought for the manufacture of decorative objects, and this quest for so-called "tortoise shell" has resulted in the worldwide depletion of the species.

The shells of green turtles are not green; it is the greenish fat of this species that gives the green turtle its common name. Green turtles (*Chelonia mydas*) grow to about 40 inches in length, and can weigh up to 300 pounds. They have a single claw on each foreflipper. Found in the Atlantic and the Indo-Pacific oceans, green turtles nest on sandy beaches during the summer months. With their close relatives, the Indo-Pacific black turtles (*C. agassizii*), the adults are the only vegetarian sea turtles, feeding exclusively on sea grasses and algae. They are the turtle of choice for soup in the Caribbean, and the greenish color of the fat has given us "green turtle soup." Not surprisingly, their numbers have been greatly reduced by hunting, and they are considered endangered around the world. In the Persian Gulf, where 1991 man-made oil fires created an unprecedented environmental disaster, green turtles are threatened not only by oil spills but also by fishermen and fishing nets. Unless immediate steps are taken, the Persian Gulf green turtles may be on the way to extinction.*

The smallest and the most endangered of all sea turtles is Kemp's ridley (*Lepidochelys kempii*), rarely weighing more than 100 pounds. They can be found in the open waters of the Gulf of Mexico and along the Gulf coasts of Mexico, Texas, Louisiana, and Florida. They nest only on the beaches of Rancho Nuevo, Mexico,

* In recent years, more and more green turtles have been found to be infected with the tumor disease known as fibropapillomatosis. It was first described in captured adult green turtles in the 1930s, and it was identified in the late 1970s in mariculture-reared green turtles at Cayman Turtle Farm on Grand Cayman, in the British West Indies. By the 1980s it began to appear in wild populations in Florida and Hawaii, where more than 50 percent of certain populations were found to be affected. We do not know the cause of fibropapillomatosis, but we may be looking at the process of extinction at work.

The green turtle, Chelonia mydas, *gets its common name from the color of its fat, not its shell. Green turtles nest on sandy beaches on the Atlantic, Pacific, and Indian Oceans, and people have been hunting and collecting their eggs for centuries. Green turtles are considered endangered around the world.*

and in the past 40 years, the number of nesting females has dropped from tens of thousands to about 400. The beaches are now patrolled by Mexican marines, but still the eggs are stolen by people or eaten by pigs and dogs. In addition, many juvenile turtles are drowned every year in shrimpers' nets in the Gulf of Mexico. The introduction of a Turtle Excluding Device (TED) into the nets has saved many turtles, but scientists fear that the numbers, estimated at only 1,300 to 1,500 animals, are already too low for the species to survive.

The olive ridley (*L. olivacea*) is widespread in the Pacific, the Indian Ocean, and parts of the Atlantic. Its range does not overlap with that of Kemp's ridley, and they may actually be a single species.

The flatback turtle (*Natator depressus*) gets its common name from the smoothness of its shell, which shows little definition of the scales on the carapace. The flatback used to be known as *Chelonia depressus* (the same genus as the green turtle), but it has now been placed in its own genus. Its shell is about 36 inches long, and adults weigh about 175 pounds. Unlike most other sea turtles, flatbacks do not migrate, spending their lives in and around Queensland, the Gulf of Carpentaria, and the Torres Straits, in shallow, soft-bottomed areas away from reefs. Flatbacks are carnivores, feeding on sea cucumbers, soft corals, and jellyfish. Like most of the sea turtles found in Australian waters, the flatback is hunted for food by Aboriginals. On the breeding islands, feral pigs often dig up and eat the eggs, but still, the flatback is the only sea turtle species in the world that is not currently considered endangered.

Alfred Sherwood Romer, the dean of vertebrate paleontologists and the author of the definitive textbook on the subject found turtles "the most bizarre of reptilian groups. . . . Were they extinct, their shells, the most remarkable armor ever assumed by a land animal, would be a cause for wonder" (*Vertebrate Paleontology*, 1966). In 1968 he wrote, "Mammals now dominate the scene. The turtles have taken no notice of us. And one may, not unreasonably, suspect that, in the far distant future, when mammals (and man) have beome as extinct as the dinosaurs, the turtles will still be plodding stolidly and conservatively along the corridors of time." This may be a credible scenario for the tortoises, but with the exception of the flatback, the marine turtles are all in trouble, and it is all our fault. We killed them in prodigious numbers for their shells and their meat, and then we stole their eggs so they could not mature and propagate. We commandeered their breeding beaches and disrupted their ancient migration routes. These ancient animals, whose armor plating enabled them to survive every predator and catastrophe for 250 million years, may not survive another thousand.

As for the tortoises, some of those that survived to the present in gigantic forms are evidently headed toward extinction much faster than their stumpy legs can possibly carry them away from it. In fact, many varieties of the great land tortoises (genus *Geochelone*) are already gone. Those that lived on the Indian Ocean islands of the Seychelles, Mauritius, Rodriguez and Réunion were all killed off by sailors by the close of the nineteenth century. The only giant tortoises left are on the little Indian Ocean coral atoll of Aldabra, and on a few of the Galápagos Islands in the eastern Pacific. Many of the subspecies of *Geochelone elephantopus* from individual Galápagos Islands are also gone, killed of by passing whalers who loaded them on their whaleships as a source of fresh meat during voyages that might last for years.

One of the most enduring discussions concerns the question of humanity's place in the evolutionary process. We are certainly products of the process, even though the fossil evidence is as poor for us as it is for the turtles, but how are we to explain extinction, a previously natural mystery that has now acquired an identifiable causative agent? Creatures of all sorts, from trilobites to ammonites, from dinosaurs to desmostylians, went extinct, perhaps by an organic evolutionary process, or perhaps as a result of an extraterrestrial catastrophe, but a new factor has now been added to the equation: *Homo destructivus.* "Civilized" man, the painter of cave walls and builder of cities, has been around for maybe 30,000 years and, in that nanosecond of evolutionary history, has so misinterpreted the biblical injunction to "have dominion over the fish of the sea, and over the fowl of the air, and over every living thing that moveth on the Earth" that it is as if we had been commanded to eliminate as many creatures as we could, lest they interfere with our dominion.

THE SEA SNAKES

It is not evident whether snakes evolved as aquatic animals and then some came out on land, or whether they are land animals, some of which returned to the sea. Traditionally, the earliest reptiles are thought to have emerged from the water, eventually to evolve into lizards, turtles, crocodiles, dinosaurs, snakes—and maybe birds. Some reptiles returned to the water: the ichthyosaurs, plesiosaurs, and mosasaurs aquired an aquatic modus vivendi, and then became extinct. The legs of these returning reptiles were modified into flippers. The terrestrial snakes, however, had no legs to modify, so if some snakes became aquatic, the major morphological change involved the flattening of the tail, to allow for more efficient movement through the water, and a modification of the breathing apparatus.

In the chapter on evolution in *Australian Snakes: A Natural History*, the biologist Richard Shine wrote (1991), "Partly because of the scarcity of good fossils, the classification of snakes is a mess. The difficulties are compounded by the fact that snakes have very simplified bodies, and there are not too many external characteristics upon which to base a classification: they look too much alike." At one time, it was believed that sea snakes were a single lineage, and biochemical studies suggested that snakes invaded the oceans several times. The egg-laying sea snakes, or sea kraits (Laticaudidae), probably arose from terrestrial elapids or perhaps from a very early Australian species. They are now found widely throughout the Pacific and have evolved flattened, paddlelike tails very similar to those of the live-bearing sea snakes (Hydrophiidae). Like the hydrophiids, the sea kraits have very large lungs (to enable them to stay under water longer), valvular nostrils (to keep out seawater), and salt-excreting glands under their tongues (to get rid of excess salt).

In 1966, commenting on the confused and inconclusive state of the study of snake evolution at that time, Alfred Sherwood Romer wrote, "In contrast to the extinction or seeming evolutionary stagnation of other reptile types, the snakes are today a group of reptiles still 'on the make.' " Minton and Heatwole (1978) wrote, "There seems to be little doubt, that [snakes] evolved in the region between northern Australia and the peninsulas of southeast Asia, a region that still harbors the greatest number and variety of species." The same authors wrote, "The snakes appeared late in the Mesozoic, and there is some evidence that, quite early in their history, they produced some huge marine species. Apparently these giant sea snakes were not very successful, for they endured but a short time and left a very scanty fossil record." According to sea snake expert Harold Heatwole, "Other snakes from the Cretaceous are known only from incomplete fossils and as no good skull material exists the most that can be said of them is that they were snakes with some characteristics intermediate between those of lizards and modern snakes" (*Sea Snakes*, 1999).

The earliest snakes had an elongate body, reduced limbs, and adaptations for chemosensory hunting. They are descended from lizards, and manifest many lizard-like characteristics. In 1997, Caldwell and Lee published a description of *Pachyrhachis problematicus* (from *pachys,* "thick," and *rhachis,* "spine"), a fossil snake that was found in the limestone quarries of Ein Yabrud, some 12 miles north of Jerusalem. Although it was certainly a snake, it had tiny hind legs consisting of femur, tibia, fibula, and tarsal bones, as well as other characteristics that provided "additional support for the hypothesis that snakes are most closely related to Cretaceous marine lizards (mosasauroids)." It was later determined that a fossil snake known as *Ophiomorphus,* also from the quarries at Ein Yabrud, was actually another specimen of *Pachyrhachis problematicus,* and though neither skeleton was complete, scientists had enough material to postulate the snake's morphology and relationships. Its heavy bones indicated that its overall density was very close to that of seawater, and it was probably a very slow swimmer. Although its head was tiny, it had powerful jaws and was believed to strike swiftly at its prey.

Then Lee, Caldwell, and J. D. Scanlon (1999) reexamined *Pachyophis woodwardi,* which had originally been described as a snake by a Hungarian amateur dinosaur expert, Baron Nopsca, in 1923,* but "the evidence was not compelling, and later workers have been reluctant to accept this view." But in their reevaluation, Lee and his colleagues believe that they have shown that *Pachyophis* was indeed a very primitive snake. The fossil was found in East Herzogovina, in the same locality as another snake known as *Mesophis nopscai,* which is now lost. The fossil of *Pachyophis* is smaller than that of *Pachyrhachis,* but its ribs are much heavier, indicating that it is not a juvenile of the latter form. The thickened bones—a condition known as pachyostosis—is characteristic of marine animals because it increases their density, which strongly suggests that this species (and also *Pachyrhachis*) were marine. "However," wrote the authors, "at the moment, whether or not all snakes went through a marine phase in their evolution remains equivocal."

Lee, Bell, and Caldwell (1999) "indicate that the nearest relatives of snakes are the Mosasauroidea, medium to gigantic marine lizards. . . . The most basal ('primitive') snakes are the Cretaceous forms *Pachyrhachis* and *Dinilysia,* along with the living blindsnakes." *Dinilysia,* discovered from the Cretaceous of Argentina, "has inspired and challenged efforts to understand snake origins. The single well-preserved skull exhibits such a blend of snakelike, lizardlike, and unique characteristics that herpetologists still argue about whether this creature was indeed a snake" (Greene 1997). It would appear that snakes and mosasaurs are "sister groups," which means that they have a common ancestor that they do not share with any other animals, and that ancestral group split into two distinct lines, one of which evolved into mosasaurs while the other became snakes. Because the mosasaurs and the early snakes were aquatic, their common ancestor must also have been aquatic.

Along with *Pachyrhachis* and *Dinilysia,* the fossil snake known as genus *Wonambi* is one of the most primitive snakes known. Wonambi (also spelled Wanambi or Wanampi) is the Aboriginal name of giant serpents that are supposed to live in sacred water holes in central Australia, including rockholes on the tops of Ayers Rock and Mount Olga. Australians John Scanlon and Michael Lee (2000) wrote that "none of these three primitive snake lineages shows features associated with burrowing, nor do any of the nearest lizard relatives of snakes (varanoids). These phylogenetic conclusions contradict the widely held 'subterranean' theory of snake origins, and instead imply that burrowing snakes . . . acquired their fossorial [burrowing] adaptations after the evolution of the snake body form and jaw apparatus in a large aquatic or (surface active) terrestrial ancestor." In other words, the earliest true snakes were probably aquatic, and the sea snakes might be their direct descendants, without having passed through a terrestrial phase. *Wonambi* was a huge snake for any period;

---

* Franz Baron Nopsca von Felso-Szilvas (1875–1933) was born into a noble Hungarian family and became fascinated by dinosaurs after he had discovered and described a new species he called *Limnosaurus.* His book on fossil reptiles, *Osteologia Reptilium Fossilium et Recentium,* was published in 1926 and established the classification of dinosaurs into the saurischian ("lizard-hipped") and ornithischian ("bird-hipped") categories still in use today. The flamboyant Nopsca (pronounced "*Nope*-sha") had a habit of dressing in Balkan peasant costume, and when Albania was freed from Turkish rule in 1913, he applied for the job of king. He served as an officer and a spy during World War I, but after the war, his estates were confiscated by the government of the newly created nation of Romania, and in 1933, reduced to poverty, he killed his male lover and then committed suicide.

the total length of the specimen described by Scanlon and Lee was approximately 16 feet.

The extinct marine reptiles known as mosasaurs are now believed to be the nearest relatives of snakes. Like snakes, mosasaurs had pterygoid (palatal) teeth and hinged lower jaws. Snakes have developed a rigid skull structure from which the highly flexible jaws are suspended. This facilitates the engulfment of large prey items, even larger than the snake's head. Harry W. Greene of Cornell University, in his 1997 *Snakes: The Evolution of Mystery in Nature*, wrote, "Several lineages of basal macrostomatans ["big-mouth" snakes] had appeared by the end of the Mesozoic and the beginning of the Cenozoic eras (65 mya); these were boa-like snakes with cranial modifications that permitted them to swallow substantially more diverse prey than are taken by Asian pipesnakes (*Cylindrophis*).... Long quadrate bones and lower jaws, enhanced movement between the tips of the mandible, and loosened connections among bones of the skull enabled these serpents to ingest bulky prey; they thus could add relatively large crocodilians, birds, and mammals to their diets." By the early Miocene (25 million years ago), the colubrid snakes had radiated so explosively that the Cenozoic (the last 65 million years) has been called "the age of snakes"—usually by herpetologists.

Since Minton and Heatwole bemoaned the lack of primitive snake fossils, fossils of their "giant marine snakes" have begun appearing, but, as with many things paleontological, newly unearthed fossils often confuse rather than clarify things. Consider *Haasiophis terrasanctus,* a particularly well-preserved fossil that was found in the same limestone quarry in the Judean Hills as *Pachyrhachis.* (*Haasiophis* was named for Professor Haas, who found the fossil in 1976; *terrasanctus* means "Holy Land" in Latin.) The *Pachyrhachis* and *Haasiophis* fossils are from the Middle Cretaceous, 95 million years ago, but the analysis of the *Haasiophis* fossil by Tchernov et al. (2000) produces conflicting conclusions from those of Scanlon et al. The fossil *Haasiophis* was about 3 feet long and also had legs, but its jaw structure suggested to Tchernov et al. that it was more closely related to the larger, living snakes of today and that both *Pachyrhachis* and *Haasiophis* were not primitive at all, but advanced snakes that had re-evolved legs. (Many living pythons retain rudimentry hind limbs, so re-evolving limbs is a possibility.) The limbs of *Pachyrhachis* and *Haasiophis* are too small in relation to body size to have had any locomotor function; they may have been used as an aid in mating, as are the hind-limb buds of pythons today.

The descent of snakes is one of the most contentious areas in vertebrate biology. Tchernov et al. are more than a little critical of the conclusions of Lee, Bell, and Caldwell and wrote, "*Haasiophis* and *Pachyrhachis* have no particular bearing on snake-mosasauroid relationships or snake origins.... Basal snakes, including basal macrostomatans, retain rudimentary hind limbs, which, however, remain much more incomplete than those of *Haasiophis.*" In the essay that introduces the 2000 Tchernov et al. paper in *Science*, Greene and Cundall criticize the methodology and conclusions of Lee, Bell, and Caldwell, who "showed their drawings and reconstructions of *Pachyrhachis* to a number of nonscientists who use 'snake' in the vernacular sense, and all identified *Pachyrhachis* as a snake rather than a lizard." Even though it had legs, the dense, heavy bones of *Pachyrhachis* suggested that it was a water dweller, and the location of the fossil in marine deposits seemed to confirm this suggestion. Because of similarities in the jaw structure, Lee, Caldwell, and Scanlon (1999) concluded that it and all other snakes had evolved from mosasaurs. In remarks published in *New Scientist* (Hecht 2000), Mike Caldwell said that "Rieppel's analysis fails to compare the legged snakes with mosasaurs, their closest relatives."

None of this discussion addresses what to many is the fundamental question about snake evolution in the first place: Why should an animal with legs evolve into one without them? Legs seem to be quite a useful bit of equipment; every other terrestrial vertebrate has them. Only the fishes are legless; the whales, to adapt to the fishes' watery environment, have lost their hind limbs completely and their forelimbs have evolved into paddles. Perhaps the sea snakes are a more advanced form of snake, for only in aquatic animals does leglessness confer an advantage—or at least, it does not handicap the animal. Snakes, terrestrial and aquatic, are unquestionably successful, but only in tautological terms: if they were not, they would not be here. (I am using "successful" here to mean "nonextinct." Many earlier forms—the nonavian dinosaurs, for example—were more successful than the snakes in that they lasted considerably longer, but our anthropocentric worldview leads us to believe that everything now living on Earth—especially us—is more successful than the legions of long-extinct creatures.)

Compared to some of the other marine reptiles, very little is known about sea snake evolution from fossils, and it is largely because some species have survived into the present that we know what we do. Like other snakes, sea snakes are legless reptiles with scales and forked tongues. To locate their prey, most snakes depend on their sense of smell and also their tongues, which absorb chemical odors from the air or water. When the tongue is retracted, odors are processed in a sensory pit on the roof of the mouth known as Jacobsen's organ. (Some terrestrial snakes also have heat-sensitive areas on the head that enable them to locate their prey in total darkness, but sea snakes do not.) Both terrestrial and sea snakes are air breathers and breathe through one lung. Except for the giant constrictors such as pythons, boas, and anacondas, where paired lungs are the rule, most snakes have lost the left lung completely. The right one has become elongated to fit the shape of the narrow body. In some sea snakes the lung extends almost the entire length of the body and consists of a long hollow tube divided into three sections: the tracheal portion, nearest the head, which is a modified windpipe; the central bronchial portion, which is the lung proper; and the posterior section, known as the saccular lung, which contains so few blood vessels that its function must be to serve as an air reservoir.

Danish herpetologist Helge Volsøe has written (1939), "Since all snakes are able to swim, it cannot cause much surprise that a distinct group of snakes, the sea snakes, have taken the sea into its possession, and have acquired such adaptations that enable them to compete with the fishes in rapidity of movement, and so subsist entirely on these creatures." Most living snakes are competent swimmers, even those whose normal habitat is in the desert. Well-known species like anacondas, water moccasins, and the common "water snakes" (*Natrix* spp.) make most or all of their living in the water. What differentiates the sea snakes from other snakes that swim? Their bodies are laterally flattened, often ribbonlike, and their tails even more so, having become modified almost into an oar that they use to scull through the water. Volsøe describes sea snakes observed in the Iranian Gulf that "move backwards with almost equal alertness as that [with] which they moved forwards. . . . This ability seems to be a new acquisition in the sea snakes, since it was never observed—as far as I know—in land snakes." While under water, the sea snake's nostrils can be shut tight with a valvelike device, and the nostrils face upward to facilitate breathing at the surface. When the mouth is closed it is sealed except for a small opening through which the tongue can protrude. Most sea snakes are viviparous and deliver their live young in the water, but the sea kraits leave the water to lay their eggs on land.

The largest family of sea snakes is the Hydrophiidae, which are closely related to cobras (Elapiidae), and are as deadly. Of the 70 odd known species of sea snakes, 53 (76 per cent) are hydrophiids. There are four species of sea kraits (*Laticauda,* meaning "broad tail"); their common name, krait, is taken from that of the terrestrial venomous elapids known as kraits that are found in scattered locations throughout southern Asia. Most sea snakes are venomous, and sea kraits are highly venomous, but even when handled, kraits rarely bite humans. Most sea snakes average between 2 and 3 feet in length, but the largest is the yellow sea snake (*Hydrophoris spiralis*), which has been measured at 9 feet, and the yellow-lipped sea snake, *Laticauda columbrina,* which has been recorded at 11 feet (Heatwole 1999).

Sea snakes use their poison to subdue their prey, which consists mainly of small fishes. Sea snakes rarely attack humans, although the venom certainly is powerful enough to debilitate or even kill a person. Most bites occur when the snakes are being removed from fishermen's nets. There are, however, occasional reports of unprovoked attacks in Asian coastal waters by the beaked sea snake (*Enhydrina schistosa*), whose venom is twice as toxic as that of the Indian cobra. Rasmussen, in his 1999 Food and Agricultural Organization report, lists three other species that are "much more aggressive": the olive sea snake (*Aipysurus laevis*), Stokes' sea snake (*Astrotia stokesii*), and the ornate sea snake (*Hydrophis ornatus*). Heatwole asks, "Why are there venomous sea snakes?" and "How did such a complicated system evolve?" His answer: "Sea snakes and sea kraits both inherited their venom and associated delivery apparatus from venomous ancestors, the elapid snakes. . . . It is likely that venom first arose in the role of facilitating feeding."

The primary prey items of sea snakes are fishes, which they swallow whole and head-first. Some snakes have more specialized tastes. *Hydrophis cantoris,* with its tiny head and long thin neck, seems designed to feed almost entirely on burrowing eels. Less-specialized sea snakes also eat eels and gobies, fish eggs, and crabs, and some freshwater species will take frogs. And what threatens sea snakes? Eagles and other predatory sea birds probably pluck them from the surface, and in Australian waters, tiger sharks have no compunctions about eating them. The greatest danger to sea snakes, however, is *Homo sapiens.* File snakes are caught in the Philippines for their skins, and numerous other Asian species are killed and used in the manufacture of exotic leather goods. After the snakes have been skinned, the carcasses generally are

The yellow-bellied sea snake, Pelamis platurus, *found only in the tropical Pacific, has been observed at sea in huge aggregations sometimes 60 miles long.*

used in the preparation of feed for poultry and pigs, although in some countries, sea snake meat is believed to be an aphrodisiac.

The geographical distribution of sea snakes is almost exclusively a function of water temperature. As Richard Shine wrote (1991), "They seem to be highly specialised fish-eaters that depend on specific tropical environments for feeding and reproduction. Fully aquatic species of snakes are almost all restricted to the tropics, perhaps because of their mode of temperature regulation. Snakes that don't emerge from the water have relatively little opportunity to select particular temperatures. They can 'bask' in the hotter surface layers of the water, or find slightly warmer or cooler pockets when they need it, but have little real control of their body temperatures."

Most sea snakes are found in the tropical and subtropical waters of the Pacific and Indian oceans, from the east coast of Africa to the Gulf of Panama. There are no sea snakes in the Mediterranean, the Atlantic, or the Caribbean. In fact, early plans for a sea-level canal in Panama evoked concern that connecting the Atlantic and Pacific oceans would allow sea snakes to enter the Atlantic. The sea kraits (laticautids) are found throughout Southeast Asia, Indonesia, and the Philippines. The most widely distributed of all sea snakes is the yellow-bellied sea snake (*Pelamis platurus*), which occurs from the east coast of Africa along the coast of southern Asia, into the Pacific northward to Japan, and across the ocean as far as Mexico. It is usually found in shallower waters, over continental shelves or in the vicinity of islands; there are very few sightings in water that is more than 50 fathoms deep. *Pelamis* is the most widely distributed of the sea snakes, but it is not the most common. That distinction belongs to the genus *Hydrophis* (*ophis* is "snake" in Greek, so *Hydrophis* means "water snake"), which includes some 24 species, and the closely related *Aipysurus*, with 8 species. Almost all of them live in saltwater, but *Hydrophis semperi* is found only in the freshwater Lake Taal in the Philippines; and the file snake (*Acrochordus granulatus*) lives in the rivers and lakes of Australia, Southeast Asia and various Pacific island groups. The file snake, which gets its common name from its granular scales, is a drab, flabby snake about 3 feet long that can remain submerged for up to five hours. It actually accomplishes gas exchange though its skin, and when submerged, its heartbeat can slow to less than one beat per minute.

Most sea snakes are solitary swimmers, but occasionally they aggregate in huge groups, for reasons that remain a mystery. The yellow-bellied sea snake has been sighted in groups of several thousand in Panama Bay, and in his 1999 book *Sea Snakes*, Heatwole mentions

> a spectacular aggregation of sea snakes in the Malacca Straits [that] was reported in 1932 by Willoughby Lowe. From the deck of a ship he observed a nearly solid mass of a species reported to be *Astrotia stokesii* in a line about three meters wide and 100 kilometers long, containing literally millions of snakes, many intertwined with each other. Such an event is certainly not a regular feature of the biology of any sea snake, and it is not known whether it was a migration or some other phenomenon.

Heatwole remarks that "there are a few air-breathing animals that have remarkable abilities to dive to great depths and remain submerged for extended periods. Among the most accomplished of these are the whales, seals, and penguins. To this list must be added the sea snakes. Some species that feed and rest on the bottom submerge to a depth of 100 meters. They can remain underwater for about two hours." Like all reptiles, sea snakes are cold-blooded, and their body temperature is usually the same as that of their surroundings. Indeed, water absorbs so much heat that a sea snake basking at the surface cannot raise its body temperature more than three degrees above the ambient water temperature.

Our understanding of snake evolution is still fragmentary. In 1999, Scanlon et al. wrote, "The origin of snakes from lizard-like ancestors was a major event in vertebrate evolution, and remains poorly understood despite much effort." Although the fossil record is too sparse to make any definitive statements (always dangerous in paleontology anyway), many paleontologists have done exactly that. After two groups of scientists examined fossil snakes from the same area, two scenarios emerged. Scanlon, Lee, and Caldwell have proposed that snakes are derived from water-dwelling lizards (mosasaurs), and emerged from the water to become the terrestrial species that we know today. As for the sea snakes, these writers say that they are simply land snakes that returned to the water and developed adaptations that enabled them to function more efficiently as aquatic animals. In their 1999 study of *Pachyrhachis*, Scanlon et al. said that it was "the most primitive snake [and] an excellent example of a transitional taxon, exhibiting most, but not all, of the derived features that are diagnostic of snakes." In their reconstruction, they show a large-bodied snake with a tiny head, minuscule hind limbs, and a short tail section.

The other group of scientists, Tchernov et al. (2000), believe that modern snakes are the descendants of small, burrowing, terrestrial lizards. Greene and Cundall (2000) in their discussion of the Tchernov et al. article wrote, "In fact, a terrestrial-to-marine transition seems more likely as it is a common theme among tetrapods—for example, modern whales evolved from a terrestrial ancestor." Thus, Scanlon, Lee, and Caldwell believe that the two species from Israel are ancestral to living snakes and that snake evolution thus began in the water. Tchernov et al. suggest that the Israeli species evolved later and were not ancestral at all to other snakes. As for the legs, which Scanlon et al. cited to demonstrate that snakes are descended from lizards, all they mean to Tchernov's group is that ancestral lizards lost their legs, became snakes, evolved into snakes with legs, and as of now have become (mostly) snakes without legs. They acknowledge that this is a less parsimonious explanation, but it nevertheless "remains a possibility, given the incompleteness of the fossil record of snakes and the recognition of multiple loss of limbs among squamates in general."

THE CROCODILIANS

Canst thou draw out leviathan with an hook? or his tongue with a cord which thou lettest down? Canst thou put an hook into his nose? or bore his jaw through with a thorn? . . . Canst thou fill his skin with barbed irons? or his head with fish spears? . . . Who can open the doors of his face? his teeth are terrible round about. His scales are his pride, shut up together as with a close seal. One is so near to another, that no air can come between them. They are joined one to another, they stick together, that they cannot be sundered. By his neesings a light doth shine, and his eyes are like the eyelids of the morning. Out of his mouth go burning lamps, and sparks of fire leap out.

—Job 41:1–32

When John Tarduno of the University of Rochester and his colleagues discovered the fossilized bones of a champsosaur on Axel Heiberg Island high in the Canadian Arctic, they saw it as a clear indication that 90 million years ago, the temperature 600 miles from the North Pole was a lot warmer. Axel Heiberg Island is now an uninhabited island in Nunavut (the eastern portion of Northwest Territories), west of Ellesmere Island. Axel Heiberg is about 16,000 square miles of snow and ice, deeply indented by fjords. In 1985, a geological survey team flying over the island observed a field of large stumps; the wood was so well preserved that it could be cut like recent lumber and readily burned. It was subsequently dated at 45 million years old, indicating an extraordinary change in climate.

For champsosaurs to inhabit the region it would have had to be very much warmer than now, because champsosaurs were cold-blooded animals that required fairly high temperatures to survive. The Late Cretaceous was ice-free, and the Arctic regions averaged 14°C (57°F) and did not go below freezing. Extensive volcanism apparently injected heat-trapping carbon dioxide into the air, which contributed to the Cretaceous global warming.

Although the champsosaurs are no more, they have the distinction of being the largest vertebrates to survive the K-T Extinction, the catastrophe that took out the dinosaurs 65 million years ago. They reached a known maximum length of 5 feet, about the size of a large monitor lizard, but their narrow snout made them look more like a modern gavial, and like the gavials they were fish eaters. The width of the skull behind the eyes suggests particularly powerful jaws. It has been suggested that they swam the way modern crocodiles and marine iguanas do, powered by sweeps of the tail, with their legs tucked in and the body undulating. Fossils of *Champsosaurus* (the nominal species) have been found in river and swamp deposits dating from the Late Cretaceous to the Eocene in Belgium and France, and also in Alberta, Montana, New Mexico, and Wyoming. They were not actually crocodilians, but they were similar in morphology and habits.

With their scaly, low-slung, heavily armored bodies, crocodilians (crocodiles, alligators, caimans, and gavials) look not unlike their prehistoric forebears. In fact, they look pretty prehistoric themselves, and it is not difficult to imagine a swamp of 150 million years ago with the eyes and nostrils of an ancient crocodile poking ominously through the surface of the water. In the 1966 textbook *Vertebrate Paleontology*, Alfred Sherwood Romer called the crocodiles the "least progressive of ruling reptiles." Something about primitiveness must work however, because they are still with us, pretty much unchanged over the last 200 million years.* As Hans-Dieter Sues of the Royal Ontario Museum has written (1989), "Crocodilians are the only living representatives of one of the most successful groups of land-dwelling vertebrates ever known—the Archosauria or ruling reptiles." Terrestrial, aquatic, or semiaquatic forms dominated the inshore and offshore environments of the Mesozoic Era, from 250 million to 65 million years ago. On the basis of a similarity in the ankle bones, they are believed to be derived from the thecodonts, which were early reptiles that dominated the Mesozoic, some 250 million years ago. Even though they bear a superficial resemblance to lizards, crocodilians are more closely related to dinosaurs (and therefore to birds) than they are to lizards. (In 1972, the British paleontologist Alick Walker suggested that birds arose from a stock of lightly-built crocodilians that took to the trees, but most paleontologists now agree that birds were descended from small carnivorous dinosaurs, not tree-climbing crocodiles.)

Crocodilians have an elongate outer-ear canal, a muscular gizzard, and complete separation of the heart ventricles. All crocodiles have a diapsid skull, meaning that it has a pair of openings on the top and sides. The teeth of crocodiles and dinosaurs are set in sockets, whereas lizard and snake teeth are fused to the jawbone. Crocodiles and dinosaurs have a deep socket in the pelvic bone for the insertion of the thigh bone, but lizards do not. A crocodile has two different walking modes; it can walk like a lizard with its legs splayed out and its belly close to the ground, or it can raise its belly off the ground and walk like a mammal, with its hind legs very nearly underneath its body. When swimming both at the surface and under water, modern croco-

---

* In his posthumously published novel, *The Dechronization of Sam Magruder*, paleontologist George Gaylord Simpson (1902–84) wrote: "It has been said by some theorists that cases like that of the crocodile, virtually unchanged for 100 million years and more, represent a failure of evolutionary force, a blind alley, or a long senescence. . . . Here was no failure but an adaptation so successful, so perfect that once developed it has never needed to change. Is it, perhaps, not the success but the failure of adaptation that has forced evolving life onward to what we, at least, consider higher levels? The crocodile in his sluggish waters has perfectly mastered life in an unchanging environment. No challenge arose."

diles use their vertically flattened tails for propulsion and steer with their hind legs while keeping their forelegs tucked in. It is probably safe to assume that the extinct marine crocodilians swam the same way.

Before the crocodiles, there were sphenosuchians (*spheno* from the Greek word for "wedge," referring to the shape of the skull; and *suchos,* from an Egyptian word for crocodile), which were big-headed reptiles, only a foot long, but the bones of the skull, neck, and ankle, as well as the vertebrae, classified them clearly as crocodilians. They lived during the Middle Triassic Period, about 225 million years ago. Among the earliest sphenosuchians were *Terrestrisuchus* and *Gracilisuchus,* both leggy, smallish land animals that probably ran down their prey the way many lizards do today. The protosuchians (including *Protosuchus*) were larger, perhaps 3 feet in total length, but they were still terrestrial. They had a mouthful of sharp teeth, including a pair of massive canines that fitted into notches on either side of the upper jaw, much like modern crocodiles. Until the rise of carnivorous dinosaurs, about 200 million years ago, the early crocs were the dominant land predators on Earth. Species like *Ticinosuchus,* found in Argentina and Brazil, were large, powerful animals, with heavy skulls and serrated, bladelike teeth.*

The phytosaurs were very common semiaquatic freshwater forms that became extinct at the end of the Triassic. Although the phytosaurs (e.g., *Rutiodon*) looked very much like crocodiles, they were only distant relatives. (The name phytosaur, which means "plant-eating lizard," was erroneously bestowed on these piscivores by an early paleontologist who thought they were herbivorous.) The nostrils of crocodilians are located at the tip of the snout, but the phytosaurs had nostrils immediately forward of the eyes, and they lacked the secondary bony palate of the crocs, which enables the latter to feed under water without flooding their nostrils.

The Mesosuchia (middle crocodiles) of the Jurassic looked not unlike today's crocodiles. Some of the 70 known genera were terrestrial or semiaquatic, but several species, such as *Metriorhynchus,* were marine. *Metriorhynchus* was an underwater hunter of fishes that was sleekly unarmored (the heavy armor of land crocs makes them largely inflexible). The body was highly streamlined and the feet were webbed and had developed more into paddles than those of the semiaquatic crocodiles. Well-preserved fossils from Germany show that the tail of these marine species had evolved into a fleshy fin that extended above the downturned caudal vertebrae—the same swimming modification achieved by the ichthyosaurs. The sea crocodiles may have even developed a small dorsal fin, also like the ichthyosaurs. Yet another similarity to ichthyosaurs is the presence in fossil marine crocodiles of a sclerotic ring, a circle of bones that compresses the eyeball, pushing the lens farther from the back of the eyeball and allowing an animal to change focus underwater. (Many living birds

---

* Serrated teeth, like steak knives, make cutting through fibrous flesh easier. Many carnivores developed these dentary modifications, including some of the early crocodilians, many of the ancient and recent sharks, some primitive whales, and some of the most terrifying meat-eating dinosaurs that ever lived, including *Tyrannosaurus rex.*

have a sclerotic ring, but no living reptiles do.) According to Neill (1971), "In marine crocodilians it may have preserved the eyeball from deformation by water pressure when the animal made a deep dive." The largest of the metriorhynchids was the 13-foot-long *Dakosaurus,* and the smallest was the slender *Geosaurus,* which did not exceed 6 feet in overall length and had a particularly large tail fin.

Another marine crocodile family was the Teleosauridae. Teleosaurids were gavial-like in appearance. Their long, narrow snouts equipped with numerous piercing teeth suggest a diet of marine invertebrates and fishes. The nominal species, *Teleosaurus,* described by Geoffroy Saint-Hilaire in 1825, probably lived in littoral (inshore) waters. This can be inferred from the fact that it was heavily armored, and true marine crocs seem to have lost their armor, to become sleeker and swifter under water. In 1992, a 22-foot-long fossil *Teleorhinus* was found at the Dickson Hardie site in Saskatchewan. Another marine croc, *Steneosaurus,* lost much of its armor by the Jurassic, and became more streamlined, again presumably for more efficient chasing of fishy prey. The forelimbs were flattened and the hind feet webbed, but the tail was not as powerful as that of the swimming crocodiles (including the living *Crocodylus porosus*), which suggested a gavial-like hunting strategy: lying in wait for fishes to swim by and then striking out with a sudden lateral movement of the jaws. Such an interpretation might explain why the teleosaurid fossils are usually found in shallow-water deposits. If living crocodilians are an example, different teleosaurids may have competed for prey and living space, and the most powerful prevailed. The metriorhynchids were faster swimmers and had more powerful jaws, and they may have contributed to the extinction of the slender-jawed teleosaurids, either by out-hunting them, or by actively attacking them (Hua and Buffetaut 1997).

Examination of the fossils suggests that the paired flippers of the marine crocodiles would have made them extremely clumsy on land, but reptile embryos have to breathe through a permeable shell, and so the eggs cannot be laid in water. Therefore, no matter how awkward they might have been, the sea crocodiles probably came ashore to lay their eggs. (The ichthyosaurs and mosasaurs, reptiles that were also equipped with paddles instead of feet, gave birth to live young in the water.) A modern analog would be the sea turtles, which, though almost completely out of their element on land, still struggle ashore to lay their eggs. Some authors believe that the metriorhynchids were too vulnerable out of the water to have laid eggs on shore, and that they must have been ovoviviparous (*ovo,* "egg," and *viviparous* for "live-birth"), meaning the embryos develop in the mother, protected by an egg membrane, and then are brought forth alive, as in some fishes, lizards, and snakes. Others think that they were oviparous, meaning that young hatch from an already fertilized egg that has been laid, as with the turtles, some sharks, and, of course, birds. (The living egg-laying mammals, like the platypus and echidnas, are also oviparous.)

Because crocodilians have an unbroken lineage that can be traced back for 200 million years and many of them look much like their distant ancestors, they are usually included with cockroaches, coelacanths, and horsehoe crabs as "creatures that time forgot." Almost everything that has ever lived is extinct (the usual figure is 99.9 percent), so the crocodilians—at least some of them—must have stumbled upon some excellent formula for survival. In *Living Fossils* (1984), a compilation of essays by authorities on various animals, Eugene Meyer contributed "Crocodilians as Living Fossils." He wrote, "It is now clear that only one group of crocodilians, the broad-nosed group, is at all stable. Species change, predators and prey change, but the common broad-nosed shape is perpetually renewed. The question arises: What makes them living fossils?" Part of his answer:

> The broad-nosed group has given rise to many offshoots, including forms with delicate narrow snouts, forms with duck-bills, and forms with strange tall snouts. Some tall-snouted species were so terrestrial that they had hooves! The broad-nosed group now appears as a central core in crocodilian evolution, able to innovate on many levels.

Meyer believes that the divergent crocodilian forms all developed in habitats in which the broad-nosed forms were rare, and thus could avoid the powerful predatory behavior of the broad-nosed alligators and crocodiles, which he says is one of the factors that has enabled them to predominate for so long. Wherever a broad-nosed species existed within a mixed crocodilian fauna, it grew to be the largest and predominant form. Meyer proposes that "combat and its avoidance select for the classic shapes and construction of broad-nosed crocodilians. Combat and dispersal are common to three processes: predation between species, cannibalism, and sexual battle":

> These processes stabilize a "central" core, often while driving innovation into side groups. Neither stability nor innovation requires that each genetic species lasts a long time. One or more of the conflicts can be given up and . . . can then change morphologically and give rise to a specialized offshoot that lives in habitats free of the ancestral forms. Crocodilians have produced such specialists again and again.

Described by Buckley et al. in 2000, *Simosuchus* did not conform at all to the crocodilian stereotype, which the authors say consists of "an elongate snout with an array of conical teeth, a dorsoventrally flattened skull, and a posteriorly positioned jaw articulation, which provides a powerful bite force." Instead, *Simosuchus* was a pug-nosed creature with a high, domed skull, and clove-shaped, multicusped teeth reminiscent of those of some ornithischian (bird-hipped) dinosaurs. "This last feature," wrote the authors, "implies that the diet of the new taxon may have been predominantly if not exclusively herbivorous." (When *Simosuchus* was described in *Discover* magazine in September 2000, it was called a "broccoli-eating croc.")

The crocodilian *Deinosuchus* ("terrible crocodile") is usually considered alongside *Tyrannosaurus rex* and *Carcharodon megalodon* as one of the most terrifying predators of all time. Found in the Late Cretaceous shales of Montana and also on the Atlantic coastal plain from New Jersey to Mississippi, these animals probably reached a length of 50 feet and may have weighed as much as 18 tons. Other ancient crocodiles, including *Sarcosuchus* and *Purrusaurus* may have been almost as large, but

The saltwater crocodile is the largest living reptile, reaching a known length of 23 feet. These reptiles live in the waters of northern Australia, Indonesia, and Malaysia, moving from island to island and sometimes venturing far out to sea. "Salties" feed on crabs, turtles, shore birds, snakes, buffaloes, domestic livestock—and, occasionally, people.

Deinosuchus is the one that gives us nightmares. Curiously, *Deinosuchus* may not be a crocodile at all. In their 1999 study, Rowe, Brochu, and Kishi wrote, "*Deinosuchus* is not the world's largest crocodile—it is one of the largest alligators, and one of at least two independent instances (the other is *Purrusaurus*) of gigantism within Alligatoroidea."

The ancient marine crocodiles had to compete with the fully marine ichthyosaurs, plesiosaurs, and mosasaurs. Those marine reptiles are only available as fossils for study, but there is a living crocodilian, which, according to Hua and Buffetaut (1997) "can, to some extent, be used as an actualistic model for reconstructions of [the marine crocodile's] physiology and ecology." Hua and Buffetaut's discussion of the saltwater crocodile *Crocodylus porosus* appears in *Ancient Marine Reptiles* (Calloway and Nicholls 1997), and they defend its inclusion by saying that it "can be considered as a marine reptile, because it seems to be able to grow in salt water, without any influx of fresh water . . . [and] the diet of *Crocodylus porosus* in a marine environment shows that it can easily feed in salt water, and its ability to swim in the open sea is well established." The largest of all living reptiles, the saltwater crocodile grows to a known length of 23 feet and may get even larger. It is also the most dangerous of all crocodiles, with a reputation as a known man eater. "Salties" occur in Southeast Asia, Indonesia, Malaysia, and northern Australia, where they move freely from island to island. They can also swim great distances in the sea, and have been found in the Solomon Islands, New Guinea, Fiji, Japan, and the Cocos-Keeling Islands, more than 600 miles southwest of Java, the nearest large island. They are not confined to salt water, appearing in freshwater rivers, lakes, and swamps, where, depending upon their size, they feed on crabs, turtles, snakes, shore and wading birds, buffaloes, domestic livestock, wild boars, monkeys—and, occasionally, people.

Ancient Marine Reptiles also devotes a chapter to a single fossil marine crocodile, *Hyposaurus rogersii,* known from the Maastrichtian through the Danian ages, 61 million to 71 million years ago (Denton, Dobie, and Parris 1997). *Hyposaurus rogersii* had an elongate, narrow snout, a long flattened tail, and light dorsal armor. This species, known from fossils found in New Jersey as well as South America and Africa, had none of the diving specializations of the metriorhynchids (streamlined shape, upper tail lobe, etc.), but it seems to have swallowed stones as ballast and propelled itself through the water by flexing its tail from side to side. The species fed primarily on fish, but occasionally ate ammonites and other mollusks. Its large, well-developed limbs meant that *Hyposaurus* could probably move around on land with a certain facility, rather like living crocodiles, but its general anatomy suggested a primarily marine existence. Unlike the saltwater crocodile, which spends most of its time in nearshore waters and only occasionally wanders far offshore, *Hyposaurus* spent most of its life offshore and ventured onto land only to lay its eggs. "Occupation of multiple habitats," wrote Denton et al., "may have enabled *Hyposaurus* (among other crocodilians) to escape from the fate that befell many marine reptiles at the end of the Cretaceous; however, their abundance diminishes significantly in the Paleocene of North America."

Whatever it was that killed off the nonavian dinosaurs 65 million years ago, it obviously did not affect the crocodilians, since they can trace their unbroken lineage from the sphenosuchians of 200 million years ago right up to the 22 species of crocodiles, alligators, gavials, and caimans alive today.* From the fossil evidence and, more important, from living species, it is obvious that crocodiles were largely unaffected by the crisis of the Cretaceous-Tertiary (K-T) boundary, which killed off the nonavian dinosaurs and various other forms as well. Crocodiles lived 240 million years ago, and crocodiles live now. Since crocodilians are known to be sensitive to temperature changes and cannot survive in a cold climate, whatever happened at the K-T Extinction could not have produced the substantial temperature drop that some theorists postulate. Why did the crocodilians survive while all the dinosaurs died off? Hans-Dieter Sues wrote in 1989:

> They belonged to a community of living beings that was little affected by what happened 65 million years ago; freshwater vertebrates (including fishes, amphibians, turtles, and crocodilians) are known to have survived the crisis much better than either terrestrial or marine communities. A possible explanation for this differential survival is that the freshwater food web was based on neither marine plankton nor flowering plants, both of which were severely affected at the Cretaceous-Tertiary boundary.

---

* Black caiman (*Melanosuchus niger*), Chinese alligator (*Alligator sinensis*), American alligator (*Alligator mississippiensis*), Cuvier's dwarf caiman (*Paleosuchus palpebrosis*), Schneider's dwarf caiman (*Paleosuchus trigonatus*), broad-snouted caiman (*Caiman lanirostris*), common caiman (*Caiman crocodilus*), Cuban crocodile (*Crocodylus rhombifer*), Morelet's crocodile (*C. moreletii*), American crocodile (*C. acutus*), African slender-snouted crocodile (*C. cataphractus*), Nile crocodile (*C. niloticus*), Orinoco crocodile (*C. intermedius*), saltwater crocodile (*C. porosus*), Johnston's crocodile (*C. johnstoni*), mugger (*C. palustris*), Siamese crocodile (*C. simaensis*), Philippine crocodile (*C. mindorensis*), New Guinea crocodile (*C. novaeguineae*), dwarf crocodile (*Osteolaemus tetraspis*), false gharial (*Tomistoma schlegelii*), gavial (*Gavialis gangeticus*).

Crocodilians breezed through the cataclysmic K-T Extinction, which eliminated some 70 percent of all living animals, but throughout their history they had never encountered anything like the biped that the Bible said was to "have dominion over . . . every creeping thing that creepeth upon the Earth." As Alistair Graham wrote in *Eyelids of Morning: The Mingled Destinies of Crocodiles and Men* (1973), "Civilized man will not tolerate wild beasts that eat his children, his cattle, or even the fish that he deems to be his." And so *Homo sapiens* has waged a war on crocodiles and alligators, slaughtering those that were perceived as a threat to personal safety and killing tens of millions for their skins. In many locations where crocodilians were considered troublesome vermin, people who killed them were seen to be doing the community a service, and if they made money in the process, so much the better. Several nations have now enacted legislation to protect their crocodilians, but many species—particularly caimans, the primary source for "alligator" leather—are still considered endangered. Some of the crocodilians made it to modern times, only to be hunted as dangerous to humans, or turned into briefcases, wallets, and cowboy boots.

BIRDS THAT SWIM

Many birds, such as ducks, geese, and swans, paddle along on the surface of the water, and some of the diving ducks actually dive beneath the surface to feed. The pelagic seabirds known as the order Procellariiformes (albatrosses, petrels, shearwaters, and so on) feed from the sea—usually by plucking food from the surface in a dive—and they are superb flyers, more agile in the air than on land or on the water. The order Pelecaniformes is a diverse group that includes the pelicans, tropic birds, cormorants, anhingas, boobies, gannets, and frigate birds, all of which can swim on the surface and dive for their food. The pelicans and gannets dive from the air, whereas the cormorants and anhingas dive from the surface of the water. (Frigate birds can catch their own food at the surface, but they prefer to steal from other birds in the air.) Gulls and terns are classified together as Charadriiformes, and they are creatures of varying habits and habitats, most of which have more than a little to do with water. The alcids (auks, murres, puffins, and so on) are flying birds that are well adapted to hunt—more accurately, to *fish*—in the water, and some of them, such as auklets and murres, actually use their wings to propel themselves underwater in pursuit of their prey.

There are a few species of birds that do not fly; the ratites such as ostriches, rheas, emus, and kiwis come to mind, but by and large, from hummingbirds to eagles, from sparrows to turkeys, the thing that characterizes birds and makes them so magical is their ability to fly. Most birds can walk more or less competently, which means that they have mastered the realms of air and land. And those that can also swim well can add water to their range of venues, making them the only animals on Earth that are able to function effectively in the air, on land, and in or on the water.

Like almost all other birds, penguins have wings, but over time these wings have become modified into flippers, which penguins use to "fly" underwater—that is, they flap their wings just as aerial birds do, but it is water rather than air that provides the resistance. Penguins are the avian analog of seals and sea lions, coming ashore to breed but swimming underwater in pursuit of their prey. Often awkward on land and utterly deprived of the aerial experience, penguins have become modified for a specifically aquatic existence. They lay their eggs and raise their chicks on land (or on ice), but they feed in the water. They have retained their wings and their powerful flight muscles, but only for use in water, a much thicker and more resistant substance than air.

Adaptation to an aquatic or semiaquatic existence occurred early in the evolutionary history of birds, and has appeared independently in several avian lineages (Kristoffersen 1999). Birds are unique among the vertebrates in that their forelimbs and hind limbs are adapted for different modes of locomotion: in most bird species, the hind limbs are designed for walking, running, or perching, whereas the forelimbs have evolved into wings, which are mostly used for flying. In some species, however, they are used for a specialized kind of swimming. Of the living birds that can move under water, the diving birds (grebes, cormorants, diving ducks) propel themselves with their hind limbs, and others such as penguins and alcids (auks, murres, puffins) use their wings to "fly" underwater.

Throughout most of the Cretaceous, the dominant aquatic birds were foot-propelled divers known as hesperornithiformes, from the Greek *hesperos,* meaning "western," and *ornis* for "bird." Fossils of these birds were found near the Smoky Hill River in Kansas. They were covered with a coat of fine hairlike feathers, which meant that they were birds.

By the late Cretaceous there was the genus *Enaliornis,* an early member of a group of highly specialized flightless, diving birds that was discovered in England. The flightless diving birds came in different sizes: *Parahesperornis* was less than 3 feet long; *Baptornis* was somewhat larger; and the largest was the 5-foot-tall *Hesperornis. Hesperornis* was very much a bird with feathers, but its upper limbs were reduced in size to the point of being useless, and its lower limbs were so specialized for swimming that it probably could not walk very well on land, if at all. This totally flightless bird still had teeth, and solid bones (bones of living birds are hollow). It had no keel on its sternum, which would have supported wing muscles, and only single small bones where the forelimbs would have been. It propelled itself through the water by kicking its powerful hind limbs, like a giant diving beetle. Most living diving birds use their wings for underwater propulsion, but *Hesperornis* used only its feet, which were webbed like those of loons or grebes and had a large, sharp claw on the end of each toe. J. T. Gregory (1951) described the beak as "long and slightly down-curved at the tip, more so than in gannets but less than in gulls and much less than in cormorants," but noted that the lower jawbone of this large, flightless, wingless bird had an intramandibular hinge, which is seen in no other creatures besides mosasaurs. These observations prompted Gregory to write an essay that he called "Convergent Evolution: The Jaws of *Hesperornis* and the Mosasaurs." O. C. Marsh's 1880 description devoted several pages to the teeth of *Hesperornis,* and in his original drawing, the skeleton is posed in an upright position like a cormorant, but there are some who believe that *Hesperornis* could not bring its legs underneath it to stand up, implying that this ungainly creature spent its entire life in the water, except perhaps to make its way out on land to lay eggs—like the sea turtles.

Hesperornis appears to have been an inhabitant of the north; most of the known fossils come from Canada, and none are reported south of Kansas. The bird lived along the shores of the large inland sea that once covered most of western North America from the Rockies to the Mississippi River. It is one of the most common fossil vetebrates from the Anderson River shales of Canada's Northwest Territories, and Russell (1967) has written that the predominance of juveniles there suggests that this region might have been a rookery. Like so many denizens of the inland seas, the hesperornithiforms became extinct at the end of the Cretaceous, approximately 65 million years ago. We know that they all became extinct because they left no descendants.

Although many penguin fossils have been found, the evolutionary history of these birds is still poorly understood. We know they go back at least as far as the Late Eocene, 40 million years ago, because even though they differ somewhat from today's 18 species, they were already penguins: upright-walking birds with reduced wings. Most penguin fossils have been found in the Southern Hemisphere, in Patagonia, New Zealand, Australia, and various Antarctic locations, and New Zealand has the largest number and the greatest variety. Most fossils consist of single isolated bones. Simpson speculated (1975): "The likelihood that connected parts of a penguin skeleton would be buried intact in seas swarming with predators and lesser consumers was clearly very slight." This may indeed be the case, yet other aquatic animals such as pinnipeds and fishes have certainly been found as almost intact fossil skeletons. Perhaps a better explanation for the paucity of penguin skeletons is that the region in which they live is not usually subjected to sedimentation, a prerequisite for the preservation of fossils.

So far, the largest fossil penguin, known only from a few bones, is *Anthropornis nordenskjoeldi* from Seymour Island in the Antarctic. *Anthropornis* ("man bird") measured 5 feet 7 inches. Another large fossil, with the descriptive name of *Pachydyptes ponderosus* ("ponderous heavy diver") may have been 5 feet 4 inches tall. For the most part, however, the extinct penguin species fell well within the size range for the living species. (The largest living penguin is the emperor, which can stand 48 inches tall.)

Where did the penguins come from? George Gaylord Simpson, a highly acclaimed paleontologist who made an extensive study of fossil penguins and wrote a book called *Penguins Past and Present* (1976), stated: "They were almost certainly derived from aerial flying birds, and their ancestors probably resembled and possibly belonged to the order Procellariformes." The change from a bird that flies in the air to one that flies under water, is, says Simpson, "radical. . . . It is hard to see how a phase of prolonged change intermediate between aerial and aquatic swimming could be viably adaptive." In other words, we have no idea how (or why, for that matter) a nonswimming flyer became a nonflying swimmer. Bernard Stonehouse, a British biologist who has edited a book (1975) on penguins, advances a theory based on his observations of pigeon guillemots flying low over the water: it would be only a short step from flying close to the water to flying in it. Fossil penguins from the Eocene were already highly specialized, wrote Simpson, which "implies a long previous history for which there is no record at all."

The oldest known penguin was found in the Waipara Greensand of New Zealand, and is known as "the Waipara bird." The Waipara Greensand, in the northern part of Canterbury Province on South Island, dates from the Late Paleocene or Early Eocene, making the Waipara bird about 55 million to 60 million years old, and therefore 10 million to 15 million years older than any other known penguin. It was a large bird, somewhere between the yellow-eyed penguin and the king penguin in height, around 36 inches. It shows "an intermediate mix of features that suggest an ancestry among volant birds but also relationships with later penguins" (Fordyce and Jones, 1990).

Further fossil evidence came from Storrs Olson (1980), who reported on "a new genus of penguin-like pelicaniform bird from the Oligocene of Washington," which he called *Tonsala hildegardae* (after Hildegarde Howard, a fossil bird specialist at the Natural History Museum of Los Angeles). Howard had earlier introduced *Plotopterum joaquinensis* from Kern County, California, and *T. hildegardae* had similar, paddlelike forelimbs, suggesting that it was a wing-propelled diver like living penguins. *Tonsala* was larger than any living penguin species except the king and emperor, and from the fossil evidence, it is assumed that it became extinct at the end of the Miocene, about 6 million years ago, roughly the same time as the giant penguins of the Southern Hemisphere disappeared. More recently, Olson and Hasegawa (1996) described several more species of Late Oligocene plotopterids (with paddlelike forelimbs) from Japan, including *Copeptryx titan* ("giant wing-oar"), which "may have been larger than any of the giant fossil penguins from the Tertiary of the Southern Hemisphere." At more than 6 feet from bill tip to tail tip, *C. titan* was the largest diving bird that ever lived. We do not know what happened to the early large penguins or the plotopterids, but they wrote, "There is a strong possibility that the disappearance of these two unrelated groups is linked with the ascendancy of seals and porpoises."

Because ancient penguins' descendants can still be found in or around the world's oceans, we are able to study the living penguins to get an idea of how their ancestors lived. The family Spheniscidae, which includes all the known fossils and living species, are flightless, aquatic birds that walk upright. The common name may come from the Celtic *pen* ("head") and *gwyn* ("white"), or from the Latin *pinguis,* which means "fat." (In English, the name was first applied to the great auk of the North Atlantic, which is now extinct.) All but the Galápagos penguin, which lives at the Equator, are confined to the Southern Hemisphere. Only a few species actually inhabit the Antarctic area. All penguins breed on land and feed at sea, sometimes making long voyages in search of fish, squid, and crustaceans. Overlapping, tightly packed feathers and a thick layer of fat insulate the birds against the cold. They can all walk upright, but they also move by crawling on their bellies, or "tobogganing" down slopes of ice or snow. Some are capable of remarkable leaps out of the water. The basic color scheme of penguins is black or dark gray and white, but some species have surprising flashes of yellow or orange. Many penguin species are threatened by overfishing, oil spills, global warming, and tourists, who trample eggs and nests in their attempts to get close to the birds.

Of the living penguins, the genus *Aptenodytes* ("wingless diver") includes two very large and very similar species. The emperor (*Aptenodytes forsteri*) is the largest of the penguins, reaching a height of 4 feet and a weight of 80 pounds. Emperor penguins have extremely dense plumage and thick fat deposits, because they live and breed on the pack ice of the Antarctic, one of the most inhospitable environments on Earth. In March, the adult birds leave the sea and trudge as much as 150 miles over the ice to reach their breeding colonies. Six weeks after their arrival, the female lays a single egg, which the male sits on as blizzards howl, temperatures drop to –80°F, and the males huddle together for warmth. After nine weeks, the female, who has been feeding at sea, now takes over the incubation chores, and the male, who has been fasting for 15 weeks, heads for the sea. When he returns, both parents take turns feeding the chick, making regular trips to the sea for food. When the chicks are about seven weeks old, they crowd together in a crèche where their parents are able to recognize them by their individualized high-pitched squeals. After the chicks reach eight months of age, their parents abandon them, and the youngsters make their way across the ice to the sea. Emperors are the only birds in the world that neither breed on land nor make some kind of a nest. They are the deepest-diving penguins, capable of reaching depths of 1,700 feet.

Second only to the emperor in size, the king penguin is similarly marked with bright yellow splashes around its head. Baby kings, in their brown fuzzy coats (actually feathers), might well be the silliest-looking birds in the world.

Second to the emperor in size, the king penguin (*Aptenodytes patagonicus*) can reach a standing height of 3 feet, and full-grown adults might weigh 30 pounds, making them among the heaviest birds in the world. (Only the nonflying ostriches, rheas, and emus are heavier than the large penguins.). The king is a stately bird with silvery-gray upper parts, a black head and chin, and a bright golden ear patch. An estimated 1,000,000 pairs breed on Antarctic and subantarctic islands. The king penguin is an accomplished diver, and has been recorded diving to 1,056 feet. King penguins aggregate in large colonies for breeding, where there is much calling and ritualized courtship behavior. When the chicks are born they form crèches. The huge chicks in their brown, fluffy suits are so different from the adults that they were once thought to be a separate species of "wooly penguin." In the early sealing days, king penguins were killed and boiled for oil, but nowadays their only predators are leopard seals and killer whales.

The penguin genus *Pygoscelis* includes three species that do not look much alike, yet they have a feature in common: Adélie, gentoo, and chinstrap penguins all have long, stiff tails that they use as props when they stand. Their habits, however, differ: Adélies and chinstraps are migratory and undergo long fasts; gentoos do not fast except when moulting. The gentoo (*Pygoscelis papua*) is the largest of the pygoscelids—after the emperor and the king, it is the third largest of the penguins. Often seen in the company of Adélies, gentoos are identifiable by their yellow feet, black-tipped, red-orange bill, and the conspicuous white patch over each eye. Gentoos stand about 30 inches high and breed in loose colonies on land, carrying out ritualized threat postures and courtship behavior. The nest is a scraped-out hollow lined with grass, stones, bones, and moss. The total circumpolar population of gentoo penguins is about 350,000 breeding pairs.

One of the "crested penguins," the rockhopper has an orange bill and bright yellow plumes that project from its eyebrows.

The Adélie penguin (*Pygoscelis adeliae*) reaches a length of about 27 inches and is easily recognized by its black back, white front, and conspicuous white eye-rings. Adélies nest in large colonies on the coast of the Antarctic continent, only rarely appearing as far north as South Georgia. Their nests are made of stones piled up in hollowed-out mounds. In the water, they swim with only their heads out, and they are capable of fantastic leaps out of the water, especially if pursued by their archenemy, the leopard seal. (The French admiral Dumont d'Urville, who explored the Antarctic in 1837, named this penguin and also Terre Adélie for his wife.)

The derivation of the chinstrap penguin's common name is a narrow band of black feathers that extends from ear to ear. Chinstraps' habits are similar to the Adélies. On South Atlantic islands and the Antarctic Peninsula, chinstraps build nests of stones, feathers and bones, and both parents take turns incubating the eggs.

The genus *Eudyptes* are the crested penguins. All six species in the genus have red or reddish-brown eyes and a crest of yellow or orange plumes. When swimming, they keep their heads and part of their backs above the water, and they "porpoise," leaping all the way out of the water, when traveling fast. The rockhopper (*E. chrysocome*) is one of the smallest of all the penguins (only the blue penguin is smaller). It is a chubby little bird with an orange-brown bill, pink feet, and a bright yellow stripe that does not meet above the bill and ends in long, bright yellow plumes that project laterally. The name comes from its habit of traveling on land by bouncing along with both feet held together. It often slides or bounces down banks and cliffs to get to the sea and then jumps feet first into the water, unlike other species, which walk or dive into the sea. Rockhoppers are the fiercest and most aggressive of all penguins, attacking almost any animal—including humans—that approaches their nesting colonies.

The erect-crested penguin (*E. sclateri*) is found in the subantarctic waters of New Zealand and Bounty Island, the Antipodes, and the Aucklands. It differs from the other crested penguins in having yellow crest feathers that it can raise upward; the crests of the other *Eudyptes* species droop. Fiordland penguins (*E. pachyrhynchus*) are found only on South Island, New Zealand. Like the other crested penguins, they are chunky birds characterized by a yellow eyebrow stripe that develops into silky plumes that droop down the sides of the nape. The species often has little white markings on the cheeks, parallel to the eyebrow stripe. They live in a thickly vegetated forest habitat, and are therefore difficult to study, but it is estimated that there are only between 5,000 and 10,000 breeding pairs.

The macaroni penguin (*Eudyptes chrysolophus*) with its striking crest gets its name from the "macaronis," British dandies of the seventeenth and eighteenth centuries who wore plumes in their hats. The largest of the crested penguins, the macaroni can stand 27 inches high and can be differentiated from the other species by the downward-pointing angle of the black coloration in the neck. As with the other crested penguins, conspicuous golden-yellow plumes project from behind the eye and droop behind the head. On their breeding grounds on subantarctic islands and the Antarctic Peninsula, the total population has been estimated a 11 million breeding pairs. The royal penguin (*E. schlegeli*) breeds only on subantarctic Macquarie Island and adjacent islets. The total population is around 850,000 pairs. The yellow-crested Snares penguin (*E. robustus*), restricted to Snares Island south of New Zealand, reaches a height of 24 inches. They forage in shallow waters near the shore and keep in contact with each other by means of short, barking calls. The population on the island is estimated at 66,000 birds.

The yellow-eyed penguin (*Megadyptes antipodes*) is monotypical, the only member of its genus. It is one of the largest and heaviest of the penguins, standing 28 inches high. The feathers of the face are pale yellow, tipped with black, and there is a lighter yellow mask around the straw-colored eyes. The long, slender bill is cream-colored with a reddish-brown tip. The yellow-eyed penguin has the smallest population of any of the species: approximately 5,000 individuals live on the southeastern coasts of New Zealand's South Island, the Stewart, Campbell and Auckland islands, and Enderby Island, where they live in an unusual habitat for penguins, a dense forest. In recent years, loggers and farmers have introduced predators such as cats, dogs, pigs, stoats, and ferrets, and the species is considered endangered.

The smallest of all the penguins is known in New Zealand as the blue penguin, and on Phillip Island, off Melbourne, Australia, where a study has been ongoing since 1967, it is known as the fairy penguin. *Eudyptula minor* stands only 17 inches tall. The monotypical blue penguins are bluish-gray on the back and have light-colored eyes. They are the only nocturnal penguins, entering and leaving the water under cover of darkness. They form rafts offshore until dawn, then come out of the surf to walk to their burrows, where they remain for the day before returning to the water to feed. Blue penguins are nonmigratory, but they do make long excursions in search of food.

The order name Sphenisciformes comes from the Greek *sphen,* "wedge-like," referring to the general shape of all penguins. The name is used for the family Spheniscidae, and also for the genus, *Spheniscus,* that includes the familiar Magellanic, Humboldt, African, and Galápagos penguins. The spheniscids all dig burrows or nest under vegetation or rock crevices, and there is a strong bond between male and female. Unlike many other penguin species, spheniscid chicks do not run after their parents demanding to be fed. The Magellanic penguin (*S. magellanicus*) is named in honor of Ferdinand Magellan, whose sixteenth-century expedition was the first to report the existence of penguins. It is the only member of the genus *Spheniscus* with two black bands across its chest, one of which loops down toward the feet and the other of which comes forward from the back to form a collar. Colonies of these penguins are found all around the southern tip of South America, from the Patagonian coast of Argentina to the coast of Chile. They are also found in the Falkland Islands. They fish in groups and have been seen hundreds of miles offshore. They lay their eggs in burrows that they dig out of sand or shingle, or sometimes, as in the Falklands, in grassy turf. If the burrows are flooded the chicks can drown. Magellanic penguins are a favorite prey species for sea lions and giant petrels. In recent years, oil spills have caused the death of thousands.

The smallest of the spheniscids, the Galápagos penguin (*S. mendiculus*), is probably the most unusual of all penguins, since unlike its relatives, which are cold-weather, cold-water birds, it breeds on hot volcanic islands on the Equator. It is a nonmigratory species and does not leave the warm waters of the Galápagos. To rid themselves of the heat of the equatorial sun, Galápagos penguins take to the water or stand on shore and pant. During the disastrous El Niño event of 1982–83, the waters around the islands warmed up so much that the small fishes that make up the penguins' normal food supply were not available, and many Galápagos penguins starved to death.

The Humboldt penguin (*S. humboldti*) is the least-known of all the members of the genus *Spheniscus.* Humboldt penguins live almost exclusively in the long, narrow band of coastal water where the Humboldt Current comes into contact with the coasts of Peru and Chile. In northern Chile, the range of the Humboldt penguin overlaps that of the Magellanic penguin, and the two species sometimes interbreed. Overfishing of the Humboldt Current has caused a substantial decrease in the numbers of penguins; not only because their food source has been depleted, but the penguins are eaten by fishermen and also used for bait.

Like their close relatives in South America, African penguins (*S. demersus*) are known as jackass penguins because of their loud, braying call. They are the only penguins native to southern Africa (though stray macaronis, kings, and rockhoppers occasionally are reported in the region). Standing about 2 feet high, this species, like the other spheniscids, is strongly marked with a black band that runs from the chest to the feet and white coloring that swoops up and over the eye, giving it the appearance of wearing a black mask. Oil spills and predation by sea lions, sharks and men have greatly reduced the population, which was estimated at over a million birds at the beginning of the twentieth century.

Although they all look like headwaiters, penguins are an enormously diversified

The great auk was a 3-foot-tall flightless bird that obtained its food in the chilly waters of the North Atlantic. Hunters killed the last pair on an Icelandic island in 1844, and they are now extinct.

group, and they live in a variety of habitats, from the lava rocks and warm waters of the Galápagos to the forests of Enderby Island to the perpetual ice and snow of Antarctica. They are awkward on shore, but supremely graceful in the water, which, it might be argued, is their natural element. They developed scalelike feathers to insulate them from the heat-absorbing water, flippers instead of wings, and a hydrodynamically streamlined shape that enables them to pass through the water almost effortlessly.

The great auk (*Alca impennis*) was a large, flightless seabird that stood close to 3 feet tall, and has been extinct since the last pair was killed by hunters on an Icelandic island in 1844. They were the original "penguins," the name being applied to the Southern Hemisphere birds because of their similarity to the great auk. Like penguins, they were strongly marked in black and white, and they stood upright with their little wings tucked into their sides. Also like penguins, they could not fly, and they dived for their food like their surviving relatives, the razorbills and the murres.

The first records of people killing great auks are from 1497, when French ships arrived at the prodigious cod-fishing grounds of Newfoundland. (It was at this time that the birds were named *pingouins,* the name that would later be applied to the upstanding, flightless, black-and-white birds of the Southern Hemisphere.) In 1535, Jacques Cartier visited an "Island of Birds" (probably Funk Island) off Newfoundland, where his crews crammed barrels with auks and collected as many eggs as they could carry. It is said that an Icelander named Latra filled a boat with great auks at Gunnbjorn Rocks on the east coast of Greenland in 1590 (Greenway 1958). The hapless birds nested on the various islands in the western North Atlantic, around Newfoundland, including the Magdalens in the Gulf of St. Lawrence, and as far south as the Gulf of Maine and Massachusetts Bay. One of the centers of the nesting grounds was Funk Island, so-called because of the overpowering stench ("funk") of the tons of dung that covered every square foot of the half-mile-long island, some 40 miles off northeast Newfoundland, about 130 miles north of St. John's. (Others have attributed the name to the smell of roasting and rotting birds.) In the eastern Atlantic, they nested at various islands off Europe, especially St. Kilda (40 miles west of Scotland), and several tiny islands off Iceland. Wherever they were found, they were harvested.

MARINE MAMMALS

At many different times in the geological past, terrestrial animals abandoned the ecological accomplishments of their remote ancestors, the air-breathing fishes, to reinvade the sea. Ichthyosaurs, plesiosaurs, crocodiles, turtles, snakes, penguins, whales, and seals, to cite a few obvious groups, have established themselves very well in marine environments. In doing so, they have mimicked the shapes of fish. However, fish have effectively invaded the land only once, 365 million years ago, and these were the ancestors of succeeding land-dwelling backboned mammals. No fish has since emerged from the water to mimic the shapes of the dominant backboned animals on land. How is it that land-dwelling creatures, though initially adapted imperfectly to living in water, can successfully compete with the marine creatures whose ancestors never left the sea, but the reverse never occurs?

—DALE RUSSELL, 1989

Marine mammals are those that spend most or all of their time in the water.* Included are the sea otters, terrestrial animals that feed only at sea; the sea lions and fur seals, quadrupedal amphibians that retain the use of all four legs for walking on land (although their feet have been modified into flippers for swimming); the eared seals, which have lost the use of their hind limbs for walking; the sirenians (manatees and dugongs), which are fully aquatic; the toothed whales (dolphins, porpoises, beaked and sperm whales); and the filter-feeding mysticetes, which feed by straining food through baleen plates that hang from the roof of their mouths. The smallest of the marine mammals is the sea otter, at 5 feet long, including the tail, and up to 70 pounds. The largest is the blue whale—the largest animal that has ever lived—which can be 110 feet long and weigh 300,000 pounds.

To a greater or lesser degree, the repatriated mammals have retained vestiges of their terrestrial, mammalian lifestyle, including hair, which only mammals have. Sea

*Because it spends most of its time on land or ice, and enters the water only occasionally, the polar bear is not usually considered a marine mammal. *Ursus maritimus* swims well, but only to get from place to place. During the past 250,000 years, it has evolved from a brown, temperate-zone, land-based bear into a white, semiaquatic, cold-weather bear, and is probably a marine mammal in the making.

otters, seals, and sea lions are thickly furred; manatees and dugongs have a sparse pelage, but have plentiful whiskers around the mouth. Dolphins and whales are hairless, but in some species there are hairs present at birth, which are quickly lost. For insulation in the heat-draining water, the cetaceans have substituted blubber for hair. Sea otters have handlike paws on their front legs, but the hind feet have become flattened and webbed, almost flippers. The four legs of pinnipeds have become flippers, and the sirenians have front flippers (some of them have fingernails), but no hind legs and a flattened, horizontal tail for propulsion. Whales and dolphins have no hind legs at all, flippers instead of forelegs, and a horizontal tail (the flukes) for propulsion. There would appear to be a seaward progression of mammals from the semiaquatic sea otters to the fully aquatic cetaceans. In fact, though all marine mammals are derived from totally distinct evolutionary groups, there are certain similarities in lifestyle and morphology, and they are considered good examples of the principle of convergence, where creatures unrelated by evolution develop similar or even identical solutions to a particular problem. As Victor Scheffer (1976) said, the return of mammals to the sea consisted "six separate launching parties from the continents to the oceans."

THE SEA COWS

If you have collected fossils to any great extent in marine Tertiary deposits of tropical or once-tropical latitudes, you have probably come across cylindrical pieces of thick, dense vertebrate bone several centimeters in diameter that someone told you were the ribs of sirenians, or seacows. You may even have learned to apply the term "pachyostotic" to such bones. Inquiring further, you probably learned that seacows (manatees and dugongs) are the legendary mermaids, that they are the nearest living relatives of elephants, and that they all replace their teeth horizontally, back to front, just like elephants do. Such is education: a random mixture of truth and error, hopefully favoring the former.

—Daryl Domning, 1999

The living sirenians are all plump, grayish, rough-skinned mammals, with a body about the size of a small cow's, rounded flippers, a broad, paddlelike tail, and a face that only another sirenian could love. They were named for the Sirens of classical Greek mythology, sea nymphs who used their beautiful songs to charm and entice everyone that heard them. In the *Odyssey,* when Odysseus comes close to their island, he stuffs the ears of his men with wax and has himself tied to the mast so he will not succumb to their "siren song." This view continued well into the Middle Ages, as shown by the *Physiologus*, a twelfth-century bestiary, in which Sirens are described as "deadly creatures who are made like human beings from the head to the navel, while their lower parts down to the feet are winged. They give forth musical songs in a melodious manner, which songs are very lovely, and thus they charm the ears of sailormen and allure them to themselves."

Somehow, the myth of the siren got mixed up with that of the mermaid, half woman, half fish. When Columbus entered the Caribbean, he noticed some strange-looking beasts floating in the aquamarine waters. In his log for January 9, 1493, he wrote "I saw three sirens [*sirena*] that came up very high out of the sea. They are not as beautiful as they are painted, since in some ways they have the face of a man."*

The earliest fossils were found in Hungary, but they are fragmentary and give little idea of what the animal actually looked like, or even whether it was aquatic. The Middle Eocene sea cow, *Prorastomus sirenoides,* was found in Jamaica in 1855, but again not enough material was available to reconstruct the animal. Additional fossil material has been found in Jamaica, but, wrote Savage, Domning, and Thewissen (1994), "This odd little sea cow has been a difficult morsel for science to digest." Nevertheless, *Prorastomus sirenoides* "is the best available approximation to a structural ancestor of the Sirenia," and was "probably an amphibious but mainly aquatic quadruped that inhabited coastal rivers and embayments. It was probably a selective browser on floating and emergent aquatic plants and, to a minor degree, on seagrasses."

In a talk given at the symposium Secondary Adaptations to Life in Water in Copenhagen in 1999, Daryl Domning said that "despite the anomalous fact that the earliest and most primitive sirenians (prorastomids) occur in Jamaica," the sirenians probably arose in the Tethys Sea, which stretched from the present-day Mediterranean east to the Arabian Sea and the Bay of Bengal. "Recent discovery of a possible prorastomid vertebra in Israel, he continued, "may be the first step toward removing this apparent artifact of paleontological sampling." The first manateelike animal, dating from about 15 million years ago, was *Protosiren,* yet it lacked the replacement teeth that enable modern manatees to eat abrasive plant material. Recently, Domning and Gingerich (1994) suggested that *Protorastamus* and *Protosiren* were quadrupedal amphibious creatures that lived along the seashores. What did they look like? Domning (1999) wrote, "The Seven Rivers [Jamaica] prorastomids, which are slightly more derived than *Prorasatomus* itself, were pig-sized quadrupeds with long trunks, and a substantial though not powerfully-muscled tail." Certain anatomical characteristics "indicate that they were more aquatic than modern hippopotami; they probably fed as well as rested in the water, but at least the earlier of them could probably support their weight on land.

In a 1916 paper, "Sea Cows, Past and Present," Frederic Lucas wrote: "It would

---

* Whatever other appellations might be applied to these animals, "beautiful" is not the first one that springs to mind. Although they bear a superficial resemblance to some of the other marine mammals such as pinnipeds and cetaceans, sirenians have no evolutionary relationships with them. In fact, paleontologists have classified them as subungulates and lumped them together with such varied company as elephants, hyraxes, and aardvarks. Like the sirenians, all these animals lack a clavicle (collar bone) and have nails or hooves instead of claws. There also seems to be a distant connection with the desmostylians, enigmatic creatures that left no descendants (see pp. 197–201). Sirenians reached their peak in the Oligocene and Miocene, but they seem to go further back than the desmostylians.

of course be one link in the chain, one step towards a four-footed animal, if we could find a four-paddled porpoise, but none has yet come to light, and here is where the sea cow comes forward with an important bit of evidence. When paleontologists were hunting in the Fayum, Egypt, for ancestors of existing elephants, they came upon the remains of a manatee, not unlike those of today, save that it possessed four well-developed paddles; and because it was so evidently the predecessor of the modern sea cows, it was named *Eosiren,*" from Eos, the Greek goddess of dawn. In fact, says Domning, "*Eosiren libyca* (Late Eocene) had vestigial hind limb bones, but the hind limbs probably did not show on the outside of the body. In no case have we actually found a complete hind foot for any of these, paddlelike or otherwise. Lucas's statement is a broadly valid inference based on the pelvic bones, which were the only relevant elements available in his time" (personal communication, 2000).

In their 1994 discussion of *Protosiren smithae,* a new species from the Late Middle Eocene (about 40 million years ago) of Wadi Hitan, Egypt, Domning and Gingerich wrote that "the new specimens serve as 'Rosetta stones' in revealing the true associations among the disparate bones described by earlier workers." *Protosiren smithae* is similar to *Protosiren fraasi,* described in 1906 by Andrews, also from Wadi Hitan; but somewhat more advanced; *Protosiren fraasi* may be "plausibly viewed as the direct ancestor" of *P. smithae.* Both species were semiaquatic, as evidenced by the well-developed fore and hind limbs, and the authors wrote that it was "safe to assume that still more primitive sirenians (*Protosiren fraasi, Prorastomus sirenoides*) were at least equally capable of terrestrial locomotion. . . . However, the legs of *P. smithae* were very short relative to its body, and it seems questionable whether the animal could have lifted its body off the ground; it may have merely slid or rested on its belly when out of the water, in a manner analogous to modern pinnipeds."

Thirty million years ago, the most widespead genus of sirenians was *Metaxytherium,* which was followed by *Dusisiren jordani,* and then *Dusisiren dewana* from Japan. *Dusisiren jordani* lived along the California coast from about 15 million years ago until the last great ice age of the Pleistocene; the youngest fossil, from Monterey Bay, is only 20,000 years old. With a cooling climate, the sea cows were moving northward, switching their diet from semitropical sea grasses to colder-water kelps. We can document the development of *Hydrodamalis* from *Dusisiren* through a continuous sequence of morphological changes, where the size of the animal increased from 6 to 8 feet to more than 30, the cheek teeth were replaced by grinding plates, and the phalanges of the front limbs disappeared altogether.

Middle Eocene sirenians have been found in North African and European fossil beds and assigned to the genera *Eotheroides, Eosiren,* and *Prototherium.* By the Middle Eocene, they had begun to lose their hind legs and to evolve the paddlelike tail fins that characterize today's living manatees and dugongs. By the early Pliocene, about 5 million years ago, manatees had colonized the Amazon basin and the Caribbean. Dugongs, which are similar in appearance and habits, were present in both of these areas, but the manatees seem to have replaced them.

When Commander Vitus Bering, sailing under the orders of Tsar Peter the Great, was wrecked on a remote island at the western end of the Aleutian chain in

Steller's sea cow was discovered in 1741 by members of Vitus Bering's crew, when his ship was wrecked on an uninhabited island off the coast of Siberia. Bering died on the island, but the survivors made it back to Russia and reported sea otters, fur seals, and these gigantic 30-foot-long manatees. Within 28 years, Russian sealers had killed every one of the ponderous sea cows, and because there were no others on any neighboring islands, they are extinct.

1741, his men were starving, so they were relieved to discover a herd of gigantic, slow-moving animals that they could kill easily: sea cows. Georg Wilhelm Steller, the naturalist on Bering's voyage, first described a new species of sea lion (*Eumetopias jubatus,* now known as Steller's sea lion); he also described a jay (Steller's jay), an eagle (Steller's sea eagle), and this animal now known as Steller's sea cow (*Hydrodamalis gigas*), one of the most unusual mammals that has ever lived.

Steller's sea cow is now extinct, but there are enough contemporaneous illustrations and descriptions to give us ample information on what they looked like and how they lived. As far as we know, *Hydrodamalis* was the only cold-water sirenian; at a length of 30 feet and a weight of 10 tons, it was also the largest. It was an overstuffed sausage of a beast, with a small head, piggy eyes, and skin that was likened to the bark of a tree. It had a forked, horizontal tail like its relative the dugong (manatees have a rounded, paddlelike tail), and its forelegs were unique in the mammal kingdom: they had no finger bones. The animal, which probably could not dive below the surface, pulled itself along the bottom on its stumps as it browsed on kelp. Instead of teeth, the mouth of the sea cow was equipped with horny plates that it rubbed together to grind plant matter into a pulp.

The skin of the sea cow was very thick and could be used for the soles of shoes and for belts, but these massive beasts were not killed for their skin, but rather for the subcutaneous fat, which could be as much as nine inches thick. In his 1745 description, Steller wrote of the fat:

> It is glandulous, stiff, and white, but when exposed to the sun it becomes yellow like May butter. . . . Its odor and flavor are so agreeable that it can not easily be compared with the fat of any other sea beast. . . . Moreover, it can be kept a very long time, even in the hottest weather, without becoming rancid or strong. In flavor it approximates nearly the oil of sweet almonds and can be used for the same purposes as butter. In a lamp it burns clear, without

> smoke or smell. And indeed, its use in medicine is not to be despised, for it moves the bowels gently, producing no loss of appetite or nausea, even when drunk from the cup.

Russian sealers killed the huge, slow-moving, oil-rich sea cows with such insensitive profligacy that there were none left by 1768. It took only 27 years for the Russian adventurers to eliminate the hapless sea cow from the face of the Earth, but they had no way of knowing that this was the last of them; they probably assumed that there were similar undiscovered islands with more sea cows. There were not.

Like whales and dolphins, sirenians are completely aquatic mammals and spend their entire lives in the water. There are three species of living manatees, the West Indian (*Trichecus manatus*), the Amazonian (*T. inunguis*), and the West African (*T. senegalensis*). The manatees and the dugong (*Dugong dugon*) have large, fusiform (spindle-shaped) bodies, flippers for forelegs, horizontally flattened tails (paddle-shaped in the manatees, forked like the flukes of a whale in the dugong), solid, heavy bones, and teeth that are highly specialized for their herbivorous diet.

Manatees are strictly vegetarians, browsing mostly on floating plants. Every day they consume 10 to 15 percent of their body weight, and a full-grown manatee can weigh over a ton. To guide the leaves into their mouths, they turn their lip pads inside out and use the bristles to tuck the food in. When they are not feeding, manatees are likely to be resting either at the bottom or just at the surface, with only the nostrils exposed. It is assumed that the manatee's sense of smell is rudimentary or absent; they keep their nostrils closed while they are submerged and only open them briefly to inhale at the surface. They never go very fast; their top speed is about 12 miles per hour, and they usually mosey along at 2 to 6 miles per hour. Manatees spend a lot of time under water; in the wild, they spend six to eight hours a day feeding.

Manatee skin is dark gray, but its coloration may be affected by various algae that grow on it. The flippers of the Florida and Caribbean species have fingernails but no fingers. Because manatees are designed to move smoothly through the water, all unnecessary protuberances have been eliminated. The ears are just pinholes, and

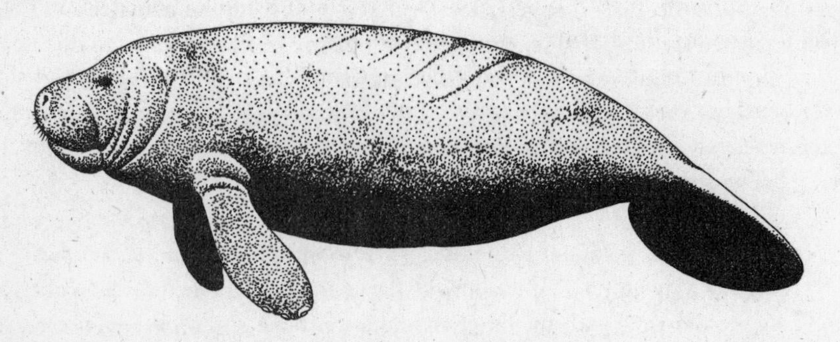

Manatees are found in Florida, the Amazon basin, and West Africa. They are placid, harmless creatures that ask nothing more than to graze on underwater foliage, but humans create problems for manatees, and they are endangered throughout their range.

the genitalia are internal. Gestation takes about 13 months, after which a 60-pound calf is born. The female nurses her calf from teats located under the flippers. (The location of the manatees' mammaries may be responsible for some of the earliest "mermaid" tales.)

Science now accepts the Florida manatee (*Trichechus manatus latirostris*) as a valid subspecies of the West Indian manatee (the other recognized subspecies is the Antillean manatee, *Trichechus manatus manatus*). The differences between the two are mostly skeletal and invisible to the layperson. The Florida subspecies is found in the state's waters year-round and during the warmer months may range north to Virginia or west to Louisiana. Although they can exist in salt or fresh water, they never wander far from shore, and they are restricted now to the inland and coastal waters of southern Florida. They favor warm waters, and in the winter they migrate to warm springs, such as at Florida's Crystal River, or the warm-water effluents of power plants. In Florida, the harvesting of manatees for food has ended, but the danger has not. Because of their desire to inhabit coastal waterways, they are in constant danger from power boats: the boats run over the manatees, and the spinning propeller slices the animal's back, often fatally. Threatened as well by development in its restricted habitat and by pollution, the Florida manatee is officially considered an endangered species. Its reproduction rate is low—one calf per adult female every three years—and there are thought to be no more than three thousand of the ponderous, blimp-like creatures left.

Unlike the Florida and Caribbean manatees, the Amazonian variety has no fingernails. Its specific name, *inunguis,* means "no nails." The Amazonian manatee is somewhat smaller than its northern relatives and can be differentiated by the presence of an irregular white patch on the belly. This species is also the only manatee that is confined to a river system and does not venture into saltwater. Its range includes the Amazon basin in Brazil (where it is known in Portuguese as *peixe-boi,* or "ox-fish"), Peru, and Ecuador, where its Spanish name is *vaca marina,* the familiar "sea cow."

Across the Atlantic, another manatee swims slowly in the coastal waters of West Africa, a virtual carbon copy of its American relative. The sirenologists John Reynolds and Daniel Odell suggest (1991) "that if a West Indian manatee and a West African manatee lay side by side, even an expert would be hard pressed to distinguish between the two animals." Their behavior and habitat are also similar, the only difference being that *Trichechus senegalensis* has a wider range, not only in Senegal but also in the littoral and riverine waters of Gambia, Liberia, Guinea-Bissau, Sierra Leone, the Ivory Coast, Ghana, Chad, Nigeria, Cameroon, Gabon, Zaire, and Angola. Like all of its relatives, the West African manatee feeds on floating vegetation and may occasionally eat clams.

Halfway around the world lives the dugong, the manatee's only living relative. The dugong is strictly a bottom feeder, and its mouth is on the bottom of its muzzle (the manatee's is on the front of its face). Its body has the same fusiform shape as the manatee's, and is slimmer; whereas the tail fin of the manatee is rounded, that of the dugong looks more like the flukes of a whale's tale. Dugongs have incisor teeth,

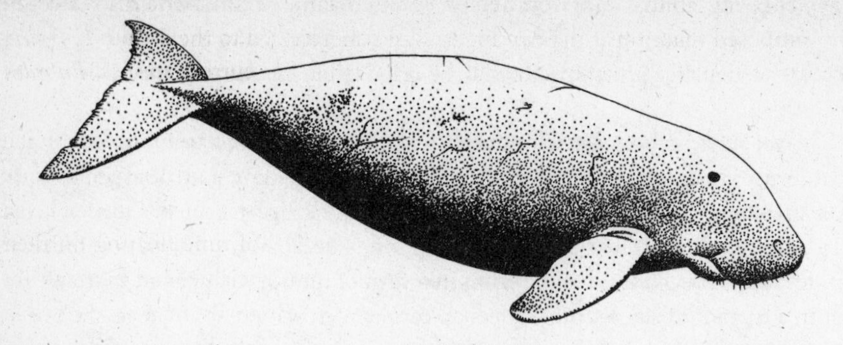

The dugong of Indopacific waters can be distinguished from the manatee of the Western Hemisphere by its down-turned muzzle and pointed, whalelike flukes.

which manatees lack, and in the males these develop into short tusks. The remainder of their teeth are molars, which are used for the grinding of plant matter.

The dugong is found in the territorial waters of 43 countries along the Indian and western Pacific oceans, ranging from Mozambique and Madagascar on the East African coast, up through the Arabian Sea and Persian Gulf to the west coast of India, through Southeast Asia and Indonesia, all along the coast of northern Australia, and around New Guinea, Melanesia, and the Philippines. With this wide a distribution, through some of the most densely populated areas of the world, the problems of the dugong are as wide and diversified as its range. Stated simply, however, the problem is this: dugong meat is said to taste like veal or pork, and throughout its range, it is hunted for food.

THE SEA WEASELS

The weasel family (Mustelidae) includes the living weasels, ermines, minks, ferrets, martens, wolverines, skunks, badgers, and otters. All of them can swim, but only the minks and otters are skilful swimmers, and only the otters have perfected a semi-aquatic existence. It reaches its peak in the sea otter, which thus can be considered a marine mammal. *Enhydra lutris* (*Enhydris* is Greek for "otter" and "water snake") come ashore to groom themselves, but they feed, mate, and sleep in the water. River otters (genera *Lutra, Aonyx,* and *Pteronura*) are better swimmers than sea otters, and they are adept at catching fish under water, which they consume on shore. Sea otters are not as agile as river otters, and feed on abalone, clams, mussels, and sea urchins, which they pluck off the bottom and bring to the surface, lying on their backs and swimming with their hindlegs while they eat.

The modern sea otter is believed to have arisen in the North Pacific at the beginning of the Pleistocene, between 1 million and 3 million years ago. There are two sea otter lineages. On the basis of an analysis of the fossil teeth, Repenning (1976a) suggested that *Enhydriodon* and *Enhydra* should be grouped together, because their teeth lost the shearing ability and were replaced by crushing teeth. Berta and Morgan (1985) proposed that one of the two lineages of sea otters led to the extinct *Enhydriodon* and the giant *Enhydritherium,* and the other led to the current species, *Enhydra lutris.*

The smallest of all marine mammals, sea otters are at most 5 feet long, including the tail. They are dark brown in color, and older animals have a grizzled appearance with a lightening of color around the head and shoulders; a bushy mustache of whiskers surrounds the nose. Sea otters spend a great deal of time floating on their backs, feeding and grooming themselves. They are highly sociable, and may raft together in a favorite kelp bed in numbers up to a thousand animals, many entwined in strands of kelp. They are among the few tool-using mammals, breaking open shellfish with rocks that they hold in their paws.

Georg Wilhelm Steller brought the news of his discovery of the North Pacific sea otters back to Russia, initiating an unrestrained hunt for these creatures, whose thick, dense fur is among the most luxurious in the world. Because they lack a subcutaneous fat layer, sea otters depend entirely on a layer of air trapped among the long, soft hair fibers. If the hair becomes soiled or impregnated with oil, it loses its insulating qualities and the animal cannot maintain its body temperature in the water. Sea otters were hunted to the brink of extinction throughout their range, from northern Japan all around the islands of the North Pacific, as far south as Baja California, but with worldwide protection, the sea otter has made a remarkable comeback. Once it was fully protected, "seed populations" were discovered in Alaska and Siberia, and the animal was reintroduced into its former range, where it is now fully protected. After a calamitous centuries-long decline, then a revival, the numbers of sea otters in California started falling again, and by 1998, only 1,937 animals were counted, a reduction of about 90 per year since 1995. Although the exact reasons for the decline are unclear, a combination of infectious disease, water contaminants, fish nets, and

The sea otter is blessed with the densest fur of any animal. In pursuit of this luxurious pelt, trappers and hunters nearly exterminated sea otters from their inshore Alaskan and Aleutian Island habitats, and only their scarcity saved them from complete extinction.

fishing pots seems to be taking its toll. There may be as many as 100,000 sea otters alive today, but oil spills also pose a constant threat. In 1989, oil from the *Exxon Valdez* killed more than 5,000 Alaskan sea otters.

Some 45 million years ago, an ancient group known as the arctoid ("bearlike") carnivores gave rise to the terrestrial mustelids, seals, and bears. Of these, the otters and pinnipeds became semiaquatic to varying degrees, and one species of bear—the polar bear, *Thalarctos maritimus*—also became adapted to spending a significant amount of time in the water. It has been suggested that the Pacific sea lions (otarioids) derive from early bears and the Atlantic seals (phocoids) derive from early otters or otterlike arctoids, but this has not been clearly proved. *Enhydriodon* probably evolved in Eurasia and dispersed to Southeast Asia (as exemplified by the living small-clawed otter, *Amblonyx*); to Africa (the African clawless otter, *Aonyx*), and to Europe and Africa (the extinct *Enhydriodon*). These are all called "crabeater otters" or "small-clawed otters" and are a distinct branch separate from the "river" (fish-eating) otters. The river otters, *Lutra canadensis,* made famous in Gavin Maxwell's *Ring of Bright Water* are found in streams and lakes of North and South America, Africa, Europe, Asia, and many Malaysian islands. Most crabeaters also live in rivers, except *Enhydriodon,* which sometimes lived in rivers, and *Enhydra,* which lives only in coastal waters.

Found in the Moss Acres Racetrack site in north central Florida, the Late Miocene *Enhydritherium terraenovae* is described as "a giant otter . . . thought to be related to both the Old World otter *Enhydriodon* and the extant sea otter *Enhydra*" (Lambert 1997). *Enhydritherium* had been previously described by Berta and Morgan (1985), but this new skeleton offered better material. Though incomplete, it included an intact skull, mandible, and virtually complete fore- and hind limbs. David Lambert of the University of Florida was able to add much to the understanding of the phylogeny of early sea otters. The animal, about the size of the living sea otter (5 feet long including tail, and weighing about 50 pounds), was less strongly bound to the sea, and probably lived in both nearshore coastal and inland freshwater environments. The resemblance of its powerful forelimbs to those of the California sea lion (*Zalophus*) suggested that, unlike *Enhydra,* it was a forelimb swimmer, and its powerful neck meant that it could also catch fish and crayfish in addition to the various mollusks, and probably did not use its hands for food manipulation to the degree that the living sea otter does: *Enhydra* spends most of its aquatic time on its back, paddling with its hind limbs, leaving its paws free. The teeth of the extinct *Enhydritherium* are heavily worn, suggesting a partial diet of extremely hard food items, such as mollusks (Berta and Sumich 1999).

The youngest crabeater otter in Europe is from England. Repenning (1983) published two records from the Arctic Ocean that are around 2.5 million years old and appear to be intermediate between *Enhydriodon* and *Enhydra.* They are younger than the specimens from California, but it would appear that they gained the Arctic Ocean from England or northwest Europe and then, as *Enhydriodon,* entered the North Pacific. There, in the North Pacific and adjacent Arctic Oceans, it evolved into *Enhydra.* There are a number of records of *Enhydriodon* from Spain, and it is possible that they could have blown across to Florida, or followed the coast or island-hopped across the North Atlantic. The marine otter (*Lutra felina*), which lives along the rocky coasts of western South America, is ecologically intermediate between the true river otters and the sea otter. Like *Enhydra,* it swims on its back at the surface, but it does not sleep in the water and comes ashore to shelter in caves at night.

At the same time that the sea otter and the sea cow were discovered—and in the same general area—Georg Steller described the "sea monkey," one of the most enigmatic creatures ever to enter the zoological (or cryptozoological) literature. It is an animal whose identity has eluded scientists since the account was first published. All Steller's other descriptions are more than adequate to identify the animals in question, and subsequent investigations have demonstrated the existence of all the other creatures he so meticulously cataloged. But what are we to make of this?

> During this time we were near land or surrounded by it we saw large numbers of hair seals, sea otters, fur seals, sea lions, and porpoises. . . . On August 10 [1741] we saw a very unusual and unknown sea animal, of which I am going to give a brief account since I observed it for two whole hours. It was about two Russian ells [6 feet] in length; the head was like a dog's, with pointed erect ears. From the upper and lower lips on both sides whiskers hung down which made it look almost like a Chinaman. The eyes were large; the body was longish round and thick, tapering gradually towards the tail. The skin seemed thickly covered with hair, of a grey color on the back, but reddish white on the belly; in the water, however, the whole animal appeared entirely reddish and cow-colored. The tail was divided into two fins, of which the upper, as in the case of sharks, was twice as large as the lower.

Steller went on to say that the only reference he could think of that might help in identifying the creature was Gesner's description in the 1587 *Historia Animalium* of something he called *Simia marina danica,* the Danish sea monkey. The animal cavorted around Steller's boat for hours, playing with strands of kelp, and finally Steller took a shot at it ("in order to get possession of it for a more accurate description"), but he missed. In *Where the Sea Breaks Its Back,* the story of Steller's voyage, Corey Ford discounts the suggestion that Steller confused the creature with a sea otter or a seal and wrote, "The simplest explanation is that the 'sea-monkey' actually existed, and that Steller saw it for the first and last time before it became extinct."

There are no fossils that might confirm the existence of such a creature, but along the rocky coasts of Maine and New Brunswick, Canada, until about the end of the nineteenth century, there lived an animal known as the sea mink (*Mustela macrodon*), which might well be compared to Steller's sea monkey. Also known as the bull mink or saltwater mink, it was about twice as large as the Eastern mink (*M. vison*) and had coarser, reddish-brown fur (Campbell 1988). By 1894, the sea mink had been hunted to extinction by trappers, and although there is no reason to assume that a comparable species existed on the West Coast of North America or the East Coast of Asia, there is an Alaskan subspecies of the common mink (*M. vison injens*) and several species of martens, sables, and other mustelids live in Siberia. In other words, if Steller had spotted the "sea monkey" off the East Coast of North America, there would be no problem with its identification.

FLIPPERS FOR FEET

> *At one time the seals were thought to be sufficiently closely related to the land living, meat-eating mammals to be placed with them in a single classificatory group, the order Carnivora. Nowadays, instead of leaving the seals as a subgroup of the Carnivora, many zoologists place them in a separate order of their own, the Pinnipedia. This is really only two ways of looking at the same set of facts, but the modern view does recognize that although seals have much in common with such animals as bears and wolves they are sharply divided from them by the considerable modifications they have undergone to suit their watery habitat—as is suggested by their name Pinnipedia, "the fin-footed ones." A member of the Pinnipedia can be recognized at a glance by its flippers and shape.*
>
> —Gavin Maxwell, *Seals of the World*

Charles Linnaeus described the pinnipeds as "a dirty, curious, and quarrelsome tribe, easily tamed, and polygamous; fat and skin useful. They inhabit and swim under water and crawl on land with difficulty because of their restricted fore-feet; feed on fish and marine productions, and swallow stones to prevent hunger, by distending the stomach" (Lowenstein 1986). In his article, Lowenstein says, "Another quarrelsome tribe, the evolutionary systematists, have been arguing about pinniped relations for one hundred years." He continued,

> The hottest issue in pinniped systematics is whether the "wingfeet" [his definition of "pinniped"] had a single common ancestor among land mammals or two different ancestors—whether they are monophyletic or biphyletic. Phylogenetic opinion has its fashions, like hair, and skirts, and medical treatments. Most fashionable marine mammalogists these days believe in two ancestors—a bearlike progenitor of the eared seals and walruses, an otterlike precursor of the earless seals. They also favor a Pacific origin for the eared, an Atlantic origin for the earless, which makes for a neat anatomical and geographical separation.

The fin-footed seals, sea lions, fur seals, and walruses are widely distributed throughout the world, from one polar ocean to the other, and at numerous locations in between. In 1922, Remington Kellogg gave a résumé of the known theories about the evolutionary history of the pinnipeds. He wrote, "The pinnipeds were well specialized in the Miocene [which began about 24 million years ago] for a pelagic life and the distribution of the fossil forms corresponds very well, in most respects, with the present distribution of their living representatives." Our records of early pinnipeds do not extend much further back than that—a time when pinnipeds already resembled their living representatives. They seem to have appeared more or less fully formed, and nobody has much of an idea about what preceded them.

By 1958, Victor Scheffer wrote that there were "alternative reconstructions" of pinneped evolution, but he didn't come to any clear conclusion. He based his reconstructions on similarities and differences that were visible to the naked eye but was able to see that this approach was limited. He wrote, "So remote are the ancestors of the Pinnipedia that little evidence upon them can be expected from the study of the comparartive anatomy and physiology of the modern forms."

The first pinnipeds in the fossil record are the family Enaliarctidae, found from the Early Miocene, about 20 million years ago (Berta, Ray, and Wyss 1989). They looked like a cross between an otter and a sea lion, with a long neck, short legs with webbed feet, and a stubby tail. The enaliarctids flourished between 24 million and 12 million years ago, along the shores of the North Pacific from Japan to Southern California. From a detailed study of a virtually complete skeleton found in central California, Berta and Ray (1990) were able to show that this harbor seal–sized animal swam by using both its hind and forelimbs, which only the walrus does today, and no living seal or sea lion does. Sea lions (otariids) and earless true seals (phocids) use only their forelimbs for propulsion in the water, but they move very differently on land; the otariids are able to use both pairs of limbs in an awkward but still efficient rocking walk that has often been described as a waddle. The earless seals cannot use their hind limbs for terrestrial locomotion, and pull themselves along with their clawed foreflippers, an awkward method of land locomotion, but one that works on ice, where many phocids live.

In their reconstruction, Berta and Ray depicted *Enaliarctos mealsi* as short-bodied and relatively long-legged, and they gave it a coat of dense fur, "since the animal was well-adapted for aquatic life." They wrote, "*Enaliarctos* was highly capable of maneuvering on land and probably spent more time near the shore than extant pinnipeds." The enaliarctids probably came ashore to eat, which Berta and Ray inferred from the fact that the animal's heterodont teeth, bladelike carnassials (shearing teeth) and crushing molars, suggested that they did not swallow their prey whole, as fully aquatic fish eaters would have done (Repenning 1976).

Living at the same time as *Enaliarctos* was *Desmatophoca,* which looked much like today's sea lions, with stronger forelimbs than hind limbs, webbed feet with elongated digits that would eventually become proper flippers, and a short tail, only about as long as the animal's skull. Its eyes were enormous, suggesting that vision was its primary sense in the water; reliance on hearing and echolocation would come later. The differences between the skeleton of a desmatophocid and an otariid are many and can also be traced through the skeletons of their ancestors, indicating that they were inherited through separate lineages (Repenning and Tedford 1977). During most of their existence, the two types of seals lived together in the Pacific.

In *Marine Mammals: Evolutionary Biology* (1999), Annalisa Berta holds that the three pinniped families (otariids, phocids, and odobenids), the bears, and the

So far, all fossil pinnipeds look very much like today's seals and sea lions. What preceded them—or, for that matter, why they took to the sea—is not known. Enaliarctos *lived between 23 million and 12 million years ago, along the shores of the North Pacific from Japan to southern California.*

mustelids have a single common evolutionary origin. She wrote, "All recent workers, on the basis of both molecular and morphologic data, agree that the closest relatives of pinnipeds are the arctoid carnivores which include procyonids (racoons and their allies) and ursids, although which specific arctoid group forms the closest alliance with the pinnipeds is debated." Irina Koretsky (1999) believes that they are diphyletic: the odobenids are derived from a bearlike animal; the phocids, from an otterlike animal. Hunt and Barnes confirmed this conclusion in their 1994 analysis of the skulls of *Pinnarctidion* and *Enaliarctos,* the earliest known pinnipeds, citing "evidence that pinnipeds are derived from from an ursid ancestor, and [that] does not support the view that pinnipeds are most closely related to mustelids."

Once upon a time in the Miocene, there was a bearlike carnivore known as *Kolponomos* that lived along the shore of the Pacific Ocean in what is now Washington and Oregon. When the first bones were discovered by R. A. Stirton (1960), *Kolponomos* was thought to be some sort of a bear-sized raccoon, but a second species (*K. newportensis*), described by Tedford, Barnes, and Ray in 1994, led to the conclusion that it was an ursid closer to pinnipeds. Stirton had written that "it evidently was adapted to live in marine waters, much like the living sea otter," and on the basis of the fossilized bones and teeth, Tedford, Barnes, and Ray were able to make some educated guesses about its lifestyle: "*Kolponomos* was probably littoral in distribution, all specimens having been discovered in nearshore marine rocks. The crushing cheek teeth would have been suited to a diet of hard-shelled marine invertebrates. The anteriorly directed eyes and narrow snout indicate that *Kolponomos* could view objects directly in front of its head. . . . Large paroccipital and mastoid processes indicate strong neck muscles that could provide powerful downward movements of the head. These features indicate that *Kolponomos* probably fed on marine invertebrates living on rocky substrates, prying them off with the incisors and canines, crushing their shells, and extracting the soft parts as do sea otters."

The history of the North Atlantic pinnipeds is incomplete. There is little fossil evidence for the transition from terrestrial to amphibian mammal. As Repenning has written,

> The oldest known North Atlantic pinnipeds are only about 15 million years old, are very well adapted to oceanic life, and are recognizable as belonging to the two subfamilies of the present seal fauna, the Phocinae and the Monachinae. They have no recorded history of transition from land carnivore characteristics to those of the marine carnivore. Nor are there any known extinct lineages of North Atlantic pinnipeds.

It is more than a little curious that the fossil record of pinnipeds other than walruses should be so sparse and in many cases nonexistent. Animals that live (or lived) along the shores of the sea seem likely to leave fossil evidence of their existence, but this has not been the case. When I asked Charles Repenning in 1999 whether he had written any more on the evolution of pinnipeds after his 1980 *Oceans* article, he said that was just about his last word because he "had run out of fossils." Not so with Irina Koretsky. From fragmentary fossil material from South Carolina, she and Albert Sanders described the "oldest known seal," which they dated from the Late Oligocene, 30 million years ago. Even though the material consisted only of parts of two femurs, Koretsky and Sanders (2000) compared these pieces with the bones of living pinnipeds and were able to show that the unnamed fossil was closest to "the most specialized phocid, the modern genus *Cystophora.*" *Cystophora cristata* is the hooded seal, found in the western Arctic. They believe that it was better adapted to the marine environment than its otterlike ancestors, and "that the material of this Oligocene seal is not sufficient to either support or falsify the hypothesis of mustelid ancestry, but it is sufficient to conclude that the ancestor of the Phocidae [uneared, or true, seals] must be sought in deposits older than the Late Oligocene."

The subfamily Monachinae (pronounced "mon-*ah*-kin-ay") includes monk seals (from the Greek *monachus,* "monk"), phocids with a subtropical distribution. The Mediterranean variety (*Monachus monachus*), found mostly along the coasts of northwest Africa, is endangered, as is the Hawaiian (*M. schauinslandi*). The Caribbean monk seal (*M. tropicalis*) is extinct, the last living specimen having been sighted in Jamaican waters in 1952. They are all chunky animals with round faces, probably accounting for their common name. Monk seal fossils are sparse (for the Hawaiian species they are nonexistent), but those of the other monachids suggest that the Caribbean species formed the root stock from which the others developed. First described by Columbus in 1494, this seal has been seen around various Caribbean islands such as the Greater Antilles, the Bahamas, and even the Florida Keys. Human beings, competing for space on the Caribbean islands, certainly contributed to the demise of *M. tropicalis,* but the overall cooling of the world's oceans over geological time has also been a factor. The Hawaiian and Mediterranean monk seals are listed as endangered, and although stringent protective measures are in place, both species will probably become extinct in the near future.

The Hawaiian monk seal is found primarily in the Leeward Islands, but they sometimes stray eastward to the main Hawaiian group. They were heavily hunted by whalers and sealers in the nineteenth century, and were disturbed again during World War II, when U.S. forces occupied Laysan and Midway islands. Recent surveys have estimated the population at about 1,400 animals. They are fully protected, but they seem to be very suscepible to attacks by tiger sharks. Of all the monk seals, this species is the best known, because it has been studied in its native habitat, and has been successfully maintained in captivity, particularly at the Waikiki Aquarium.

Also classified in the subfamily Monachinae are the Antarctic seals, including the southern elephant seal (*Mirounga leonina*), the crabeater (*Lobodon carcinophagus*), the Ross seal (*Ommatophoca rossi*), the Weddell seal (*Leptonychotes weddelli*), and the leopard seal (*Hydrurga leptonyx*). Each of these seals has evolved differently, and their habits and apppearance are substantially different. For example, the southern elephant seal is, with its northern relative, the largest of all living seals, the males reaching a length of 20 feet and a weight of 10,000 pounds. Females are considerably smaller, and do not reach more than 11 feet in length. The southern elephant seal is found north of the pack ice in the Southern Ocean, on various subantarctic islands, and on the Valdés Peninsula in southern Argentina. Elephant seals get their common name from the large, inflatable proboscis of the males, through which they make a variety of gurgling and growling noises, especially during the breeding season. Males fight viciously for dominance over a harem of females. When not hauled out on their breeding grounds, elephant seals spend most of their time at sea, where they dive to

Twenty feet long and a weighing 2 tons, the male elephant seal has a "trunk" like its namesake. There are separate species in the Northern and Southern Hemispheres.

prodigious depths, well over a mile, in search of the fish and squid on which they feed.

Unlike most of the other phocids but like the otariids, male elephant seals are much larger and heavier than females. Until recently, their destinations after leaving the breeding beaches were completely unknown, but attaching radio transmitters to individuals has allowed their migrations to be tracked. Their wanderings were surprising enough—they appear to spend most of their time at sea, which is why they were not found at haul-out sites other than breeding beaches—and their feeding dives astonished everyone even more. Brent Stewart and Robert DeLong affixed time-depth recorders to six adult male elephant seals off San Miguel Island, California, in February 1991, and recorded dives that lasted up to 77 minutes, with a maximum recorded depth of 5,105 feet, "the deepest yet measured for air-breathing vetebrates." The greatest recorded depth for a southern elephant seal, as recorded by Hindell and colleagues in 1992, was 3,936 feet. Sperm whales and northern bottlenose whales are believed to dive as deep or even deeper, but so far the deepest dive recorded for a bottlenose is 4,765 feet. Hooker and Baird (1999), who affixed suction-cup radio tags to bottlenose whales off Nova Scotia in 1996 and 1997, wrote that "the combined evidence leads us to hypothesize that these whales make greater use of the deep portions of the water column than any other mammal so far studied." A sperm whale "possibly" dived to 6,560 feet, but the average dives recorded by Watkins et al. in 1993 were between 1,300 and 2,000 feet. As noted by Malcolm Clarke (1976), the maximum depth of a sperm whale dive was not actually recorded but was inferred from a known bottom depth of 10,000 feet at a spot where a sperm whale was caught whose stomach contents included a freshly consumed bottom-dwelling shark.

Marine mammals that dive to great depths to feed usually do so in pursuit of prey. Because sperm and bottlenose whales have long been hunted by whalers, their stomach contents have been closely examined, and it is known that squid make up a large proportion of their diets. From "stomach lavage," whereby the stomach contents are pumped out without killing the animal, it was learned that northern elephant seals feed primarily on squid, but also take octopuses, starfishes, crabs, and assorted fishes (Antonelis et al. 1987). The ability of various unrelated marine mammals to dive to incredible depths in pursuit of food is another example of convergence, and it also exemplifies the profound modifications that occurred since the first terrestrial mammals returned to the sea.

However they evolved, the pinnipeds are with us in force today, and in some places, such as the Antarctic, the *only* native mammals are pinnipeds and cetaceans. There are both phocids and otariids there, occupying different niches and performing different jobs in the complex ecology of this vast icy wasteland. Officially, the Antarctic continent is the landmass that encompasses the South Pole, but "the Antarctic" means the entire region, including the islands of South Georgia, Heard, Bouvetøya, the South Shetlands, the South Orkneys, and the South Sandwich Islands. The continent itself is roughly circular, with the Antarctic Peninsula curling in the direction of the tip of South America. With an average elevation of 9,000 feet and a maximum height of 16,000 feet at the Vinson Massif in the Ellsworth Mountains, it is the highest continent in the world. It is also the driest and coldest. The continent's ice sheet represents 90 percent of the world's ice and more than half of its fresh water. There are no trees or terrestrial mammals in Antarctica, but penguins abound, seabirds are extremely numerous, and whales and seals used to be so abundant that their presence encouraged whalers and sealers to explore this region before anyone else.

The fur seals and sea lions are members of the eared seal family, Otariidae. There are eight species in the fur seal genus *Arctocephalus* ("bear-headed"), mostly found south of the Equator, but also on Guadalupe Island off the coast of Baja California, and in the Galápagos Islands. All southern fur seals have a sharply pointed snout, and many of the Southern Hemisphere sea lions have shorter, stubbier muzzles. The species of southern fur seals, differentiated by minor skeletal differences and geography, are the Galápagos (*A. galapagoensis*), Guadalupe (*A. townsendi*), Juan Fernández (*A. philippi*), South American (*A. australis*), South African and Australian (*A. pusillus*), New Zealand (*A. fosteri*), Antarctic (*A. gazella*), and subantarctic (*A. tropicalis*). In recent years, the depleted populations of fur seals have rebounded, probably because of the cessation of hunting, and also because the baleen whales have been so reduced in numbers that more food is avaliable for the seals. (There are no significant differences between fur seals and sea lions; the fur seals were so named because they were hunted for their fur.)

Also known as the Pribilof fur seal, the northen fur seal (*Callorhinus ursinus*) is the northern version of *Arctocephalus*. It occurs in the Pribilof island chain, which extends from Alaska to Kamchatka, and has been known to venture as far south as Santa Barbara, California, in the eastern North Pacific and Honshu, Japan, in the western. The large males (8 feet long and 600 pounds) arrive first on the breeding grounds, and then the females arrive to form into harems. Once hunted mercilessly for their luxurious pelts, the northern fur seal was "harvested" in the Pribilof Islands under the North Pacific Fur Seal Convention of 1911, an agreement between the United States, Japan, Canada, and Russia. This treaty has now

The northern fur seal, Callorhinus ursinus, *found only in the North Pacific, was hunted to near extinction in the nineteenth century by Russian, Canadian, Japanese, and American sealers.*

run out, and no nation is legally involved in a seal hunt. Like the Steller's sea lion, with which it shares much of its habitat, the population of northern fur seals has declined seriously, perhaps because of food limitations.

Although it is commonly seen performing in circuses and oceanariums as a "trained seal," *Zalophus californianus* is actually the California sea lion. It has visible ear flaps and hind flippers that can be rotated forward for locomotion. Males are much larger and heavier than females, reaching a weight of 600 pounds and developing a prominent "sagittal crest," or forehead bulge. As its name implies, the California sea lion lives in and around the waters of California and Mexico, but a subspecies was found in the Galápagos and another was found in Japan before it became extinct.

Steller's sea lion (*Eumetopias jubatus*) is the largest of all the eared seals, and the males attain a much greater size than the females. Bulls can weigh as much as a ton, whereas females do not exceed 700 pounds. Found only in the North Pacific, Steller's sea lion breeds in the Pribilofs, the Aleutians, and the coast of North America as far south as Southern California. These animals are extremely wary of people and are difficult to approach on land. Until the 1980s, the estimated population was some 300,000 animals, but in recent years, for reasons that are not understood, the population has declined at an alarming rate. As of 1997, the National Marine Fisheries Service has listed Steller's sea lion as "endangered."

Sometimes known as the "hair seal," the Australian sea lion (*Neophoca cinerea*) is one of the few species that is nonmigratory. These animals spend most of their lives on or near the beach where they were born. As with all of the eared seals, males are considerably larger and heavier than females. They are renowned for their ability to move about on land, and one individual was found 6 miles from the sea. In the waters of South Australia and Kangaroo Island, Australian sea lions are the preferred prey of great white sharks. Seal hunters massacred this species during the eighteenth and nineteenth centuries, and it is believed that no more than 5,000 are left.

The New Zealand sea lion (*Phocarctos hookeri*), also known as Hooker's sea lion, is found mostly on the subantarctic islands of New Zealand, but it is sometimes seen on South Island as well. The males are much larger than females, and can be 10 feet long and weigh 800 pounds. In addition, bulls are dark brown and the females are silvery gray. Like many of the Southern Hemisphere fur seals and sea lions, *Phocarctos hookeri* had been hunted to the brink of extinction by the end of the nineteenth century.

The southern sea lion, also known as the South American sea lion, inhabits both coasts of southern South America, from Cape Horn to Peru on the west coast and as far north as Uruguay on the east coast. Male *Otaria flavescens* are distinguished by their short, upturned muzzle, massive head, and neck with a mane that gives them their common name. These animals are known as *lobos marinos,* "sea wolves," in Spanish. The bulls can weigh as much as 750 pounds; the females are much smaller, lighter in color, and more graceful. Like most of the pinnipeds, this species is primarily a fish eater, and it can often be seen in shallow inshore waters. Southern sea lions were hunted extensively by sealers in the nineteenth century, and although they are now protected, the populations are still declining.

With its thick mane and upturned muzzle, Otaria flavescens *shows why the first visitors to South America called it a sea lion.*

In the classification of pinnipeds, the walrus (*Odobenus rosmarus*) is usually placed between the otariids and the phocids because it shares some characteristics with both groups. Nevertheless it represents a separate lineage, and is not considered closely allied with either. The odobenids are also believed to have derived from the otterlike *Enaliarctos,* but many changes had to occur before the fish-eating *Enaliarctos* became the shellfish-eating ancestral walrus *Imagotaria.* Its canine teeth were large, but were nothing like the 3-foot-long tusks of the modern walrus, and the back teeth had not evolved into the shell-crushers of today's odobenids. *Imagotaria* was probably transitional between the sea lions and the walruses, and probably fed on both fish and shellfish. The primitive walruslike pinnipeds evidently flourished and became greatly diversified in the Late Miocene and Early Pliocene times (15 million to 10 million years ago); Repenning (1976b) writes that "there are more named species of [fossil] odobenids than any other group of pinnipeds." Walruses with tusks have been found in deposits in Baja California, and although there are no Caribbean fossils, walruses are believed to have entered the Atlantic by passing through the Central American Seaway and moving up the Atlantic coast, as evidenced by fossils in South Carolina and Maryland.

There are two subspecies of living walrus, the Atlantic and the Pacific, both inhabitants of the moving pack ice and rocky islands of the Arctic. Walruses are extremely gregarious and often cluster together in large groups. Males are larger than females, and both sexes have tusks. The largest bulls can weigh 3,500 pounds. Walruses are the only mollusk-eating pinnipeds. Feeding underwater, they use powerful suction to eat the clams and other mollusks that make up their diet. They may employ the tusks as "guide rails" as they suck up clams from the muddy bottom, but they do not dig with them. They do occasionally pull themselves along on the ice with their tusks, however. Walruses have been exploited for their ivory tusks, tough hides, and blubber oil for thousands of years, and although most commercial hunting has ceased, native peoples still kill them in large numbers for their ivory, which they carve into trinkets for the tourist trade.

The phocids (earless seals) are believed to be descended from a line of carnivores separate from the otariids, but there are very few fossils to establish their ancestry. Considered the most "primitive"—manifesting the characteristics of its ancestors—is the bearded seal (*Erignathus barbatus*), a Northern Hemisphere species whose long and plentiful whiskers are responsible for its name. Males grow larger than females and can reach a length of 12 feet, but 10-footers are more common. Throughout their ice-defined range, bearded seals are hunted for their meat, blubber, and skin, but they were never the object of a major sealing industry because they are usually solitary animals and do not assemble in large herds the way many other pinnipeds do. They have a circumpolar Arctic distribution, from Labrador and Greenland to Siberia and Alaska, and as far south as northern Norway. Bearded seals are very vocal, and they are among the few mammals that actually "sing."

The common seal (*Phoca vitulina*), also known as the harbor seal, is a small, stocky animal that lives throughout the temperate and Arctic waters of the Northern Hemisphere and has the widest distribution of any pinniped. Some populations are nonmigratory, breeding and feeding in the same area throughout the year, but others may migrate hundreds of miles. Like all the other phocids, the harbor seal uses its

There are two subspecies of walrus, the Atlantic and the Pacific, both inhabitants of the moving pack ice and rocky islands of the Arctic. Males are larger than females, and both sexes have tusks. The largest bulls can weigh 3,500 pounds.

hind flippers for propulsion in the water, but on land, it hitches itself along using only its foreflippers, which are equipped with sturdy nails. Harbor seals eat almost anything they can catch, which is mostly fish. They occasionally raid and ruin fishermen's nets, and they are killed for that reason, as well as for their meat and highly prized fur.

Probably the most strikingly colored of all pinnipeds is the ribbon seal (*Histriophoca fasciata*), found only in the Arctic North Pacific, from Alaska to Northern Japan. The population of the Sea of Okhotsk has been estimated at about 250,000 animals. They are creatures of the ice, and show a remarkable indifference to the approach of predators, including polar bears, arctic foxes, or humans. Ribbon seals are born white, but after a couple of weeks the white coat—known as the lanugo—is shed. The "ribbons" that characterize the adults do not appear until the animals are between two and four years old.

Creatures of the Northern Hemisphere ice pack, ringed seals (*Phoca hispida*) have a completely circumpolar distribution. They are the smallest of the earless seals, rarely reaching 4.5 feet in length. Females give birth to 2-foot-long pups in caves or tunnels in the snow, a most unusual habit for seals. Throughout their range they are hunted by Eskimos, who use the meat, blubber, and skins, and they are the favorite prey of polar bears, who wait for them to poke their heads out of breathing holes in the ice and then swipe them out and kill them. There are several subspecies of ringed seals, distinguished primarily by distribution.

Harp seals (*Pagophilus groenlandica*) are 6-foot-long seals found throughout the Northern Hemisphere ice pack from Canada and Greenland to Siberia. They are born white and fluffy, and it was for these "whitecoats" that the Newfoundland sealers, notoriously, killed them with clubs. Within a month the pups begin to develop the characteristic coloration of the adults. Harp seals are extremely aquatic and spend much of their lives in the water, where they display particularly effective diving skills.

The hooded seal (*Cystophora cristata*), an animal of the North Atlantic and Arctic seas, is named for an inflatable sac that occurs only in mature males. Unlike the proboscis of elephant seals, which is inflated by muscular action, the hooded seal's sac is inflated with air to form the characteristic hood. These seals can also blow a bright red, balloonlike structure out of one nostril, which accounts for their other name, bladder-nosed seal. This behavior may be used to intimidate subordinate males, and it has also been observed when humans get too close. Hooded seals reach a length of about 8 feet and a weight of 800 pounds.

The Southern Hemisphere is the home of four phocid species whose ranges often overlap but whose habits do not. The leopard seal is an active carnivore, whereas the Weddell and crabeater seals have been likened in their feeding habits to penguins: they are fish eaters that dive deeply after their prey. Named for its spots as well as its predatory habits, the leopard seal (*Hydrurga leptonyx*) is found throughout the Antarctic and surrounding island groups. Leopard seals are active carnivores, feeding on krill (which they strain from the water in the manner of crabeater seals), fish, squid, other seals, and penguins—particularly Adélies, which they devour by grasping the birds in their powerful jaws and shaking them out of their skin. Unlike many other seals and sea lions, leopard seals are solitary, and it is rare to encounter more than one at a time. Female leopard seals are larger than males, unusual in pinnipeds. The only animal powerful enough to take on a leopard seal is a killer whale.

With its fat body, small head, and benign expression, the Weddell seal (*Leptonychotes weddelli*) is easily recognized. These seals probably live farther south than any other mammals, and they spend a great deal of time under the ice, breaking through it, with specially developed canine teeth, to breathe. Weddell seals (named for the Antarctic explorer James Weddell) are among the most accomplished divers of all pinnipeds. They can remain submerged for well over an hour, and descend to almost 2,000 feet in pursuit of fish or squid.

Named for James Clark Ross, who explored the Antarctic from 1839 to 1843, the Ross seal (*Ommatophoca rossi*) has a proportionally small head and well-developed vibrissae (whiskers). It is streaked rather than spotted, and it has longitudinally oriented stripes on the neck and throat. Essentially solitary, it inhabits the pack ice all around the Antarctic. Its vocalizations consist of clicks, gurgles, and a sound that has been described as "chugging."

Their name suggests that crabeater seals feed on crabs, but they don't. *Lobodon carcinophagus* (*carcinophagus* means "crabeater") feeds primarily on small shrimp, the same krill that is eaten by baleen whales. Crabeaters have specially designed teeth that enable them to strain these small organisms from the water. They inhabit the pack ice of the circumpolar Antarctic. Although the actual estimates differ, there seems to be little doubt that the crabeater is the most numerous of all pinnipeds, and perhaps the most abundant large mammal in the world (excluding, of course, *Homo sapiens*). Various studies have produced figures that range from 15 to 30 million crabeaters. Because of the remoteness of their habitat, crabeaters have never been hunted commercially, which may account for their populousness.

THE DESMOSTYLIANS

More than 9 million years ago, some strange mammals of the Pacific rim became extinct after inhabiting the basin for about 21 million years. They may have looked like hippos, sea lions, stumpy reptiles, or maybe some combination that we cannot envision. They may have carried their weight on their toes or on the soles of their feet. They may have walked upright, splayed out their legs like a lizard, waddled like a sea lion, or they may never have come ashore at all. They may have fed on sea grass like modern manatees; on plants and algae in tide pools; or they may have paddled out to sea and dived for their food on the shallow bottom. They were almost certainly herbivores. At least one of them was attacked by a shark. The only thing we can say with certainty about the desmostylians is that we will never see their like again.

Somewhere in the ancestry of the desmostylians there lurks the shade of a still earlier mammal whose affiliations are unclear. Less than 3 feet tall, it was a lightly built, piglike animal found in the Upper Eocene deposits of Africa, from Senegal to Egypt. Its name, *Moeritherium,* is derived from Lake Moeris in Egypt. It had a thick, round body, and although it stood upright on all fours, it is believed to have been partially aquatic, because all the fossils have been found in deltaic deposits. Many believe that *Moeritherium* gave rise to the proboscideans, animals with trunks (from the Greek *proboscis,* for "nose" or "snout"). McKenna (1975) has grouped it in the Tethytheria, a group that includes the sirenians, proboscideans, and the desmostylians. (The Tethytheria were named for the very ancient Tethys Sea, which circled the Earth at about the Tropic of Cancer in Paleozoic times, and the Greek *therio,* "beast.")

Paleontologists are still arguing about the form of the desmostylians, and whether they were terrestrial, aquatic, or both, and they also disagree about the relationships and descent of these unusual mammals. They agree, from the fossil evidence, that the desmostylians lived in the North Pacific during the Oligocene and Miocene epochs. Having gone to their extinction without leaving any identifiable descendants, they are the only extinct order of marine mammals. From fossil skeletal evidence, we know that the adults were about the size of a rhinoceros, with splayed toes that resembled those of a hippopotamus and may have been webbed. The front limbs are considerably larger than the hind ones, as they are in most modern seals and sea lions.

The cheek teeth in *Desmostylus* look like small columns bound together, and are responsible for the name, which comes from the Greek *desmo* meaning "bond" or "chain," and *stylos,* a pillar or column. The tusklike canines projected forward at such an angle that they seemed ideal for prying shellfish from rocks, and though some authors have suggested that they fed on mollusks or other benthic invertebrates, there is no evidence to support such a claim. Barnes, Domning, and Ray (1985) wrote, "a molluscan diet has been suggested for *Desmostylus* . . . but this idea has not gained acceptance," and in their 1986 discussion of the desmostylian lifestyle, Domning, Ray, and McKenna wrote, "Certainly we see nothing in the dentitions . . . that would suggest a diet radically different from that of proboscideans, hippos, pigs, or other plant-eaters. . . . If we visualize primitive desmostylians as feeding on benthic grass and other marine plants along North Pacific shores, a major potential adaptation that suggests itself is subtidal feeding." Charles Repenning believes that the teeth "resemble a hay rake for scooping up marine grasses for food. . . . The blunt worn ends are unsuitable for prying an abalone holding fast to a rock" (personal communication). With the exception of the walrus, which has special anatomical adaptations for sucking clams from their shells, no pinniped or sirenian eats mollusks.

There are complete skeletons for *Desmostylus* and also for the genus *Paleoparadoxia,* but other genera are known only from scattered jawbones and teeth. In Miocene deposits all along the eastern Pacific, from Baja California to Unalaska in the Aleutians, the teeth of the desmostylid known as *Cornwallius* have been found. As recently as 1986, a new genus named *Behemotops* was described from a massive tusk, lower jawbones, and teeth that were found off the Oregon coast and on the Strait of Juan de Fuca, on the north shore of the Olympic Peninsula in Washington. The *Behemotops* fossils date from the Late Oligocene, about 30 million years ago, and because they are the oldest and most primitive of the desmostylids, they are consid-

Reconstruction of the desmostylian Paleoparadoxia, *based on fossil skeletal material from North America and Japan. The desmostylians might have been fully aquatic or semiaquatic, but the bones do not tell us how it lived. Take a good look at the teeth.*

ered ancestral to the later forms. Domning, Ray, and McKenna named the genus for the biblical Behemoth, which means "a great beast." *Behemotops emlongi* was named for the late Douglas Emlong, a well-known fossil collector from this region, who found the fossils of the species that now bears his name, as well as all of the material of the other species.

In 1965, Charles A. Repenning supervised the reconstruction of a skeleton of the appropriately named *Paleoparadoxia*—because scientists aren't sure how it lived—that he had found during the excavations in preparation for the construction of Stanford University's linear accelerator in Palo Alto, California. Repenning's much reproduced drawing shows the skeleton positioned like that of a sea lion, but with its forelegs arranged in such a way that it appears to be walking on the back of its hands (modern sea lions walk on the *palms* of their hands). In a later analysis, Repenning decided that it probably didn't walk on land, and possibly it never came out of the water at all. The skeleton exhibited at the Stanford Linear Accelerator is mounted in a swimming position*, with its large, possibly webbed, forefeet paddling in polar bear fashion; thereby avoiding the question of how it might have walked on land, because it probably didn't.

Along with the Stanford *Paleoparadoxia,* Repenning found many shark's teeth,

---

* Actually, it's a plaster cast, as is common in many museum exhibits. Nowadays, casts are usually made of fiberglass or epoxy, but many of the older ones are plaster painted to look like the real thing. The real fossils, too important to be drilled or otherwise mangled in the construction of a exhibit skeleton, are usually stored in museum collections so they can be studied by scholars. The real bones of Stanford's *Paleoparadoxia* are in the University of California Museum of Paleontology at Berkeley. In some museums, however, many actual fossils are carfully mounted and are visible to scientists and museum visitors. This is the case at the American Museum of Natural History in New York, which has one of the largest fossil collections in the world.

some of which he identified as belonging to *Isurus planus,* a large mako common in Pacific Miocene deposits. Both of the hind legs were broken and one tooth was embedded in a vertebra, which might have meant that the mammal was killed by the shark, but, as Repenning wrote, "There was no way of knowing if the animal had died first and was then scavenged." The rear legs of this individual had already been broken in some other form of accident before it died (Repenning and Packard 1990).

There are eight whole paleoparadoxian skeletons in the world, six in Japan and two in the United States, but because casts have been made from the bones and distributed to other museums, there are more desmostylians on exhibit than there are actual skeletons. The Stanford *Paleoparadoxia* bones are in the museum at Berkeley; casts are on exhibit at the Linear Accelerator Visitor Center and the Tokyo Natural Science Museum. A cast of *Paleoparadoxia tabatai* from Japan is on exhibit in the American Museum of Natural History in New York, where the label reads, "*Paleoparadoxia* is an advanced member of an extinct group of tethytheres called desmostylians. The name refers to the paradoxical features of the skeleton which make it difficult to interpret how the animal lived. *Paleoparadoxia* may have been primarily aquatic but it may also have used its 'pigeon-toed' feet to haul itself out on land as sea lions do." (The museum's label writers had a rather unusual idea of "pigeon-toed," because this animal is shown with its feet splayed *outward* in a stance not unlike the forefoot position in otariids when they walk on land.) A cast of the Japanese specimen is also on display in the Natural History Museum in London.

In his 1984 "Skeletal Restoration of the Desmostylians: Herpetiform Mammals," Norihisa Inuzuka of the University of Tokyo discussed in detail the previous restorations of *Desmostylus* and *Paleoparadoxia,* and then reassembled the fossil skeleton of the holotype of *Desmostylus mirabilis,* which was found at Keton on Sakhalin Island in 1933. This skeleton had already been the subject of two previous restorations, and in a complicated and technical essay, Inuzuka explains why all the restorations are wrong and then laboriously reconstructs the Keton skeleton with "the limbs in a lateral position like amphibians or reptiles." A photograph of this restoration shows the animal slung low to the ground, with its forelegs bent at a lizardlike angle (it was for this reason that he decided to call the desmostylians "herpetiform mammals"), and its weight on the tips of its splayed toes. Repenning believes that Inuzuka's restoration is the best so far, but the lizardlike arrangement of the forelegs is impractical and unnecessary, because Repenning says that the animal may never have left the water.

In 1985, L. Beverley Halstead, then at the University of Reading in England, took issue with Inuzuka's "herpetiform" posture, and submitted a critique to the journal in which Inuzuka's paper had appeared. Halstead wrote, "Contrary to every previous restoration, he proposed that the upper limbs were not held vertically as in all other mammals, but rather they projected horizontally and laterally in a basically reptilian manner." Halstead visited Inuzuka's laboratory at the University of Tokyo, and together they examined and photographed the bones of the Utanobori *Desmostylus* and put them through the motions that Inuzuka said showed that the animal stood with its legs held like those of a reptile. Halstead points out several perceived discrepancies in Inuzuka's interpretation, and quotes a sentence of Inuzuka's 1984 paper ("The desmostylian posture with limbs stretching laterally seems to be inefficient for support of weight or terrestrial locomotion, but is however, extremely stable") to show that Inuzuka actually agrees that such a posture is unlikely, and then writes, "The important point here is not merely that the reptilian posture is inefficient for the support of weight, but in the case of *Desmostylus* the bulk of the animal could not be physically supported on dry land in such a posture." (In a 1996 study, Inuzuka estimated that the living animals were about the size of adult white rhinos or hippos, which can weigh over 2 tons.) In the same issue Inuzuka responded with a point-by-point refutation of Halstead's critique, and ended up exactly where he started, saying that *Desmostylus* was a very strange large mammal that stood with its legs splayed like those of a reptile.

The first known desmostylid was *Desmostylus hesperus,* described in 1888 by Othniel C. Marsh from a skull and lower jaws found in Alameda County, near San Jose, California. Based on his correspondence with paleontologist Henry Fairfield Osborn, Marsh decided it was some sort of a sirenian. When another specimen was found in Japan in 1902, Yoshiwara and Iwasaki believed theirs was a kind of proboscidean. Osborn (1902) placed the California and the Japanese specimens in the monotypic genus *Desmostylus,* which he believed to be either a sirenian or a proboscidean. Subsequent authors reassigned the desmostylians to such diverse orders as Monotremata (platypuses and echidnas), Marsupiala (kangaroos and opossums), and Perissodactyla and Artiodactyla (odd- and even-toed hoofed animals), but they were usually assumed to be some sort of protosirenian. (The discovery of *Moeritherium* did nothing to alleviate the taxonomic discord, since it too seemed to be either an ancestral proboscidean or a sirenian, that may or may not have had affinities with *Desmostylus.*) In his 1953 "Diagnosis of the New Mammal Order Desmostylia," Reinhart defined the order Desmostylia as excluding their possibly too-distant relatives.

It has been suggested that the desmostylids—which probably deserve the oft-employed appellation "bizarre"—are somehow affiliated with proboscideans and sirenians, "but," as Barnes et al. wrote in 1985, "the details of these inter-ordinal relationships have yet to be clarified." In a recent analysis, Domning, Ray, and McKenna (1986) suggest that the desmostylians were amphibious herbivores that fed on marine algae and that the earlier genera depended to a large extent on plants exposed in the intertidal zone. The uncertainty about the desmostylids is exemplified by the name *Paleoparadoxia,* which can be translated roughly to mean "ancient paradox." These peculiar animals are in that very special limbo reserved for creatures whose affiliations, morphology, and lifestyle have been only provisionally deduced, and often only imagined.

THE CETACEANS

Because of their perfected adaptation to a completely aquatic life, with all its attendant conditions of respiration, circulation, dentition, locomotion, etc., the cetaceans are on the whole, the most peculiar and aberrant of mammals. Their place in the sequence of cohorts and orders is open to question and is indeed quite impossible to determine in any purely objective way.

—George Gaylord Simpson, 1945

Beginning with some carnivorous (or herbivorous) mammals that reentered the water, the cetaceans evolved into an enormously varied group that includes, with some of the larger sharks, the top predators of the seas. There are only 77 or 78 kinds of whales and dolphins, and over time they have filled some niches already occupied by sharks and colonized others that were previously vacant. Some cetaceans live over deep water, and some can even dive miles below the surface, but all must spend much of their lives near the surface, because their aquatic development has persistently retained one critical link to their terrestrial forebears: they all must breathe air. There are no bioluminescent cetaceans (there are no bioluminescent mammals of any kind), but there are cetaceans that have been categorized by some as the most highly developed animals on Earth. Evolution does not move toward more complexity, or perfect the designs for living, since it is obvious that many of the species long extinct were as complex and "advanced" as any alive today. But when we consider the diversity of the cetaceans that now range in size from 5 to 100 feet, and include little fish-eaters; coastal or riverine creatures that are exquisitely attuned to sound; deep-divers that can hunt squid in the inky, icy depths; fast-moving ocean rangers that live in huge schools; and 100-ton giants that have achieved a body mass attainable only by creatures rendered neutrally buoyant by the displacement of the water they live in, we can get an inkling of the incredible subtlety and richness of the evolutionary process. From a terrestrial hoofed hyena, a miscellanea of sleek water-dwellers has evolved, breathing from the top of their heads, communicating with one another in ways that we have not been able to decode or understand, living in a watery world so alien to us we that we can barely enter it without benefit of artificial aids, such as face masks, snorkels, scuba gear, or submersibles.

Charles Darwin was as puzzled as anyone else on the subject of whale evolution. He read of a bear swimming through the water with its mouth open, catching insects, and from that came to the rather surprising conclusion that bears could, over time, develop into whales. In the first edition of *The Origin of Species* he wrote, "I can see no difficulty in a race of bears being rendered, by natural selection, more and more aquatic in their structure and habits, with larger and larger mouths, till a creature was produced as monstrous as a whale." He removed the bear story from later editions, but he seemed to regret having done so, and later wrote that he saw "no special difficulty in the bear's mouth being enlarged to any degree useful to its changing habits."

Recognizing that whales had managed to develop an entirely unique feeding mechanism, Darwin also reflected on the development of baleen, the filtering plates that hang from the upper jaws of some whales. In his response to the objections to the theory of natural selection as elucidated by St. George Mivart, a clergyman who generally supported Darwin's theories but was unhappy about the absence of God in the explanations, Darwin suggested that baleen had arisen from some earlier, non-baleen structure. Mivart then wrote that if the baleen "had once attained such a size and development as to be at all useful, then its preservation and augmentation within serviceable limits would be promoted by natural selection alone. But how to obtain the beginning of such a development?" Darwin answered by comparing the baleen to the lamellae [strainers] in the beak of a duck: "Whales, like ducks, subsist by sifting the mud and water."* He then discusses at some length the beaks of various ducks and geese, to show how the variations might have developed, but then makes an unfortunate leap into a region totally unfamiliar to him, the physiology of whales. He writes, "The Hyperoodon bidens is destitute of true teeth in an efficient condition, but its palate is roughened, according to Lacépède, with small, unequal points of horn. There is, therefore, nothing improbable in supposing that some early cetacean form was provided with similar points of horn on the palate, but rather more regularly placed, and which, like the knobs on the beak of a goose, aided it in seizing or tearing its food."

The "Hyperoodon bidens" is the bottlenosed whale (*Hyperoodon ampullatus*), and the males do have teeth (*bidens* means "two teeth"). Lacépède was wrong when he says that the species is "destitute of teeth." Darwin therefore based his entire argument for the development of baleen on a misreading of Lacépéde, or more likely, a willingness to see in lumps on the palate the precursor of baleen plates.

In a much later, much more sophisticated analysis, Carl Zimmer came up with a similar conclusion. In *At the Water's Edge* (1998) he wrote, "A living cetacean could be a model for this final step in the transition to baleen whales: Dall's porpoise, a six-foot-long, white-flanked cetacean that lives in the North Pacific, is a toothed whale, but its minuscule teeth are completely surrounded by a horny set of gums that it uses to grip squid or fish. On a microscopic scale, this hard tissue is almost identical to baleen. Perhaps some early toothed mysticete 30 million years ago might have followed a similar path, developing gums as tough as nails for capturing some particular prey." Teeth are among the hardest (and sharpest) of bones, and it is extremely un-

* In fact, most baleen whales do not sift mud and water. Only the gray whale, probably unknown to Darwin, sifts organisms from the mud. The bowhead whale, the primary subject of Darwin's disquisition, uses its long baleen plates to strain small organisms from the water as it swims through them with its mouth partially open, allowing the water to pass out through the sieve created by the baleen plates while trapping the food items in the hairy fringes. This is a different procedure than that employed by the rorquals (blue whale, fin whale, minke), which take in a mouthful of food and water, close the mouth, and with the tongue, force the water out through the filter of the baleen, trapping the food items in the fringes.

likely that gum tissue would replace them as a device for capturing slippery prey.

Like many writers, Zimmer tries to explain evolution as if it were a rational process with recognizable goals, but it is difficult, if not impossible, to determine why certain animals evolved the way they did. In 1980, Stephen Jay Gould wrote, "Evolutionary biology has been severely hampered by a speculative style of argument that records anatomy and ecology and then tries to construct historical or adaptive explanations for why this bone looked like that or why this creature lived here. These speculations have been charitably called 'scenarios'; they are often more contemptuously, and rightly, labeled 'stories' (or 'just-so stories' if they rely on the fallacious assumption that everything exists for a purpose). Scientists know that these tales are stories; unfortunately, they are presented in the professional literature where they are taken too seriously and literally." Bjorn Kurtén (1924–1988), a paleontologist, wrote in an essay published in 1991, "Evolution is like poetry; it explodes into forms and colors that are always new, without any overlapping goals or meaning, always individualistic, always shaping and reshaping; it creates, it plagiarizes, it annihilates."

The three major groups of cetaceans, the baleen whales (Mysticetes), the toothed whales (Odontocetes), and the extinct whales (Archaeocetes), have more similarities than differences. They all have (or had) dense ear bones, space around the bones for fat deposits, air sacs to isolate the ear from the skull, a long palate, and nostrils located on top of the snout. They all have a long body and a short neck (often with the cervical vertebrae fused), reduced or completely absent hind limbs, paddle-shaped front limbs, and a long tail designed to move up and down. All the living whale species are completely aquatic and perform all their activities in the water, including feeding and giving birth to live young. In recent years, more ancestral "whales" have been identified. Some of them appear to have been at least partially aquatic and some were terrestrial, and this has led to a certain degree of confusion and controversy about how whales developed.

The oldest whale fossils are from the Middle Eocene, 49 million years ago. Though it cannot be proved that whales did not exist before then, it is currently believed that whales arose around that time, descending from terrestrial mammals known as mesonychids (also considered to have been the ancestors of modern ungulates). The recent anaylsis of postcranial fossil material (Zhou, Sanders, and Gingerich 1992; O'Leary and Rose 1995) has clearly shown that mesonychids were hoofed, omnivorous running animals designed for endurance rather than speed. In 1968, Leigh Van Valen, then a research associate at the American Museum of Natural History in New York (and now at the University of Chicago), wrote that "only two known families need to be considered seriously as possibly ancestral to the archaeocetes and therefore to the ancestral whales. These are the Mesonychidae and Hyaenadontidae."

Briefly, mesonychids (meaning "middle claw") were carnivorous mammals with hooves, and hyneodontids were creodonts, only distantly related to modern carnivores. Although it seems counterintuitive, mesonychids are considered strongly rooted in the ancestry of whales. The mesonychids lasted for about 20 million years, from the Middle Paleocene to the Oligocene. They are among the most enigmatic of fossil mammals, not only for their unexpected appearance in the genealogy of whales, but also for the confusion about what they actually were. Edward Drinker Cope published a description of an early mesonychid called *Pachyaena* in 1884, saying that its hind legs were higher than its forelegs; it was a good swimmer; and its teeth appeared to be designed for catching fish—all of which were probably incorrect. Cope also said that when it walked, its soles rested flat on the ground—the type of gait known as plantigrade and seen today in bears. But when Leigh Van Valen examined the *Pachyaena* fossils in the 1960s, he realized that they were archaic ungulates, not carnivores.*

The mesonychids are now classified as ungulates, but they did not behave at all like today's herbivorous ungulates. The mesonychids were *carnivorous* ungulates that resembled big-headed dogs or skinny bears. Probably the best-documented is *Mesonyx,* about the size of a wolf with a disproportionately large skull equipped with powerful jaws and teeth. Where canids have claws, the mesonychids had hooves, one on each digit. They ranged in size from the fox-sized *Hapalodectes* to the bear-sized *Pachyaena,* all the way up to *Andrewsarchus,* a 13-foot-long mesonychid with huge crushing and tearing teeth—the largest carnivorous mammal that ever lived on land.

Andrewsarchus mongoliensis, "the giant mesonychid of Mongolia," described by Henry Fairfield Osborn in 1924, was first thought to be a huge omnivorous pig. (Osborn was a paleontologist and later president of the American Museum of Natural History.) When the skull reached the American Museum in New York, the vertebrate specialist W. D. Matthew immediately recognized it as a gigantic version of *Mesonyx.* In his awed description, Osborn wrote, "This is the largest terrestrial carnivore which has thus far been discovered in any part of the world.** The cranium far surpasses in size that of the Alaskan brown bear (*Ursus gyas*), which, when full grown, weighs 1,500 lbs; in length and breadth of skull, *A. mongoliensis* is double *Ursus gyas* and treble the American wolf (*Canis occidentalis*)."

The first thing you notice in the reconstructions of the mesonychids is the size of the head. *Mesonyx* was about the size of a wolf, but its head was as big as a bear's. The gigantic skull was probably the reason that Osborn thought it was a much larger animal than it actually was. The mesonychids were probably hyenalike in their behavior, scavenging everything they found, from dead mammals to dead fish. Somehow, these terrestrial scavengers that looked like big-headed hyenas metamorphosed into short-legged semiaquatic animals that spent half their time in the water and half out. The carnivorous, terrestrial mesonychids are usually placed in the ancestry of whales, but they do not reside there comfortably.

In his 1998 summary of the paleobiology of cetaceans, Philip Gingerich ob-

---

* A hoof is a modified claw that consists of a dorsal, scalelike plate, known as the unguis (Latin for "fingernail") and a softer ventral plate, the subunguis. The unguis is usually curved so that it encloses the subunguis laterally, but the ventral portion of the digit—the part the animal walks on—is not covered or enclosed.

** When Osborn described *Andrewsarchus mongoliensis* as the "largest terrestrial carnivore which has thus far been discovered," either he or the proofreaders left out a critical word. It was (and still is) the largest *mammalian* terrestrial carnivore, but the dinosaur *Tyrannosaurus rex,* which had been described by Osborn himself in 1905, was 40 feet long compared to *Andrewsarchus*'s 13. Where the mesonychid's skull was 35 inches long, that of *Tyrannosaurus* was 52 inches long, and the skull of *Carcharodontosaurus,* another gigantic carnivorous dinosaur, was even longer. The skull of the sperm whale, the largest carnivore that has ever lived, can be 18 *feet* long.

Today's whales are believed to be descended from land mammals that looked something like this reconstruction of Mesonychus.

served that there are conflicting theories about the rise of whales. Although the details are far from clear, said Gingerich, "most authors now accept as a working hypothesis Van Valen's idea that Mesonychia gave rise to Archeoceti." Because of the similarity of mesonychid teeth to those of primitive whales, it is accepted that the mesonychids are ancestral to whales and dolphins. Of the early cetaceans, Rice (1998) wrote, "Some of them are so similar to mesonychids that it is difficult to decide whether to call them mesonychids or cetaceans." Philip Gingerich (1998) compares the form of mesonychids to dogs, and writes, "Mesonychians are usually interpreted as solitary carrion feeders and scavengers that spent many of their waking hours trotting in search of dead animals and were best able to chew flesh after it was partly decomposed. This is plausibly the kind of animal from which the archaeocetes evolved."

Like the process it describes, the story of the evolution of whales occurred in stages. In the first stage, all accounts of cetacean evolution were devoted to legless whales. The first-found of the archaeocetes, as they came to be known, *Basilosaurus* reached a length of 70 feet and had a small head with a mouthful of sharp teeth. It might also have had horizontal tail flukes and a dorsal fin, and in 1990 evidence was found that it *did* have small hind legs. The fossil record seemed devoid of any creature with legs that could have become a whale; all evolutionary biologists knew was that there were extinct whales like *Basilosaurus,* and then there were fossil whales that were directly referable to living cetaceans. Whales and dolphins are descended from land mammals, and they have the anatomy to prove it. They are warm-blooded, have lungs that they use to breathe air, they suckle their young, and have rudimentary facial hair. In addition, the flippers of cetaceans contain the characteristic five-fingered *manus,* and many large whales have small hind-limb bones deep within their bodies.

Because it was obvious that whales were descended from terrestrial mammals, and it was equally clear that they could not have developed spontaneously, the absence of any transitional forms made cetacean evolution the most enigmatic and troublesome of studies. Nigel Bonner, writing in 1980, said, "The origin of cetacea, like so many orders of mammals, is shrouded in mystery. The earliest remains that can be recognized as derived from cetaceans appear in the middle Eocene, some 40 million years ago. These animals, the archaeocetes [like *Basilosaurus*], so differ from existing whales that it sems unlikely that they can be regarded as ancestral to either of the two surviving lines, the toothed whales (Odontoceti) and the whalebone, or baleen, whales (Mysticeti)."

Basilosaurus, the earliest fossil whale, was discovered in Louisiana in 1828 by Richard Harlan. At first it was thought to be a lizard (the name means "king of the reptiles," from the Greek *basilikos,* "royal"). Nine years later, Richard Owen realized it was a mammal and renamed it *Zeuglodon* in reference to its double teeth, from the Greek *zeugos* for "pair," but the rules of scientific nomenclature dictate that the first name has priority, no matter how misguided. This early whale reached a length of 70 feet, but it was considerably less bulky that the leviathans of today. As reconstructed from numerous fossils found in Louisiana, Mississippi, Alabama, Australia, and Egypt, *Basilosaurus* was a stretched-out teardrop of an animal, with a proportionately tiny head, greatly elongated vertebrae, powerful foreflippers, and hip bones that supported tiny, fully formed hind legs. It had a mouthful of differentiated teeth, cone-shaped in the front of the jaw for catching its prey, and multicusped molars with double roots ("zeuglodont") in the rear of the jaw.

The first *Basilosaurus* specimens, found and inappropriately named by Harlan, were believed to have no hind legs. A second species (*Basilosaurus isis*), found by Gingerich, Smith, and Simons (1990) in Egypt, was equipped with tiny complete hind legs, but they were so small that they must have been useless. Puzzled by the tiny, inadequate limbs, Gingerich and his colleagues wrote, "Hind limbs of *Basilosaurus* appear to have been too small relative to body size to have assisted in swimming, and they could not possibly have supported the body on land." In their discussion of the hind limbs, Gingerich and his colleagues suggested that "the hind limbs of *Basilosaurus* are most plausibly interpreted as accessories facilitating reproduction." Annalisa Berta (1994), however, wrote that they "could just as reasonably be interpreted as vestigial structures without a function."

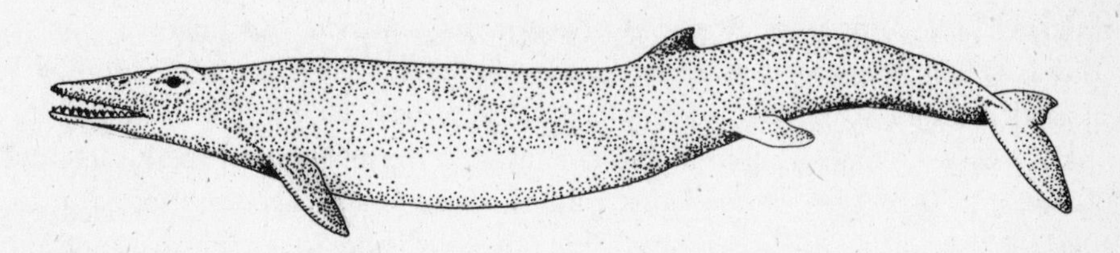

Because cetacean dorsal fins have no bones, they usually do not fossilize. This reconstruction of the 40-million-year-old Basilosaurus *has a dorsal fin, and also the little hind legs that were found in a recently discovered fossil.*

Reduced hind legs now appear to be an artifact of the evolution of four-legged animals to an aquatic existence, since they appear not only in early cetaceans but also in the marine reptiles such as ichthyosaurs, plesiosaurs, and mosasaurs. There are a few recorded instances where living whales were found with what have been described as "vestigial" hind legs, including a bowhead, a fin whale, a humpback, a sperm whale, and a striped dolphin. These atavisms are usually cited as evidence that whales are descended from terrestrial mammals—which they surely are, but the tiny bones embedded in the muscle of the pelvic region seem a much stronger argument than protruded hind limbs. After all, mammals are occasionally born with anomalous mutations—if a two-headed calf is born, nobody argues that cows are descended from animals that had two heads.

Some 50 million years ago, the Tethys Sea stretched from the Mediterranean east to the Arabian Sea and the Bay of Bengal, thus separating Africa and India (which was still an island) from Eurasia. It was probably in this area, some 40 million to 50 million years ago, that the condylarth mesonychids began to colonize the coastal fringes and swamps. Condylarth means "knuckle joints," and condylarth mesonychids are a primitive group of herbivorous and carnivorous mammals that includes the mesonychids. Some ancestral marine mammals survived the great K-T Extinction that saw the end of the dinosaurs and the marine reptiles such as mosasaurs some 65 million years ago, and these mammals might have exploited the niches abandoned by these predators. The archaeocetes lived in a world of tropical vegetation, swamps, and advancing seas. They had elongated bodies and were mainly aquatic, although there is some evidence that they might have been able to leave the water. The first archaeocetes were heterodont, meaning that their incisors, cuspids, and molars were differentiated. They had dense ear bones, long palates, nostrils on top of the nasals (nasal bones), space around the ear bones for fat deposits, and air sacs to isolate the ears from the skull. The body was elongated with a short neck (or none at all), a long tail, and reduced hind limbs. The forelimbs were paddle-shaped (retaining the bones of the mammalian hand), and there was evidently a point of flexion in the caudal vertebrae, allowing for up-and-down as well as side-to-side movement. They were probably not particularly deep divers, and may have occupied the shallower inshore waters.

Between 38 million and 25 million years ago, the archaeocetes became extinct and were replaced by the squalodonts (shark-toothed; the probable ancestors of the toothed whales) and the protomysticetes, aetiocetids, and cetotheres. The most primitive of the odontocetes still had heterodont dentition, but this was soon replaced by long rows of similar teeth, the condition known as homodonty. The squalodonts were shark-toothed cetaceans whose behavior may have resembled that of killer whales. These 12-foot-long cetaceans were the ancestors of the Eurhinodelphinidae, long-snouted, homodont creatures whose fossils have been found in the Early and Middle Miocene deposits of Europe, North and South America, and Australia. They were about 6 feet long, approximately the size of some of today's smaller dolphins. The lower jaws of *Eurhinodelphis* were fully equipped with teeth, but the uppers, which extended far beyond the lowers, were toothless for that portion that overhung the lower and could contact no other teeth, unlike the modern river dolphins, such as *Platanista, Inia,* and *Pontoporia,* where the teeth fill the jaws to the tips and are obviously used for grasping prey. The teeth were simple pegs like those of modern dolphins, not serrated like those of the squalodonts. Its earbones suggest that it had the ability to echolocate, so it is not unlikely that *Eurhinodelphis* fed the way long-snouted dolphins do today, by chasing down and seizing fishes. The toothless overhang, however, remains a puzzle. What good are teeth if there is nothing to oppose them in a bite? Several popular authors (such as Dixon et al. 1988) have suggested that this early dolphin used its upper jaw the way a swordfish does, slashing at its prey to kill or stun it, but this is hard to picture.

In a 1998 study, Gingerich wrote, "We now know, thanks to the fossil record, that the modern orders Mysticeti and Odontoceti have a fossil record extending back to the Oligocene, and that they are thought to have diverged from each other sometime in the late Eocene or early Oligocene, no more than 40 million years ago." The oldest known whale fossil identified to date is *Himalayacetus,* described by Bajpai and Gingerich in 1998. They found only a partial jawbone with teeth, but the animal is said to be an archaeocete because its molars closely resemble those of the mesonychids, and there is a small canal in the mandible that suggests underwater hearing. *Himalayacetus,* which is 50 million years old, was found in the Simla Hills of northern India, which lay under the Tethys Sea when India was an island off the Asian continent. *Himalayacetus* was probably seal-like in size and habits and is believed to represent the beginning of the transition from terrestrial mesonychid to aquatic archaeocete. It is, however, more than a little difficult to postulate its complete life history and affinities from a single jawbone.

When a fossil later named *Pakicetus* was first discovered in Pakistan in the Himalayan foothills in 1979, it was hailed by paleontologists as the missing link between land mammals and whales. Since all that was found was a part of the skull, the appearance of the rest of the animal was not obvious, and although Gingerich and his colleagues named it and described it in 1983 as an ancestral whale, there was no way of telling whether it had hind legs, flippers, or flukes. Its teeth were not unlike

Originally conceived as a sort of giant, long-jawed otter, Pakicetus, *the ancestral whale from Pakistan, has now been reconfigured as a long-legged terrestrial creature related to pigs, goats, and camels.*

those of the giant mesonychid *Andrewsarchus,* and it was probably a completely terrestrial animal. The find also included earbones, and since they were similar to those of a known archaeocete, it was concluded that the animal had been a protowhale, because it could probably hear well under water—a characteristic that would later come to define the cetaceans. Some 50 million years ago, the region where it was found was under water, so even though the skull fragments were found in conjunction with the fossils of terrestrial mammals, Gingerich and his colleagues concluded that it might have been somewhat aquatic. The latest reconstructions have it looking like a seal with a long tail, or perhaps an otter. There is now almost complete unanimity on *Pakicetus*'s place in the lineage of cetaceans, and since its discovery, more fossil whales have been found in the same location.

Found later, but believed to have lived at the same time, is the archaeocete *Nalacetus,* named by Thewissen and Hussain (1998) for the Urdu word *nala,* which means "seasonably dry riverbed," where the fossil was found in the Kuldana Formation of Pakistan. As shown by measurements of the teeth (all that was found of this species was a mandible with fragmentary teeth), *Nalacetus* was intermediate in size between the other pakicetids, *Ichthyolestes* and *Pakicetus,* and, like them, dates from the Early Eocene, approximately 49.0 to 49.5 million years ago. It is therefore considered among the oldest of all archaeocetes. In O'Leary and Uhen's 1999 review of the origin of cetaceans, they say, "The fragmentary fossil *Nalacetus* was found to be the most primitive cetacean, based on improved understanding of character transformation in the dentition of cetaceans."

Indocetus ramani was found by Gingerich and colleagues after the *Pakicetus* fossils but again placed earlier in the sequence. Only the leg bones were found, but they were clearly from an early whalelike animal, and their large size encouraged Gingerich et al. (1993) to erect the new genus *Indocetus* and place it between *Pakicetus* and *Basilosaurus.* They speculated "that *Indocetus,* like *Pakicetus,* entered the sea to feed on fish, but returned to land to rest and to birth and raise its young."

Under closer inspection (Gingerich et al. 1995), *Indocetus* was renamed *Remingtonocetus harudiensis* (for the American cetologist Remington Kellogg), a species for which the skull was already known. *Remingtonocetus* had originally been described by Kumar and Sahni in 1986 from fossils found in western Kutch on the India-Pakistan border. The remingtonocetids are 49 million to 43 million years old and are characterized by a narrow, elongated skull with correspondingly long and narrow mandibles filled with sharp teeth. Kumar and Sahni wrote that the overall dentition of *Remingtonocetus* was well adapted for a dominantly carnivorous food habit. Its premolars probably served both purposes.

Fordyce and Barnes (1990) described the "bizarre" remingtonocetids as "a short-lived archaeocete clade characterized by a long, narrow skull and jaws with cheek teeth placed relatively far forward of the eyes." Because *Remingtonocetus* and *Platanista,* today's Ganges River dolphin, are both shallow-water inhabitants with long, narrow, tooth-filled jaws, it seems not unreasonable to compare them. It is believed that the remingtonocetids died out without leaving any descendants, and the living dolphins, which live in the turbid, murky waters of the Indus and Ganges rivers, have greatly reduced vision, and specialized bony processes in the skull that enhance their echolocating abilities.

Again in Pakistan, in 1992, Hans Thewissen, a former student of Gingerich's, found a skull, ribs, and leg bones of a fossil that appeared to him and his colleagues, Madar and Hussain, to be an animal that might fall between the mesonychids and *Pakicetus.* The skull of *Ambulocetus natans,* "the walking whale that swims," was huge and the jaws powerful; it had strong forelegs, flipperlike hind legs, and a long, thick tail. Thewissen, Hussain, and Arif (1994) wrote that it "was the size of a male of the sea lion *Otaria byronia* (approximately 300 kg)," but its long tail accounted for much of its 12-foot length. The forelimbs were similar to those of a sea lion, and in the 1994 reconstruction it is shown with its forelimbs arranged splayed to the sides or even slightly tailward, as in sea lions.

A careful examination of the leg bones and vertebrae *Ambulocetus natans* suggested to the authors that it was not a fast swimmer and that it probably swam by moving its webbed hind feet up and down through the water like a sea otter. It didn't look much like a whale—in the reconstruction drawing in Thewissen et al. (1996) it looked like a giant web-footed shrew. Other illustrators have portrayed it as a sort of furry crocodile (rather as though the head of a crocodile had been affixed to the body of an otter), and it probably did feed crocodile fashion by ambushing its prey in the shallows. Also like a crocodile, its jaws and teeth were powerful, and its head was disproportionately large.

But it was clearly an important transitional form. As Thewissen and his colleagues noted:

> This kind of revolutionary change is common in the long history of life, as new morphologies were typically acquired early in the diversification of ma-

Ambulocetus *("walking whale") doesn't look very much like a whale, but paleontologists believe that today's cetaceans are descended from a creature that looked like this.*

jor clades. However, few of the transitional forms that arose are preserved in the fossil record. The origin of whales has become an exception to this rule; it is better documented than most other morphological transitions. As such it forms one of the best examples of evolutionary change across a strongly constraining environmental threshold.

The discovery of a whale that walked prompted many scientists to reexamine the very definition of the term "whale" as used in evolutionary biology. The living odontocetes and mysticetes are clearly whales: All are vertebrate mammals that are fully adapted to a marine existence. They do not walk because they have no legs and never leave the water. They all have flippers in place of forelegs and no visible hind limbs, and they propel themselves through the water by flexing their horizontal flukes up and down. They breathe air through single or paired blowholes on top of the head, and they give birth to live young that they nurse under water. (The only other completely aquatic mammals are the sirenians, the manatees and dugongs, which fulfill all of these criteria except they breathe through nostrils at the front of the head. They are descended from a completely separate line of mammals.)

When it comes to the modern whales' ancestors, definitions become more murky. In an article in *Science* that accompanied Thewissen, Hussain, and Arif's 1994 discussion of the first "walking whale," Annalisa Berta wrote:

> Whales can be defined in several different ways, emphasizing either possession of certain characters . . . or with respect to certain ancestry. . . . Character-based definitions are problematic. For example, how can whales be defined as lacking hindlimbs, since some [prehistoric] whales (for example, several archaeocetes) possess them? Another problem arises considering that discoveries of ostensible whales occur quite frequently with new combinations of characters making it difficult to decide whether they are whales following a strictly character-based definition. A more reasonable solution is to use a phylogenetic definition, that is, one based on a common ancestry. In the previous example, because archaeocetes are more closely related to modern whales than they are to mesonychids, *Ambulocetus* is a whale by virtue of its inclusion in that lineage. Evidence to support the inclusion of *Ambulocetus* in the whale lineage comes from derived characters it shares with modern whales. . . . Although its relationship with other whales is uncertain, *Ambulocetus natans* is a whale.

Almost all archeocete fossils have come from Asia. In the same subhimalayan area that produced *Pakicetus,* Gingerich and several Pakistani paleontologists unearthed a partial skeleton of a protocetacean they named *Rodhocetus,* for the Rodho Formation in the Punjab of Pakistan where the fossil was found. The skeleton measured approximately 6.5 feet long, but with the tail, the animal would have reached 10 feet. The structure of the vertebrae and pelvis indicated that *Rodhocetus* could walk on land, but there were also some characteristics, particularly the unfused vertebrae, that suggested a spinal flexibility usually associated with aquatic movement. No forelimbs were found and the terminal caudal vertebrae were also lacking. Also, the femur was considerably shorter than the corresponding hind-leg bone found in *Pachyaena,* the mesonychid believed to be ancestral to the cetaceans, leading the authors to write that they could not "access the possible presence of a caudal fluke, but it is reasonable to expect the development of a fluke to coincide with shortening of the neck, flexibility of the sacrum, and reduction of the hind limbs first observed in *Rodhocetus.*"

In the Cross Quarry of Berkeley County, South Carolina, workers found fossil fragments of a specimen that has not been named and is referred to as the "Cross Whale," provisionally placed in an intermediate position between *Rodhocetus* and *Protocetus,* the next stepping stone on the path. Another American archaeocete was found in Louisiana that is larger than *Rodhocetus* and *Protocetus.* It also has no scientific name, so it is "the protocetid whale of Uhen" (Williams 1998.)

In a 1998 review of the fossil marine mammals of the Sulaiman Range of Pakistan (the origin of *Rodhocetus* and *Remingtonocetus*), Gingerich et al. wrote,

> Protocetids appear to have been the most active swimmers, and to have been the ancestors of basilosaurid archaeocetes and through dorudontine basilosaurids, the ancestors of later modern cetaceans. The third stage of specialization of archaeocetes involved streamlining the body by shortening the neck and reducing the hind limbs, with the development of cetacean-style swimming locomotion involving "caudalization" of the lumbus and sacrum associated with dorsoventral oscillation of a heavily muscled tail.

When *Protocetus* (found in the Egyptian desert) came along 8 million years after the pakicetids, it was much more whalelike, with paddlelike forelimbs, functional hind limbs, and tail vertebrae that suggest the musculature necessary for supporting flukes. This creature had a long snout and teeth arranged in a zigzag pattern in the front of the jaws. Whereas the nostrils of *Pakicetus* and *Rodhocetus* were at the tip of the snout, those of *Protocetus* had begun to move rearward. Eventually, this modification would enable whales and dolphins to inhale and exhale without slowing down, but it is difficult to imagine the advantage of a slight transitional rearward movement of the breathing passages, unless the evolution of this character was somehow "designed" to become advantageous in the distant future. *Protocetus* had dentition that suggested that it chewed its food before swallowing, dense tympanic bullae (ear bones), and "fat windows" in the lower jaw like the ones known to be used for transmission of sound in modern dolphins.

One of the more difficult adaptations to explain in cetaceans is the horizontal tail flukes. The flukes consist of collagenous tissue attached to the numerous short caudal vertebrae and intervertebral disks. In simple terms, the flukes represent a unique, flattened appendage attached to the tail bones that is used for propulsion. As with many of the modifications that enabled cetaceans to return to the water, it is difficult to understand how natural selection could have produced a horizontally flattened instrument that is the whale's sole means of propulsion.*

Since the flukes have no bones, they do not lend themselves to fossilization, so there can be no record of half-formed flukes—if half-formed flukes could have served any useful purpose. They are not modifications of the hind legs, since many whales have vestigial hind leg bones, and some of the protocetaceans may have had reduced hind legs *and* tail flukes. In an analysis of the origin of cetacean flukes, F. E. Fish wrote (1998), "Although we understand the evolution of terrestrial locomotion because of available skeletal remains and footprints, no such record exists for swimming by cetaceans as no fossilized imprints of the flukes have been unearthed and the sea leaves no tracks."

What was the purpose of even a *slightly* flattened tail in a land animal? In their description of *Ambulocetus,* Thewissen and his colleagues wrote, "Like modern cetaceans, it swam by moving its spine up and down, but like seals, the main propulsive surface was provided by the feet. As such, *Ambulocetus* represents a critical intermediate stage between land mammals and cetaceans." Stephen Jay Gould, in his 1994 essay "Hooking Leviathan by Its Past," translates this leg motion into the development of flukes, concluding that the discovery of *Ambulocetus*'s swimming technique of " 'forward motion supplied primarily by extension of the back and subsequent flexing of the hind limbs' [was] directly responsible for the development of flukes." Thewissen et al. actually wrote, "*Ambulocetus* shows that spinal undulation evolved before the tail fluke. . . . Cetaceans have gone through a stage that combined hind limb paddling and spinal undulation, resembling the aquatic locomotion of fast-swimming otters." From that sentence, Gould somehow finds that "the horizontal tail fluke . . . evolved because whales carried their terrestrial system of spinal motion to the water." Gould apparently bases his conclusion on the propulsive actions of otters. He writes that they move in the water by powerful vertical bending of the spinal column in the rear part of the body: "This vertical bending propels the body forward both by itself (and by driving the tail up and down) by sweeping the hind limbs back and forth in paddling as the body undulates." In fact, otters move by paddling with their legs, not their tails, although there is an attendant flexion of the spine that causes the tail to undulate. Besides, no matter how or how much the spine undulates, flukes still are not modified legs but a modified *tail*—an adaptation that evolved separately in two kinds of marine mammals, the sirenians and the cetaceans. Cetaceans and sirenians were the only animals that ever developed flattened horizontal tails as a means of propulsion. This modification has turned

---

* There is, of course, a semiaquatic mammal with a greatly flattened tail that doesn't use its tail at all for propulsion. The beaver (*Castor canadensis*) uses its webbed hind feet for propulsion, and its paddlelike tail may be used as a rudder in the water, or as a prop when the animal sits upright on land to cut down a tree. Beavers also slap their tails on the surface to express alarm or fear. Like that of the beaver, the tail of the platypus (*Ornithorhynchus anatinus*) is also flattened, and furred where that of the beaver is naked, but the platypus also swims by paddling with its webbed feet.

out to be efficient for fast-swimming cetaceans, but a flattened, horizontal tail has provided the manatees and dugongs with little power and even less undulating flexibility.

After the important discovery of *Ambulocetus natans* in 1994, Thewissen and Fish (1997) tried to figure out how it moved on land or in water, and what this meant to the future development of whales. After a careful analysis of its structural morphology, they wrote that it probably did not paddle with four legs, did not move its tail up and down, and did not flex its pelvis. The length of the tail implied that "a fluke at the distal part of the tail would have a poor lever arm and would not be an efficient hydrofoil." And in any event, however *Ambulocetus* may have locomoted, Thewissen and Fish were careful to point out, "This does not imply that *Ambulocetus* was ancestral to all later cetaceans" or that "all early cetaceans swam similarly." *Ambulocetus,* a four-legged amphibian mammal, walked on land and probably swam like an otter, but it vouchsafed us little additional information on the swimming motions of the early or later whales.

In 1983, a fossil protocetid was discovered during excavations of Eocene deposits at the Georgia Power Company's Alvin W. Vogtle generating plant in Burke County, near Augusta. The three skeletons were the first formally described protocetid cetaceans from North America; all other species originated from either Africa or the Indo-Pakistani region (Hulbert 1998). The most complete of the specimens (designated as the type) consisted of a nearly complete skull, much of the left jawbone, 23 vertebrae, 12 ribs, and both halves of the pelvic bones. The vertebrae were similar to those of *Zygorhiza* but were appreciably smaller, and some of the double-rooted teeth (like those of *Basilosaurus*) were specialized for shearing. The whale was named *Georgiacetus vogtlensis,* in honor of the state (it was Georgia's first whale fossil) and the Vogtle power plant. It was about 40 million years old, and is the oldest known whale whose pelvis did not articulate with the sacral vertebrae, meaning that it could move its hind legs independent of the flexion of the backbone, an important step toward a fully aquatic existence. Basilosaurids are fairly common in the southeastern United States, but *Georgiacetus* is "the youngest, most aquatically adapted protocetid for which postcranial and skeletal elements are associated with a skull," according to Richard Hulbert, who published the first description. He wrote, "*Georgiacetus* neatly (for the most part) fills the morphologic gap . . . between early protocetids (such as *Rodhocetus*) and basilosaurids." The find was particularly important, because it showed that the archaeocetes were not restricted to India and Pakistan.

In 1994, Mark Uhen of the Cranbrook Institute in Bloomfield Hills, Michigan, was examining the U.S. National Museum's collection of fossil cetacacans when he found a collection of bones that had been collected in Natchitoches Parish, Louisiana, and had not described before. He named this new species *Natchitochia jonesi.* Unlike *Georgiacetus,* it had a direct connection between its sacral vertebrae and pelvis, which meant that it was not particularly flexible, and probably swam by kicking its hind legs.

The archaeocete genus *Dorudon,* named in 1845 by Robert Gibbes of South Carolina, from jawbones, cranial fragments, and caudal vertebrae of a specimen found "in a bed of Green sand near the Santee Canal," was smaller than *Basilosaurus,* but similar in shape and habits. Having hydrodynamic streamlining, a shortened neck, reduced hind limbs, and a powerful tail but lacking the strangely elongated vertebrae of the basilosaurs made them what Gingerich (1998) called "good candidates for the ancestry of modern whales." Another, similar species known as *Dorudon atrox* (previously known as *Prozeuglodon atrox*) flourished in the shallow seas that covered what is now northeast Africa some 40 million years ago. One of the many skeletons collected from the marine deposits at Wadi Hitan in the deserts of western Egypt was particularly well articulated, and it was prepared for exhibit at the Exhibit Museum of Natural History of the University of Michigan at Ann Arbor. Under the supervision of Mark Uhen, who studied under Gingerich and wrote his doctoral dissertation on *Dorudon atrox,* a mold of the fossil was made and the 19-foot-long skeleton was put on exhibit in the museum.

Like *Basilosaurus, Dorudon* had a streamlined shape, paddles for forelimbs, and tiny hind limbs, but it was only 20 feet long to *Basilosaurus*'s 70, and fossils of both species have been found together. Another early archaeocete was *Zygorhiza,* serpent-like in form and found on the Atlantic coast of North America. When Barnes and Mitchell published reconstructions of *Basilosaurus* and *Zygorhiza* in 1978, they gave both species tail flukes that are not notched, and therefore look not unlike the flukes of a beaked whale. They wrote:

> In most modern mammals with tails modified for aquatic propulsion, the tails are expanded in the transverse plane. There is clear-cut evidence for horizontal caudal flukes on the tail based on analogy with modern cetaceans. This analogy depends on the abrupt change in length versus diameter and the change from flat-faced to round-faced vertebral centra within the tail. . . . Our restoration shows that while the osteological evidence only shows the length of the base of the caudal flukes, if a wide caudal fluke of about the same proportions as that on a large balaenopterid whale is restored in a basilosaurine, it makes a reasonable appearing propulsive organ, even though the base of the fluke is much shorter.

In other words, we don't know whether the basilosaurines had flukes, but because modern whales propel themselves this way, and this was a certainly a whale, we'll put flukes on our reconstruction. (Not a word about how the tail might have grown these attachments, however). The beaked-whale analogy is even stronger in these reconstructions when we look at the body proportions of these basilosaurines, particularly *Zygorhiza.* The authors have added a dorsal fin (probably for the same reason they added flukes (it is useful for swimming, and most modern cetaceans have them), so that the animal looks quite like a beaked whale, but with more teeth. With this body plan, *Zygorhiza* looks as if it might have been the ancestor of the beaked whales, but there is not enough material on primitive ziphiids to make the necessary comparisons.*

Almost all living beaked whales are deepwater species, and it is not unreasonable to suggest that their ancestors were too. Deep-ocean sediments, although plentiful, do not yield many fossils, largely because they are so inaccessible. It is probably the exception rather than the rule that the body of a marine mammal is preserved at all, because those that die in the open sea are likely to be fragmented by scavengers like sharks, and the most propitious circumstances for fossilization occur when an animal is stranded or dies close to shore and is washed up on the beach. (Over the course of their development, continents separated and moved around, and some were submerged, but by 40 million years ago, at the time the first whales appeared, the continents were arranged much as they are today.)

Some paleocetologists believe that modern odontocetes and mysticetes had different origins, and they cite the symmetrical skull, vestigial femur, paired blowholes, and no teeth as characters that only the mysticetes possess. Others, like Lawrence Barnes, Leigh Van Valen, and David Gaskin (quoted in Barnes 1984), believe that both groups of living whales had a common archaeocete ancestor, and according to Barnes, "When we look at fossil cetaceans of the Oligocene, about 25 [million] to 30 million years ago, we find odontocetes and mysticetes that are remarkably similar to each other and to archaeocetes. They represent exceptionally good intermediate stages." An example cited by Barnes is *Aetiocetus* from the Oligocene of Oregon, which had many features of baleen whales, such as a loosely articulated lower jaw, but also had a full complement of mammalian teeth. Many cetologists disagree, however, and believe that *Aetiocetus* was the ancestor of the mysticete whales. In her 1994 "What Is a Whale?" Annalisa Berta wrote, "Archaeocetes are a 'scrapbasket' group of extinct Eocene whales that together with the two living whale lineages, the toothed whales (Odontoceti) and the baleen whales (Mysticeti), comprise the mammalian order Cetacea."

In 1883, William Flower, a famed British cetologist, published a discussion he entitled "On Whales, Past and Present, and their Probable Origin," in which he suggested that cetaceans were the direct descendants of hoofed mammals known as ungulates. When Flower's thesis was published, it was generally derided, because anyone could see that whales and dolphins bore no resemblance to cows, horses, and pigs, but in recent years the idea has not only been revived, but it is now accepted by at least a portion of whale evolutionists. Those who follow the traditional route of classifying animals on the basis of similarities in bones and flesh are inclined toward the mesonychians as the ancestors of whales, since there appears to be a trace-

* In a 1989 essay on the beaked whales, J. G. Mead wrote: "The evolution of the entire family of beaked whales is unclear. They suddenly appear in the lower Miocene, and are well represented in the upper Miocene. One of the problems with dealing with fossil ziphiids is that usually the only material available are the remnants of rostra. It still remains for a system to be worked out whereby the information present in the rostral fragments can be utilized. It is my firm impression that the Miocene was the heyday of ziphiid evolution, and that the family has slowly decreased in diversity since."

able pattern leading from these terrestrial carnivores through the protocetids, the basilosaurids, and right up to the modern whales and dolphins.

Recently, however, this line of thinking has come under attack. Molecular biologists, using new techniques of DNA analysis, have suggested that whales are actually artiodactyls (even number of digits), and are closer to hippos, camels, and pigs than they are to the hyenalike mesonychids. Hans Thewissen, who discovered the walking whale *Ambulocetus,* has said that the teeth of the pakicetids are not as highly specialized as those of the mesonychids, thus the mesonychids came later—and it is unlikely that whales are descended from a group that came later. Thewissen edited a book, *The Emergence of Whales* (1998), that includes an article by Milinkovitch, Bérubé, and Palsbøll unequivocally titled "Cetaceans Are Highly Derived Artiodactyls," in which they said that "the close relationship between cetaceans and ungulates (especially artiodactyls) was strongly supported by basically all molecular data published to date, from immunological studies, to analysis of mitochondrial and nuclear amino-acid and DNA sequences." Living ruminants appear to be the closest living relatives of whales (the mesonychids, of course, are extinct), but the fossil record has not revealed anything resembling a common ancestor. If the mesonychid connection is rejected, it might be that the whales are descended from an artiodactyl ancestor, and that hippopotamuses are their closest living relatives.

Also in *The Emergence of Whales* is a contribution by the molecular biologist John Gatesy entitled "Molecular Evidence for the Phylogenetic Affinities of Cetacea." It starts: "Recent phylogenetic analyses show more conflict than compromise between molecules and morphology." Gatesy then reviews the previous phylogenetic and molecular hypotheses, which show whales related to everything: mesonychids, artiodactyls, perissodactyls, lagomorphs, insectivores, rodents, carnivores, and subungulates (the hyraxes, sirenians, and elephants). In molecular analyses, DNA and tissue samples are taken from living animals and, through an elaborate computer program, checked for matches—the more matches, the closer the animals are said to be on the molecular level. Because these tests cannot be conducted on fossils, they are useful only for living animals, and therefore the mesonychids cannot be figured into any of these calculations. Various authors have used the similarity of some mesonychian characteristics to those of ancestral whales to combine the two groups into a single clade known as "Cete."

The idea that whales and hippos share a common lineage—sometimes called the "whippo" hypothesis—may be obvious to some molecular biologists, but the paleontologists who rely on the interpretation of fossils have a different idea. In a letter to *Science* (1999a), Maureen O'Leary wrote, "Hotly debated areas of genuine inconsistency do persist between molecular and morphological phylogenetic analyses of cetaceans—in particular, on the issue of artiodactyl monophyly," and in an interview in *Science News* (Monastersky 1999d), O'Leary said, "I think that's an unfair dismissal of the paleontological data, and it's just unwise not to consider the fact that the fossil record is in contradiction to the molecular data." Artiodactyls have an ankle bone called an astragalus, which enables them to flex the foot up and down and avoid twisting it while bounding over rough terrain, but mesonychians have no such bone. If an astralagus could be found in a whale, ancient or modern, it would confirm the whippo hypothesis, but such a bone has not been found in any of the fossil cetaceans, and modern cetaceans are notoriously deficient in ankle bones. According to Monastersky's report (1999d), "Gatesy and O'Leary are just starting out on a massive phylogenetic project that will combine genetic information with morphological data describing anatomical structures in fossils and living species. Once they put all the details into a database, a computer program will sort out the possible family trees relating the different animals." But O'Leary says, "There are some intricate problems with combining the molecular and morphological data into one analysis."*

In July 2000, Trisha Gura wrote an article for *Nature* that she called "Bones, Molecules . . . or Both?" in which she noted: "Systemacists have long classified even-toed hoofed mammals (artiodactyls), as a group, or 'clade,' that could be further broken down into four extant subclades—camels and llamas; cattle and deer; pigs and peccaries; and hippopotamuses. Modern cetaceans—whales, dolphins and porpoises—don't have toes, but early fossil whales did have even-numbered appendages. The teeth of ancient cetaceans look almost identical to those from an extinct group called mesonychids [so] morphologists have classified the cetaceans and mesonychids together as a sister group to the Artiodactyla."

After comparing certain molecular markers, however Japanese workers concluded that hippos are more like whales than other artiodactyls, which suggested that the artiodactyls were not a natural group. Said Maureen O'Leary, "We might be able to say that carnivorous whales were derived from some fully herbivorous animal like ancient hippos." Summing up the differences, Hans Thewissen said, "We just don't know enough to determine who is right. . . . From the morphological side we have to be smarter about how we pick the animals we analyze and go out and get better fossils."

There is a living odontocete in which the dental arrangement is so bizarre that if the animal were known only from fossils, it would certainly be regarded as one of the weirdest creatures that ever lived. Imagine a small whale with only two canine teeth in its mouth, one of which is tiny and does not erupt and the other of which is a spiral ivory horn that grows through the upper lip, sticks straight out in front of the whale, and may be 8 feet long. It is, of course, the narwhal (*Monodon monoceros*) of Arctic waters, long known to Inuit hunters as a source of meat and blubber and to Europeans as a source of the fabulous ivory tusks, which have been carved into walking sticks, used to build the coronation throne of Denmark, made into drinking cups because they were said to render the drinker immune to poison, and were somehow responsible for the myth of the unicorn. (The horns of the unicorns in the unicorn

* In an August 6, 2000, review in the *New York Times Book Review* of Tattersall and Schwartz's *Extinct Humans,* Richard Dawkins commented on the "whippo" hypothesis: "A group of molecular taxonomists has recently claimed that hippopotamuses are closer cousins to whales than they are to pigs. This is an astonishing assertion. A world in which it was true would be meaningfully different from the world in which zoologists had hitherto thought they were living. . . . But if the molecular claim about hippos and whales is upheld, it will be as though we had suddenly discovered that humans are closer to bushbabies than they are to chimpanzees. Now that would really be revolutionary. It would mean a difference in the real world."

The fantastic Odobenocetops, *with the face of a walrus, the body of a dolphin, and a single tusk that extended rearward, lived some 3 million years ago.*

tapestries at the Cloisters in New York and the Musée de Cluny in Paris look exactly like narwhal tusks.)

A *fossil* whale from the Pliocene of Peru is even stranger than the narwhal. It has been christened *Odobenocetops*; *Odobenus* is the walrus, and means "tooth walker," and *cetops* means "whale-like" in Greek, so *Odobenocetops* is a "cetacean that walks on its teeth." It was found in the Pisco Formation of Peru in 1990 and was described by Christian de Muizon, a paleontologist who was then director of the French Institute for Andean Studies in Lima. (Muizon is now director of research for the National Center for Scientific Research in Paris.) His original report (1993a) sounded more like instructions for building a jabberwocky than a description of a fossil whale skull.

Muizon told of a skull that was found that had sockets for backward-pointing tusks, no other teeth, and muscle scars that indicated a prehensile upper lip that was probably used for suction feeding. This would be a pretty good description of a walrus skull, except this wasn't a walrus, it was a whale. ("The structure of the face and basicranium indicates is was a delphinoid cetacean.") The first fossil consisted only of a single skull but it was so unlike any other cetacean or pinniped skull that Muizon placed it in its own family, the Odobenotocetopsidae. He speculated on its lifestyle: "It is likely that *Odobenocetops* fed upon benthic invertebrate fauna (bivalve and gastropod mollusks and/or crustaceans) abundant in the Pisco formation at Sacaco. When searching for food on the bottom with the body in an oblique position, the dorsally oriented orbit would have eventually provided a fairly good anterodorsal binocular vision and compensated for the poor echolocation ability." As in the living walrus, the tusks possibly served a primarily social function, and were not used as digging implements.

In describing *Odobenocetops,* even a scientist like Muizon has difficulty keeping the amazement out of his voice as such an incredible creature begins to take shape out of the rocky shales of the Pisco Formation. Of the 1990 find, he used the words "startling" and "exceptional" (1993a) and when three new specimens were uncovered in 1991, he referred to them as belonging to "the astonishing walrus-convergent delphinoid cetacean from the Pliocene of Peru" (1993b). Because of some differences in skull morphology, Muizon placed one of the new fossils in a separate species, which he called *Odobenocetops leptodon* ("slender tooth"). The skull of *O. leptodon* had both tusks in place; the right one was 52 inches long and the small left one was 9.95 inches long. Another skull of *O. peruvianus* was also found with two small tusks, and because only male narwhals have the long spiraling tusk, Muizon hypothesized that "this second skull . . . is probably that of a female." Even though we can watch living narwhals, we are still not sure what the males do with that tusk. Is it a weapon? A feeding device? A sound conductor? A secondary sexual characteristic? Muizon writes that "it is commonly admitted that its role is social." If we can't figure out what a living whale does with its tusk, imagine the problem trying to figure out what a fossil whale did with a tusk that points in the opposite direction. Muizon: "Considering the orientation and length of the large tusk of *Odobenocetops* (observed in *O. leptodon* and inferred in *O. peruvianus*), this tooth must have been very fragile and unlikely to have been used to exert force (in fighting or digging). The possible sexual dimorphism in size in the tusks of *Odobenocetops peruvianus* confirms the interpretation given in a detailed review of their possible functions which pointed to a nonviolent social role."

The "astonishing walrus-convergent delphinoid cetacean from the Pliocene of Peru" raises many questions, not only about the descent of whales, but about evolution and extinction as well. With its single, backward pointing tusk, *Odobenocetops* looks very much like a failed experiment. Indeed, since it exists only as a fossil, it could be argued that it is an experiment that didn't work because it was too weird or too impractical. But we have to look no further than its suggested living relative, the narwhal, to see a comparable experiment that also shouldn't work but does. Such creatures do not spring up overnight. It took generations of minor modifications to reach the form these animals had when they died. If evolution is defined as the less well adapted making room for the better-adapted, then how can we explain bizarre detours such as *Odobenocetops,* where it looks as if the hapless creature was designed for inefficiency and predestined for extinction?

The evolution of whales is a complex subject, and there are many questions still unanswered. Nevertheless, the basic chronological sequence is known. Philip Gingerich, probably, since his discovery of *Pakicetus,* today's best-known paleocetologist, summed it up succinctly in a popular article written in 1997:

> What history tells us is that the teeth of whales changed first, before their ears or limbs. The transition to whales started when the mesonychians went into the water to feed, possibly first on dead or dying fish, and later on healthy fish. The ears changed next, as hearing replaced vision as the dominant sensory mode, enabling archaeocetes to communicate and find their prey. Archaeocetes, however, never developed the sonar or echolocation found in modern odontocetes. The next and final modification involved the

reduction of the sacrum and pelvis for tail-powered swimming, which enabled archaeocetes to disperse widely in the world's oceans.

In whales, as in no other living mammals, the trend has been toward gigantism. Just as the blue and fin whales are the giants of the mysticetes, the sperm whale is the largest of the odontocetes, and in bulk if not length it is the largest predatory animal in history. Sperm whales are almost as enigmatic in the fossil record as they are today. A late Miocene physeterid (sperm whale) was *Ferecetotherium,* found in the Caucasus and originally described as an archaic mysticete (Mchedlidze 1984), but the published drawings clearly show a creature that resembles a sperm whale, with peglike, homodont teeth and a large panbone in the lower jaw. *Ferecetotherium* is probably from the Late Oligocene, between 23 million and 30 million years ago. Fossils have also been found in Early Miocene deposits (22 million years ago), and by that time, the fossils show characteristic sperm whale features, such as a deep cranial basin; large, tusklike teeth; and an enlarged earbone. It is likely that the early physeterids were deep divers like their descendants and were probably teuthophages (squid eaters) as well. Primitive sperm whales had upper- and lower-jaw teeth, but members of the sole surviving genus (*Physeter*) have visible teeth in the lower jaw and none in the upper jaw, suggesting some adaptive advantage to losing half of your teeth. Of today's toothed whales, most of which feed on squid, only the sperm whale uses one set of teeth, but it actually has two—those in the upper jaw remain unerupted throughout the whale's life.

The sperm whale has the largest brain of any animal that has ever lived, but this does not give any clear information on its intelligence—difficult enough to assess for humans, let alone for animals that cannot take IQ tests. But the "encephalization quotient," or EQ, developed by Harry Jerison (1973), provides a ratio of an animal's

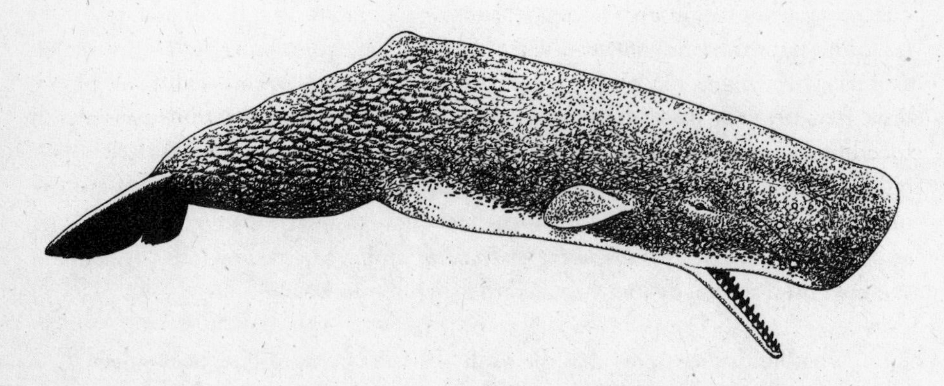

From its central role in Moby-Dick, *the sperm whale is probably the most familiar of all the great whales. But many aspects of its life remain hidden, such as how it catches the vast multitudes of squid required to sustain a 60-foot-long, 60-ton warm-blooded mammal; how deep it can dive (at least a mile, maybe more); what the purpose is of the great oil reservoir in its titanic nose; and, most intriguing, why the sperm whale needs the largest brain of any animal that has ever lived.*

brain weight to its body weight. Not surprisingly, given who designed the test, the animal with the highest EQ is *Homo sapiens,* but immediately below him or her is a river dolphin known as the tucuxi (*Sotalia fluviatalis*). Then comes the Pacific white-sided dolphin (*Lagenorhynchus obliquidens*), the common dolphin (*Delphinus delphis*), and the bottlenose (*Tursiops truncatus*). A couple more dolphins, and then *Australopithecus afarensis,* the fossil hominid known as Lucy, and well below her, the chimpanzee. More recent studies (Roth 1999) have pointed out that really small animals actually have the largest brains relative to their body size: shrews have brains that can equal 10 percent of their body weight, whereas the blue whale, the largest animal that has ever lived, has a brain that constitutes 0.01 percent of its body mass. If we look at the number of convolutions in the cerebral cortex, humans come out near the top of the scale, but animals with really large brains, like whales, dolphins, and elephants, have even more convoluted cortices. Convolutions seem to be more a measure of brain architecture than intelligence.

To determine brain size, an endocast is made by filling the brain case in the skull of a fossil. Measuring the endocasts of fossil whales, we find that *Pakicetus* had a brain the size of a walnut and that of *Ambulocetus* wasn't much larger, so we have to figure out how to make the quantum leap from a dull-witted protocetacean to *Tursiops,* an animal that many dolphin trainers say is smarter than they are. (Though brain size does not directly correlate to intelligence, animals with small brains relative to body size are usually considered to be less intelligent than those with larger brains.) Echolocation doesn't provide the explanation, since bats also echolocate, and they are not known for their high EQ (or IQ, for that matter). The brain of an adult sperm whale weighs about 20 pounds, as compared with a human's, which weighs about 3 pounds. The blue whale can weigh 300,000 pounds but it has a smaller brain than *Physeter macrocephalus.* Sperm whales may use their large brains for complex communications (among other things), but we need to find out why (never mind *how*) cetaceans such as dolphins and sperm whales developed these enormous brains.*

Of all the modifications for life in the ocean (development of flippers, blubber instead of hair, migrating nostrils, etc.), the giant brain is the most difficult to account for. After all, the other marine predators, like sharks, large bony fishes, and crocodiles, are not known for their problem-solving skills. Of course, the argument could be offered that these mammalian interlopers developed large brains in order to compete with their dumber rivals, but this suggests a determinism usually absent from evolutionary theory.

The abundance of recently discovered early cetacean fossils—Thewissen refers to a "cascade"—may give us some idea of the changes that were required for a terrestrial mammal to become an aquatic one, but we still have no idea *why* a group of animals

---

* It has long been known that sperm whales communicated with one another by means of a complex series of "click trains," composed of bursts of high-frequency sounds. Now Andreé and Kamminga (2000) have identified rhythmic signatures of individual sperm whales, which they believe the whales use to identify themselves as part of the general cacophony of sperm whale sounds. These signatures are "structured individual sequences which are precisely organized in complete, coherent structures." André and Kamminga do not use the word "sentences," but the implication is clear.

would leave one environment for another. In *The Emergence of Whales*, Thewissen wrote:

> A question that cannot be answered satisfactorily is why cetaceans took to the water. It has been suggested that they took to the water to take advantage of a plethora of resources that had gone untapped since the extinction of the Mesozoic marine reptiles. This explanation is too simplistic. The earliest cetaceans lived in or near freshwater, and it is unlikely that they profited from extinctions in the nearshore marine realm. Lack of competition for food is also an unlikely reason for the subsequent shift to the seas. The earliest marine cetaceans did not live like modern cetaceans, but resembled crocodiles ecologically. Crocodiles were not greatly affected by the K-T extinctions, and they are abundant in the same deposits that yield the earliest cetaceans. If the early cetaceans were generalized feeders, they must have suffered considerable competition and predation from crocodiles.

THE LIVING WHALES

Fossils are often difficult to read, but unlike *Archaeocetus, Pakicetus,* or *Himalayacetus,* which have been extinct for tens of millions of years, a diversity of cetaceans exists today in all the oceans of the world, and in many of its rivers as well. As with the pinnipeds, we can see in them the "end product" (so far, anyway) of cetacean evolution. In many cases, the family tree of these animals contains fossils that bear a strong resemblance to the living forms, so we can make well-educated guesses about how the early cetaceans lived. How and when the branching that produced the mysticetes and odontocetes occurred and whether there is more than one trunk to the tree is a matter of some conjecture, but the two groups of living cetaceans are believed to have diversified some 40 million years ago. The archaeocetes of the Upper Eocene (*Basilosaurus, Dorudon,* etc.) looked fairly modern, possibly with horizontal tail flukes, but they were proportionally longer and skinnier, with paddlelike flippers and tiny hind legs. They may have had a fleshy dorsal fin, since this seems to be a primitive feature of most modern cetaceans, and only a few living species lack this fin, which is believed to be a stabilizing aid in swimming.*

Odontocetes and mysticetes have diverged considerably since the Oligocene. We know that the baleen whales were derived from toothed ancestors, not only from the paleontological evidence (fossil mysticetes with teeth) but also because teeth can be found in mysticete embryos and are absorbed as the fetus develops the whalebone

---

* Of the living cetaceans, the right and bowhead whales (Eubalenidae) are sleek-backed and have no dorsal fin whatever. The common name of the right whale dolphins is derived from their finless similarity to the right whales. Both of the Monodontidae, the narwhal (*Monodon monoceros*) and the beluga (*Delphinapterus leucas*), lack dorsal fins. The finless porpoise (*Neophocaena phocaenoides*) is so named because it has no fin on its back. The gray whale (*Eschrichtius robustus*) and the sperm whale (*Physeter macrocephalus*) have no fin as such, but both species have a series of "knuckles" on the back, the most prominent of which is located where a dorsal fin might be.

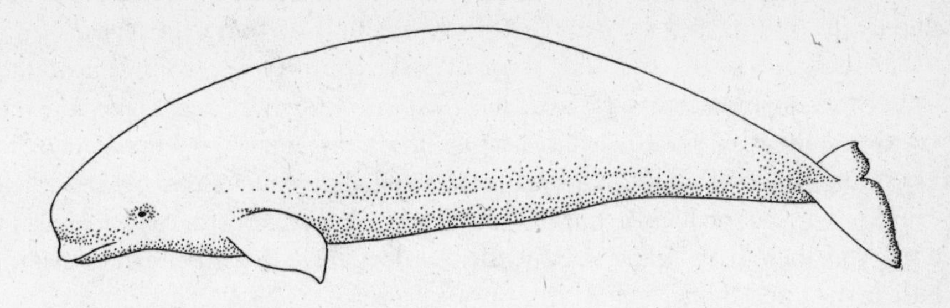

The beluga (from the Russian word for "white") is found only in the high Arctic. Belugas are among the noisiest of whales, making such a succession of whistles, squeals, barks, and snores that the old whalers used to call them sea canaries. These 16-foot-long whales do well in captivity and are often seen in oceanariums.

that is characteristic of the order. As the teeth disappeared in the whalebone whales, they proliferated in most of the odontocetes, particularly in the pelagic dolphins, some of which have greatly increased numbers of teeth. The common dolphin (*Delphinus delphis*) has 250 teeth, more than any other mammal. Of the Monodontidae, the beluga has a normal complement of teeth, but the narwhal, its only living relative, has only one tooth—the spiraling, ivory tusk of the male, actually the upper left canine. Deborah Kulu (1972) wrote, "With the exception of the specialized 'ramming tooth' now restricted to the male, the teeth of *Monodon* seem to be in the process of atrophy."

A further indication of the separation of the toothed whales from the baleen whales is that toothed whales can echolocate, whereas baleen whales cannot. "It seems reasonable to postulate," wrote David and Melba Caldwell in 1972, "that evolving toothed whales retained the useful original pathways for sounds in the lower frequencies which were significant in their original terrestrial environment but evolved new pathways to process the higher frequency ranges associated with echolocation." Although their terrestrial ancestors did have a sense of smell, cetaceans do not. Their eyes are well developed, but the poor light-transmission quality of water imposes severe limitations on vision. The cetaceans therefore developed an enormously effective hearing apparatus, and are more dependant on hearing than any other mammals except for bats. Cetaceans inherited the structure of their ears from their terrestrial forebears, but the ear had to be adapted for underwater hearing.

When we are under water, sound seems to come at us from all directions simultaneously, because it is conducted not through the usual channels of pinna (outer ear) and auditory meatus (ear hole), but through the bones of our heads. Cetaceans have adapted to be able to identify the direction from which a sound is coming, and the way they accomplish this is that their ears are isolated from each other and are not attached to the skull. Surrounding the ear bones are air sinuses, which are filled with a foam of gas bubbles and an oil-mucus emulsion, which provides acoustic isolation for each ear. The isolation of the ears from each other and from the skull allows the animal to hear with each ear separately, and it can calculate the source of the sound by the minute difference in the time the sound takes to reach each ear.

Not only do whales hear exceptionally well, but they also complement their hearing with sophisticated high-frequency sound production, and taken together, these two abilities result in what is perhaps the most amazing component of the cetacean repertoire: echolocation. This astonishing ability allows the toothed whales to send out bursts of high-frequency sounds, listen directionally to the returning echoes, and then analyze the size, density, and quality of the substance that the echoes have bounced off.

Most of the tests of echolocation that have been conducted with the usually amenable bottlenose dolphins show that they can differentiate between a one-fourth-inch and one-eighth-inch ball bearing; locate a swimming fish a football field away; and easily locate and retrieve tiny objects on the bottom of the tank, even if their eyes are covered. (Most dolphins and the larger toothed whales are known to echolocate, but few of them have been so extensively tested in controlled circumstances.) Of the extraordinary sophistication of the dolphin's echolocation capabilities, Carl Zimmer wrote (1998): "If a dolphin is blindfolded and presented with a pyramid at different orientations—or a cube, or a rectangular block—it can identify the different views as belonging to the same object. Researchers have had dolphins look at intricate constructions of plastic pipes with their eyes, and then successfully pick them out from a pipe lineup while blindfolded. Both experiments show that a dolphin not only perceives objects as sharply as we see, but can build abstract, multidimensional representations that are not tied to a particular sense."

It is extraordinary enough that dolphins can echolocate, but it is almost impossible to envision the evolutionary pathways that led from awkward, slow-moving, shoreside scavengers like *Mesonychus* or *Andrewsarchus* to *Tursiops truncatus,* the bottlenose dolphin, the large-brained underwater wizard that may have a more sophisticated communication system than we do. In a 1988 article, Shannon Brownlee cites what she calls the "Norris notion," an unpublished thesis of Ken Norris's that early dolphins restricted their noisemaking to periods when their nostrils and ears were above the water. This led to the completely anomalous situation for mammals where the animals would produce sounds from their noses (i.e., blowholes) instead of the larynx, where most other mammals generate sounds. Even an elephant's trumpeting, which certainly emerges from the animal's nose, comes from the larynx. Furthermore, according to Norris's thesis, dolphins also needed to hear under water, so they developed the (again anomalous) system of hearing through pockets of fat that are located in the lower jaw, transmitting the sounds to the inner ear. Sounds generated in the blowhole emerge through the "melon," a large lozenge of fat that lies just under the forehead and in front of the nasal passages; thus, the dolphin broadcasts and focuses sounds from its forehead. So the dolphin, such a familiar animal, is actually very, very weird. Along with all those other unmammalian cetacean characteristics, such as a fishlike body, a horizontal tail, a hairless body encased in blubber, and only one functional pair of limbs, it hears through its lower jaw and speaks through its forehead. Like most evolutionary history, Norris's ideas about the development of dolphin sound transmission and reception cannot be proved, but, according to Brownlee, they are "almost certainly right."

The living odontocetes are divided into four suborders, the most primitive of which is the family Platanistidae, which includes the Indian river dolphins (*Platanista* spp.), the franciscana (*Pontoporia monotypical*), the Amazon River dolphin (*Inia*), and the Chinese river dolphin (*Lipotes*). Except for the franciscana, a 5-foot-long dolphin that lives in the coastal waters of southern Brazil, Argentina, and Uruguay, these are all freshwater inhabitants which depend largely on echolocation because in rivers their vision is reduced. In the case of the Ganges River species, their eyes are virtually nonfunctional. Visibility in turbid rivers is usually less than in the ocean, so mammals that live in fluvial or littoral environments have developed senses to compensate for their reduced vision. The Indian freshwater dolphins have suffered major population losses as a result of hunting and the building of dams in their native rivers, and in the Indus River system, their numbers have been reduced to the point where they are among the rarest dolphins in the world. In addition to their acutely sensitive hearing, some river dolphins also hunt by touch. The inhabitants of Indian rivers are often seen swimming on their sides with one flipper dragging along the bottom, and the Amazon River dolphin (known locally as the *boutu*) has bristles on its snout that are believed to serve a tactile function. Because of their proximity to human habitation, all the river dolphins are considered endangered, but the Indian and Chinese species have been so reduced in numbers by hunting and accidental trapping in fishing nets that they may become extinct soon enough that our children will regard them, like *Ambulocetus,* as creatures of the past.

Classified together because of similarities of bone structure, the beluga and the narwhal are the only members of the cetacean family Monodontidae. They are toothed whales, but only one of them has a normal complement of teeth. The beluga has teeth in its upper and lower jaws, but the male and female narwhal have only one tooth between them: the spectacular ivory spike of the male. As with many peculiar living forms, we do not see just how odd an animal the narwhal really is. If a small fossil whale were found with a single, gigantic tooth sticking straight out of its upper jaw and no other teeth at all, it would surely be regarded as some sort of an evolutionary "sport" that could not possibly survive. We have obviously entered the story of narwhal evolution somewhere in the middle, or perhaps even close to the end, but there is no fossil evidence to document the rise of such a strange appendage, the function of which is not known even today, and has no analogs in any other living whale species.

The Delphinidae is a large and varied family that includes the better-known smaller dolphins like the bottlenose and the common dolphin, and also some cetaceans that are actually dolphins but—at least in English—have acquired such misleading names as pilot *whale,* false killer *whale,* and killer *whale.* The 36 species of delphinids constitute the largest and most diverse family of marine mammals, and they have radiated to fill many ecological niches. Some are feeders on small fishes and squid, others chase larger prey items, and others, like the misnamed killer whale, the ocean's apex predator, feed on everything from squid and fishes to seals, sea lions, and

The most widely distributed of all cetaceans, the killer whale is found in the high polar latitudes of both hemispheres and in almost every ocean in between.

smaller dolphins and porpoises, and, because they hunt in packs, prey on even the largest whales.

Most cetaceans have a dorsal fin of some sort; it is believed that the vertical fin on the back is useful to the animal, or it would have disappeared long ago. (The fact that most efficient swimmers, including ichthyosaurs, sharks, and fishes, also developed dorsal fins supports this argument.) Only a few cetaceans lack this fin, among them the right whale dolphins. The northern right whale dolphin (*Lissodelphis borealis*) is perhaps the slimmest of all dolphins; a 10-foot-long specimen would weigh only 180 pounds. Unlike its southern relative, *L. borealis* is mostly black, except for a white hourglass pattern on the ventral surface and white-tipped upper and lower jaws. It is found in the North Pacific from southern California to Alaska, Siberia, and Japan, but is difficult to see because of its unobtrusive silhouette and its low-angled leaps from the water. The southern version (*Lissodelphis peronii*) is found in a circumpolar distribution throughout the Southern Ocean; it is a 7-foot-long, elongated teardrop of a dolphin, its sleek silhouette uninterrupted by a dorsal fin. Because of its sharply demarcated white face, Herman Melville called it the "mealy-mouthed porpoise."

The word "porpoise" is derived from the Latin *porcus pisces* ("pig-fish"), and even now, among the local names for the harbor porpoise are "puffing pig" and "herring hog." *Phocoena phocoena* (pronounced "fo-*seen*-a") is the commonest cetacean in European waters, and is also found in the inshore waters of the western North Atlantic and the North Pacific. It is commonly encountered around river mouths, and has gone upstream in various European rivers, including the Seine, the Thames, and the Danube. Among the smallest of cetaceans, the harbor porpoise rarely reaches 5 feet in length, and mature animals do not weigh more than 140 pounds. They are frequently entangled in fishermen's nets, and they have been heavily affected by organochlorine pesticides and heavy metals.

The vaquita (*Phocoena sinus*), also known as the cochito, is a close relative of the harbor porpoise, but it is restricted to the Sea of Cortez (Gulf of California) between Baja and mainland Mexico. It may be the smallest of all the cetaceans, at a maximum length of 4.5 feet. It is similar in coloration to the harbor porpoise, although the mouth-to-flipper stripe is not as distinct. The dorsal fin is more sharply curved. Directed hunting and entanglement in gillnets in the totoaba fishery have seriously reduced the population of vaquitas, and with only a few hundred remaining, they are among the most endangered of all marine mammals. Other porpoises are Burmeister's porpoise (*P. spinipinnis*) found in the coastal waters of eastern and western South America, from Uruguay around Cape Horn to Peru, where it is often trapped in fishermen's nets; the spectacled porpoise (*Phocoena dioptrica*) previously known only from southern South American waters, but there are now several records from the New Zealand region, suggesting a circumpolar distribution; and the Dall's porpoise (*Phocoenoides dalli*) of the North Pacific, from California up through Alaska and Siberia, and south to Japan.

The terms "porpoise" and "dolphin" are sometimes used interchangeably, but for the most part, "porpoise" refers to members of the Phocoenidae, including the harbor porpoise (*Phocoena phocoena*), the vaquita (*P. sinus*), the spectacled porpoise (*Phocoena dioptrica*), Burmeister's porpoise (*Phocoena spinipinnis*), the finless porpoise (*Neophocaena phocaenoides*), and the Dall's porpoise (*Phocoenoides dalli*). Most of the other delphinids are commonly known as dolphins, except of course the delphinids that are referred to as whales. Porpoises have spade-shaped teeth, and dolphin teeth are usually conical.

Since the first appearance of the beaked whales (Ziphiidae) in the Middle Miocene, there has been a radiation of these peculiarly specialized animals into several groups, including the giant bottlenose whales (*Berardius*), the goosebeak whale

Similar in appearance to the harbor porpoise, the vaquita, also known as the Gulf of California porpoise, is one of the rarest animals in the world. Mexican fishermen have accidentally caught so many in their nets that the little porpoise is hovering on the brink of extinction.

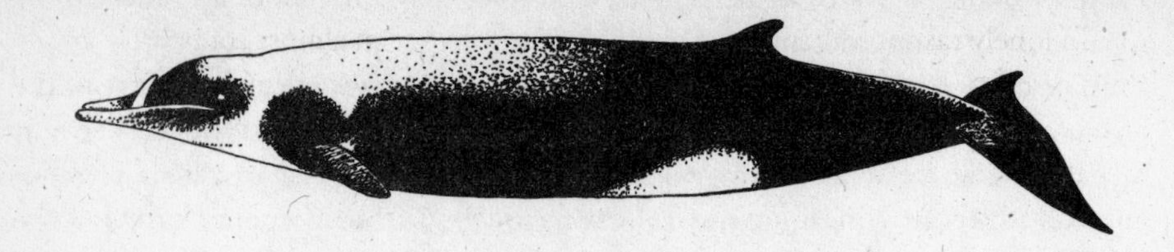

Layard's beaked whale has one of the strangest dental arrangements in the world. Only the males have teeth, and as they mature, the teeth grow and curl over the upper jaw, making it almost impossible for the animal to open its mouth. How can it eat if it can't open its mouth?

(*Ziphius*), Shepherd's beaked whale (*Tasmacetus shepherdi*), the northern and southern bottlenose whales (*Hyperoodon*), and the 13 known species of the genus *Mesoplodon.* Little is known about these animals in life; they are surely the least-known large animals in the world, and even less is known of their evolutionary history.

The beaked whales are anomalous in almost any grouping, because they seem to be unrelated to either the whales or the dolphins, but form their own family, the Ziphiidae, characterized by a spindle-shaped body, protruding upper and lower jaws, no central notch in the tail fin, and teeth only in the lower jaw of the males. Also known as Mesoplodonts ("middle-tooth," from the location of the teeth in the lower jaw), very little is known about their habits, and some species are represented only from animals that have stranded. In cases where the stomach contents have been examined, it was seen that beaked whales feed mostly on squid. (In *Mesoplodon layardii,* the straplike teeth of adult males grow out of the lower jaw and then arch over the upper jaw so that the animal can no longer open its mouth, apprently making it impossible for the animal to eat at all.) In recent years, two unsuspected new species have been described, the Peruvian beaked whale (*Mesoplodon peruvianus*) in 1991, and Bahamonde's beaked whale (*Mesoplodon bahamondi*), in 1995. For many years there have been reports of a mysterious beaked whale species that was sighted at sea—usually from a distance in Indo-Pacific waters—and conformed to the description of no known species. Some thought the sightings might be the least known of the beaked whales, *Indopacetus pacificus,* recorded from only two skeletons and never seen in the flesh. But the mystery beaked whale seemed larger than *Indopacetus* was known to be, and appeared to some observers to be some sort of bottlenose whale (*Hyperoodon* or *Berardius*), yet these animals are mostly inhabitants of the higher latitudes, and are not known from the Indo-Pacific. In 1999, Pitman et al. published a paper in *Marine Mammal Science* in which they looked at all the possibilities (and all the photographs), and tentatively concluded that there might be a heretofore unknown species of tropical bottlenose whale.*

* An anomaly within anomalies, both male and female Shepherd's beaked whales (*Tasmacetus shepherdi*) have functional teeth in the upper and lower jaws, and the males have a pair of larger teeth at the tip of the lower jaw. Because the beaked whales are descended from animals with teeth in both jaws, *Tasmacetus* is regarded as the

THE SPERM WHALES

In lonely taxonomic splendor are the sperm whales, an almost completely anomalous group, consisting of *Physeter* (*Physeter macrocephalus* means "big-headed blower") and the puzzling little kogiids, the pygmy and dwarf sperm whales. The cachalot is like no other animal on Earth, not only because of its vast size (a full-grown bull can be 60 feet long and weigh as many tons; females are two thirds that size), but also because it is constructed so strangely that the term *sui generis* could have been coined with *Physeter macrocephalus* in mind. The only thing that resembles a sperm whale is another sperm whale.

Sperm whales are the only great whales with a single nostril at the front of the head; all the other large whales are mysticetes, with paired blowholes located much farther back. In the sperm whale's head there is a huge reservoir of clear oil, and this may be used to focus and resonate sounds. Inside the nose—the largest nose in history—there are valves, sacs, and tubes, and also a unique organ known as the *museau du sange* ("monkey's muzzle"), which consists of a pair of horny "lips" that are believed to be involved in sound production. The nose also contains what the whalers referred to as the "case," a reservoir for the oil, and the "junk," which is a spongy, oil-filled construction directly below the case. The function of all these elements is unknown. We also do not know what use this animal makes of its 20-pound brain—the largest and most convoluted of any animal that ever lived—but it probably has to do with the processing of sounds employed during hunting and communication. Sperm whales are among the deep-diving champions of the mammalian world, able to dive to 10,000 feet and hold their breath for an hour and a half. They are primarily teuthophages (eaters of squid), but it is not known how they are able to catch enough of these elusive cephalopods to maintain their high metabolism.

It is conceivable that the teeth of sperm whales serve no practical function at all. Up to the age of about 10 years, sperm whales have no teeth, but they manage to feed themselves after they are weaned at about the age of two. More significant, there have been numerous instances of sperm whales found with their lower jaws so twisted, deformed, or broken that they couldn't possibly have used their jaws or teeth to feed, and yet they were well fed and otherwise healthy. This suggests a feeding technique that has nothing to do with teeth.

If the sperm whale does not use its jaws and teeth to capture its food, what does it use? One of the earliest theories was propounded by Thomas Beale, a British surgeon who shipped aboard a whaler in 1831. He could not imagine how "such a large and unwieldy animal as this whale could ever catch a sufficient quantity of such small animals, if he had to pursue them individually for his food," and suggested that the whale descends to a certain depth, where he "remains in as quiet a state as possible, opening his narrow elongated mouth until the lower jaw hangs down perpendicularly." The jaws and teeth, wrote Beale, "being of a bright glistening white colour . . .

---

most primitive (least specialized) of the ziphiids. *Tasmacetus* has never been seen alive, and is known only from specimens that have stranded in New Zealand, South Australia, Argentina, and Chile.

seem to be the incitement by which the prey are attracted, and when a sufficient number are in his mouth, he rapidly closes the jaw and swallows the contents."

In 1963, a short article appeared in a Soviet publication entitled "The Whale—An Ultrasonic Projector." Bel'kovich and Yablokov suggested that the sperm whale might be able to use its great nose somehow to project sounds loud enough to stun its prey. Other Russian scientists followed up on this idea, and in 1972, A. A. Berzin wrote of the sperm whale: "If the animal dives to the bottom, it swims with wide open mouth. When mobile squid and fish are discovered, the ultrasonic beam narrows and focusses on them, then its frequency sharply increases and the prey is stunned and then seized." In 1983, Ken Norris and Bertel Møhl published their hypothesis that odontocetes could debilitate their prey with sound. Theories about how the whale maintains a position of neutral buoyancy and waits for a school of squid to swim within range did not explain how a large animal like the sperm whale could obtain the 2,000 to 3,000 pounds of food per day that would be required to sustain it. The "sonic boom" hypothesis not only explains how the cumbersome sperm whale hunts and captures the swift cephalopods but also answers many other questions that had heretofore been as problematical as the feeding technique. Sonic debilitation of prey also seems to be a technique used by various dolphin species. Bottlenoses, which are the most closely observed of all dolphins in captivity and in the wild, have been recorded emitting loud bangs, which somehow cause fishes to die or at least to become seriously disoriented. Rachel Smolker (a student of Norris's), who spent years observing the inshore bottlenoses of Monkey Mia, Western Australia, described her experiences:

> *BANG!* I was "plugged in," headphones on, tape recorder rolling, hydrophone draped over the side of the boat, when I was nearly knocked overboard by a painfully loud sound. The needle on the tape recorder meter spiked into the red "overload" zone. The dolphins were feeding on a school of fish. Besides the crack, there was intense buzzing and clicking of all eight dolphins echolocating. Then another *BANG!* Just at the same moment, a silvery fish hurtled a couple of feet into the air and then landed on the water surface. Instead of swimming away, it lay on its side, as if stunned. A dolphin snapped it up and gobbled it down. *BA-BANG.* Two little splashes of water happened simultaneous with these bangs, and from one, a fish came flying out. Just as before it landed on the surface, stunned, before being gobbled up by a dolphin.

Despite the exaggerated stories, most of the squid consumed by sperm whales are considerably smaller than the giant squid (*Architeuthis*). From the examination of the stomach contents spilled on the decks of whaling ships, we know that the whales eat lots of squid. Finds of 5,000 to 7,000 beaks per whale are not uncommon, and one Soviet scientist found 28,000 beaks in the stomach of a single whale, indicating a feeding session in which 14,000 squid were consumed. (The beaks of squid come in pairs, an upper and a lower mandible.) But even when the echolocating capabilities of odontocetes were understood, there was a piece missing from the puzzle. The first cetological acousticians simply assumed that the whales found the squid by listening to their echoes and then dashed around gobbling them up. Upon reflection, this did not appear to be a terribly efficient method of hunting, especially considering the speed and maneuverability of the prey, and also its inherent unwillingness to be eaten. The sperm whale has massive peglike teeth in its lower jaw, so if the whale chased down its prey and snagged it in its jaws, the squids ought to have shown some evidence of having been bitten, but they didn't. After examining the stomach contents of a large number of sperm whales, the Soviet cetologist A. A. Berzin wrote that "neither the teeth nor the lower jaw need to participate in obtaining food and in the digestive process." Moby Dick is described as "having a sickle-shaped lower jaw," and there are documented instances (Nasu, 1958; Spaul, 1964) of sperm whales with lower jaws so deformed or encrusted with barnacles that they couldn't use their teeth if they wanted to. Sperm whales have even been captured that were completely blind (Beale 1839), all of which suggests a method of prey capture that has nothing to do with seeing or biting. Until someone actually observes a sperm whale hunting, that leaves us with the whales bombarding their prey with sonic blasts.

If the sperm whale debilitated or killed its prey with sound, it would go a long way to explaining the unusual construction of the jaws: the whale could use its tooth-studded lower jaw pincer fashion to pluck the floating squid out of the water or off the bottom, which would account for the lack of tooth marks on the prey. If the squid floated to the bottom, the sperm whale might plow its lower jaw through the sediment to pick them up, which would explain the strange items occasionally found in sperm whale stomachs: Japanese whalers in the vicinity of the Aleutian Islands reported stones, sand, crabs, glass buoys, a coconut, and deep-sea sponges. It would also account for the occasional entrapment and drowning of sperm whales in undersea telegraph cables. While plowing through the sediment, the whale might accidentally have become entangled in a loop of cable, or it might even have mistaken the cable for the tentacle of a giant squid. The deepest recorded entanglement was 620 fathoms, or 3,720 feet (Heezen 1957).

The sperm whale lives in structured family groups led by a dominant female in an arrangement not unlike that of elephants. The large bulls spend most of the year in high polar latitudes, visiting the family groups only during the breeding season. The babies, 14 feet long and weighing 2 tons when born, are nursed by their mothers for five years. The peatlike substance known as ambergris forms in the intestinal tract of some sperm whales, but we do not know why. In groups that can number two to 50 animals, sperm whales swim in all the world's oceans, but their migration routes and home territories (if there are any) are still poorly understood. Although they were heavily hunted during the eighteenth, nineteenth, and twentieth centuries, sperm whales are now protected throughout the world, and the population is believed to be more than 1 million.

The evolution of whales, with all its stops and starts, and its connections with such diverse creatures as mesonychids and hippos, has led us (or rather, has led them) to the sperm whale—one of the most unusual animals that has ever lived. Its history, wherever and however it began, took it into the water, and then, at least for feeding bouts, into the abyss. *Physeter macrocephalus* has retained its air-breathing proclivity, but hardly anything else about it is standard mammalian issue. Oh yes, it has the requisite bony skeleton—a straight vertebral column, ribs, and flippers with what look like hand bones in them (thus confirming its terrestrial origins)—and the regulation suite of innards, including lungs, heart, stomach, liver, intestines, kidneys, etc. Its great, sleigh-shaped skull is madly asymmetrical, for reasons that might have to do with sound production—or might not. Its nose, the largest nose in history, is fitted out with an array of equipment that enables it to perform acts that no other animal on Earth can perform: it can hold its breath for an hour, dive to 2 miles down, and produce a cacaphony of sounds that may be used for communication, echolocation, prey debilitation, or some other things that may not have occurred to us yet. It has scrimshaw pegs for teeth—only the tusks of elephants and walruses have been more used for art—that erupt only in the lower jaw, a Y-shaped affair that fits into a concavity in the upper jaw. It eats fast-swimming, water-breathing squid that it catches in total darkness while holding its breath.

Kogia breviceps and *Kogia simus* are the pygmy and dwarf sperm whales. The derevation of the genus name *Kogia* is unknown, and the same might be said of the animals themselves. They are both small compared to their huge relative, *Physeter:* the pygmy can be 13 feet in length, and the dwarf grows no longer than 8.5 feet. They resemble the larger sperm whale in the asymmetry of their skulls, in the presence of a spermaceti organ in their squarish heads, and in their underslung lower jaws, but there the resemblance ends. Both little species have dorsal fins, proportionally larger in *K. simus* than in *K. breviceps,* and where the blowhole of *Physeter* is an S-shaped affair at the left front end of the massive nose, the blowholes of the kogiids are above the eyes, and only slightly off-center. In contrast to the peglike ivory teeth of *Physeter,* its little cousins have narrow, sharp-pointed teeth, and *K. simus* has a few

The dwarf sperm whale and the pygmy sperm whale are miniatures of their larger cousin, the great sperm whale. However, the junior-sized sperm whales have a prominent dorsal fin, and they both have an inexplicable pattern of "gills."

teeth in its upper jaw as well. Both little sperm whales have mysterious "bracket marks" behind the head, which look very much like gills. It is very difficult to imagine what evolutionary purpose such marks would serve. What possible benefit could accrue to a whale that resembled a fish or a shark? Could it be mimicry?*

THE BALEEN WHALES

The earliest known mysticete, or baleen whale, is *Aetiocetus,* from the Upper Oligocene (23 million years ago) of Oregon. The fossil was found by Douglas Emlong, the collector who also unearthed the remains of the early desmostylian *Behemetops.* Although this species had teeth, the structure of the skull is otherwise typical of primitive mysticetes. Emlong wrote, "It is very unlikely that this animal would possess so many mysticete features if it were not directly in mysticete lineage." The family Aetiocetidae appears to provide the root stock of all subsequent mysticete families, including the cetotheres (now extinct), the Balaenidae (right whales and bowheads), and the Balaenopteridae, which includes all the rorquals: the blue, fin, sei, Bryde's, and minke whales. The small skull of the Eschrichtiidae (gray whales) is a primitive feature that may suggest an earlier origin.

Some explanation seems required when discussing the vast size that has been achieved by some of today's whales. *Basilosaurus* was a long, skinny creature that may have reached a length of 70 feet, but today's blue and fin whales can exceed that, and though the blue whale is long, it is anything but skinny; an adult female was weighed at 150 tons. Many marine creatures were larger in earlier times. There were crocodiles and turtles much larger than their living descendants, and although they left no heirs, some of the ichthyosaurs and mosasaurs were also giants. Freed from the restraints imposed by gravity, aquatic animals can reach a huge size, but why the large crocodilians and monster turtles died out while the whales got larger and larger is one of the historical mysteries of life in the sea.

It would seem that the Atlantic gray whale was already disappearing before the Basques helped it along its journey. The Pacific population is divided into eastern and western groups, and the gray whales of the western Pacific—those that wintered along the shores of Japan, China, and Korea—are also gone. Two out of the three known populations of gray whales (the Atlantic and the western Pacific) have been eliminated in the last five hundred years.

---

* Mimicry is a difficult concept to justify in terms of natural selection. How is it possible for totally different animals to have achieved the same shape, coloration, and size? Or for a creature to develop an inexplicable resemblance to a unrelated species, as with the dwarf sperm whales' "gills"? The leafy sea dragon, a sea horse, has evolved a costume of frilly protuberances that make it disappear in the weeds and coral near which it lives. The same is true of the sargassum fish, which looks almost exactly like the sargassum weed in which it lives. Occasionally we read that these creatures grew to look like leaves or plants *so that they could fool predators,* but this suggests a determinism that Darwinists say is completely lacking in the evolutionary process. Richard Dawkins (1996) sees no problem at all with mimicry, writing that "it is enough to say that we are sure this resemblance, and other examples of 'mimicry,' are not accidental. They are either designed or they are due to some process that produces results just as impressive as design." His final answer? "That magnificently non-random process which creates an almost perfect illusion of design. . . . Darwinian natural selection."

There used to be a population of Atlantic gray whales, but they are extinct. The only viable population of gray whales today is found in the eastern North Pacific, where they make an annual round-trip migration of 13,000 miles from the Bering Sea to Baja California—the longest migration undertaken by any mammal.

In the eastern Pacific, however, there are still gray whales. The world's only surviving population of these whales lives off the coast of western North America, making a 13,000-mile round trip from the Bering and Chukchi seas to Baja California—the longest migration of any mammal. In fact, there are probably more of them now than there were before 1857, when Captain Charles Melville Scammon told his fellow whalers about his discovery of their breeding grounds in Baja California. With its mottled grayish appearance and "knuckles" where other cetaceans have a dorsal fin, the gray whale reaches an adult length of 40 to 50 feet, has an overhanging upper jaw, and is usually covered with barnacles and whale lice. (The whales are born black and acquire the barnacles and lice with age; these produce the whales' scars and scratches.) Gray whales feed by scooping up mouthfuls of mud and tiny crustaceans from the bottom. After forcing the mud and water out through their white, sievelike baleen plates, they swallow the crustaceans. The gestation period for gray whales is a year; the young are conceived in the lagoons of Baja, and then born there 12 months later, after the mother has made a complete round trip.

Once thought to be nearly extinct because of heavy whaling pressure, the population has now stabilized at about 22,000, and is protected in Canadian, American, and Mexican waters. Gray whale watching, particularly off the coast of California and in the lagoons of Mexico, is enormously popular. Soviet whalers take about 175 gray whales annually for feeding captive minks, and the Makah Indians of Washington were recently granted permission to catch 5 gray whales per year in recognition of their traditional practices. The only "great" whales ever maintained in captivity were gray whales: "Gigi" was captured in Scammon's Lagoon in Baja California in March 1971 and was kept for a year at Sea World in San Diego before she was released. In January 1997, "J. J." was found sick and rolling in the surf near Point Loma, California, and she was also taken to Sea World for rehabilitation and exhibition. When she was released in February 1998, she weighed 19,000 pounds and was 31 feet long.

The word "rorqual" comes from the Norwegian *rörhval,* meaning "grooved whale," and is used for any of five baleen whale species that are characterized by long pleats on the underside of the throat. In descending order of size the rorquals are the blue, fin, sei, Bryde's whale, and minke whale. They are long, graceful animals that eat small organisms, such as krill and herring, by catching them in the hairy fringes of their relatively short baleen plates. All except the sei whale feed by swimming through a school of prey with their capacious mouths opened wide—often greatly distending the throat pouch as well—and taking in a huge mouthful of food and seawater. All the rorquals were hunted commercially, and the larger ones now exist in drastically reduced populations. Only the minke, which was once considered too small to hunt, has increased in numbers, because the larger rorquals, its competitors for food, were eliminated.

For sheer bulk, the blue whale (*Balaenoptera musculus*) is the largest animal ever to have lived on Earth. The biggest specimens have been over 100 feet long, and weighed 150 tons. Blue whales feed on shrimplike creatures that they ingest by the millions. They take in and expel huge mouthfuls of water, trapping the food items on the fringes of their baleen plates and then swallowing them. Blue whales were hunted to near extinction in the Antarctic by twentieth-century commercial whalers, and although hunting them has been illegal since 1966 and they are fully protected around the world, the current population of around 5,000 animals does not appear to be increasing.

Second only to the blue whale in size is the fin whale (*Balaenoptera physalus*), which can reach a maximum length of 80 feet and a weight of 50 tons. Finners are less robust than blues, and they are considerably faster and more graceful. The fin whale (also known as the razorback) gets its common name from its high dorsal fin, much larger than that of the blue whale. Fins are among the fastest whales, having been clocked at 25 mph. Their lower jaws are black on the left side and white on the left, making them the only consistently asymmetrically colored animals in the world. Moreover, the baleen plates on the forward part of the mouth match the coloration of the jaws; black on the left and white on the right. The reason for this curious coloration—if there is one—is unknown. They are found throughout the temperate and subpolar waters of the world and are probably the most common of the large whales. After the whalers had decimated the blue whale population of the Antarctic, they turned their harpoons on the fin whales and killed them in astonishing numbers. For example, in the period between 1946 and 1965, Antarctic whalers killed 417,787 fin whales, an average of 20,889 per year. During this period, fin whales were also being hunted in the North Atlantic and North Pacific, so these numbers are only part of the total. Although their numbers were severely depleted, they are now protected throughout the world.

At a maximum length of over 60 feet, the sei whale (*Balaenoptera borealis*) is the third largest of the rorquals. The name comes from the Norwegian *seje* (pronounced "say"), the coalfish that appeared off the coast of Norway every spring along with these whales. The sei can be distinguished from the fin whale because it lacks the white on the fin's right lower jaw. Where other rorquals such as the blue and fin whales take in mouthfuls of small organisms and trap them by expelling the water through the fringes of the baleen plates, the sei whale is a "skimmer," swimming through schools of krill with its mouth open, allowing the food items to become trapped in its silky baleen fringes. After the Antarctic whalers had practically wiped out their larger relatives, they began to kill these whales. What may have been a pre-exploitation population of 300,000 around the world has been reduced to perhaps 75,000.

One of the smallest of the groove-throated whales is Bryde's (pronounced "Brew-dah's"), which was named for Johann Bryde, who was the Norwegian consul to South Africa in 1913 when the species was first described. *Balaenoptera edeni* reaches a maximum length of 50 feet and a weight of 30 tons. It closely resembles the sei whale, but where the sei has a single ridge on the top of the upper jaw, Bryde's whale has three ridges. Bryde's whale is the only rorqual that does not migrate to high polar latitudes, and is sometimes referred to as the "tropical whale."

The minke (pronounced "minky") whale (*Balaenoptera acutorostrata*) is the smallest of the rorquals, reaching 30 feet and 10 tons. Until around 1870 this species was known as the "little piked whale"; it is said to have got its current common name from a Norwegian whaler named Meinecke who supposedly mistook a minke whale for a baby blue whale. There are Northern Hemisphere and Southern Hemisphere minkes; the northern type is characterized by a broad white stripe on the flipper, a feature often lacking in the southern version. Along with the larger rorquals (the blue, fin, and sei whales), southern minkes come to the Antarctic to feed in the summer. When the large species became too scarce to hunt economically, whalers went after the minkes. Even though commercial whaling has ended, Antarctic minkes are still hunted by the Japanese for "research," and the Norwegians and Icelanders continue to take a couple of hundred every year. The minke is the only Antarctic whale whose numbers have increased; as the larger species were removed from the food chain, the minkes proliferated.

The humpback has grooves on its throat, but otherwise it does not bear much resemblance to the other rorquals. Easily recognized by its long flippers and lumpy dorsal fin, *Megaptera novaeangliae* is one of the best-known of the baleen whales, but rare nonetheless. There are (or were) populations in most of the Earth's oceans, and because they are slow swimmers and breed in inshore waters, humpbacks were among the first species to be decimated when commercial whalers moved into any new area. From an estimated pre-exploitation population of perhaps 250,000, there are probably no more than 15,000 left in the world. Southern Hemisphere humpbacks migrate north from their Antarctic feeding grounds to calve in South African, Australian, and New Zealand waters; the northern populations feed in Alaska and New England and breed in Hawaii and the Caribbean. Humpbacks are the object of whale watchers in New England, Hawaiian, Alaskan, and Japanese waters. Their spectacular feeding behavior, where many whales corral small fishes underwater and then lunge open-mouthed to the surface to engulf them, has been the subject of many films. Humpbacks are the whales that "sing"; their eerie, repetitive vocalizations have been recorded and analyzed since the early 1970s, when Roger and Katy

In addition to its unusually long flippers—15 feet long in a full-grown adult—the humpback whale has developed the ability to sing complex "songs" that change every year. Why they change their songs, and how they actually make the eerie, haunting sounds, are unanswered questions.

Payne made *Songs of the Humpback Whale*, the best-selling animal recording of all time. The humpbacks in Australia sing completely different songs than those in New England. It is believed that only the males sing, but the mechanism by which they produce these whoops, yawps, moans and gurgles has not been identified. Although they were long considered "gentle giants," it has recently been observed that male humpbacks fight viciously for females, butting each other with their barnacle-encrusted heads, and slashing with their flippers.

The right whale got its name because it was the "right" whale for the early whalers to kill: right whales were slow swimmers, and they had plentiful blubber, which meant that they floated when they were killed. In addition, they had extremely long baleen plates, sometimes up to 10 feet in length, which were used in the manufacture of milady's "whalebone" corset stays and skirt hoops. The right whales of the Bay of Biscay were probably the first whales ever hunted commercially. Around A.D. 1000 the Basques began the industry that would last for another millennium and threaten almost all of the world's large whale species. Some cetologists believe there are two distinct species; the northern (*Eubalaena glacialis*) and the southern (*E. australis*), but they are similar enough to be discussed here as a single species. They are thickset, heavy animals with no dorsal fin, that have been measured up to 60 feet in length and weighed at as many tons. With their long baleen plates, they feed by "skimming" through shoals of small crustaceans, allowing the water to enter through the opening between the two "sides" of baleen plates and to pass out through the plates while the food items are trapped in the inner fringes. Right whales are born with "callosities," hardened patches of light-colored skin, on their heads, the patterns of which remain constant throughout their lives, making individual identification of the whales possible. Because of their inshore breeding habits, right whales were often among the first hunted when settlers arrived in a new area. The pre-exploitation

The bowhead, named for its arching mouth, was hunted to near extinction in the eastern Arctic by European whalers in the seventeeth century. When another population was found in 1848 off the north shore of Alaska, they too were decimated, mostly by American whalers.

population has been estimated at 200,000 animals, but almost as soon as they were discovered, they were eliminated from the waters of Cape Cod, Alaska, South Africa, Japan, Australia, and New Zealand. Only the lack of arable land and settlers in Patagonia saved the right whale population there. Some 750 right whales breed in the protected bays of Peninsula Valdéz in southern Argentina. The North Atlantic right whale, which breeds off the coasts of Georgia and Florida and feeds in the Gulf of Maine and the Bay of Fundy, numbers only a couple of hundred individuals, making it one of the rarest large animals in the world. No more than 5,000 remain in all the world's oceans.

The bowhead (*Balaena mysticetus*) is a heavy-bodied whale with a head that can account for a third of its 60-foot length and the longest baleen plates of any whale; they can be 15 feet long. Also known as whalebone, baleen was used during the seventeenth, eighteenth, and nineteenth centuries in the manufacture of corset stays, skirt hoops, and buggy whips, and its oil fueled the lamps and tanneries of Europe. Dutch and British whalers hunted the bowhead to extinction in the eastern Arctic around Greenland and Spitsbergen. About 7,000 bowheads live in the Bering Sea and off the North Slope of Alaska, where they are hunted by the Inuit under a strict quota system established by the U.S. government.

In their return to an aquatic existence, the whales experienced profound anatomical modifications. They lost their fur, external ears, and hind legs, and acquired a fishlike form and orientation, a thick coat of insulating blubber, a horizontal tail, and the world's most sophisticated auditory system. To make breathing easier, their nostrils migrated to the top of their heads, and they developed remarkable swimming and diving skills, generally exceeding those of the fishes that often serve as their prey. Probably the most unexpected improvement was the expansion and enlargement of the cetacean brain, which afforded whales cerebral capabilities that still puzzle the other mammals with exceptionally large brains, the readers of this book. The pinnipeds, too, underwent major changes, although the fossil record is less revealing than that for whales. From being upright terrestrial quadrupeds they became flippered, low-slung, densely furred swimmers with aquatic skills that are unsurpassed—except by some dolphins. Not that it was supposed to, but evolution does not seem to have improved the manatees and dugongs very much; as they were 20 million years ago they are still: blimplike, slow-moving, slow-witted creatures of warm shallow waters, endangered throughout their range, and with one of the more significant man-induced extinctions in their family tree. Perhaps the most surprising thing about the sirenians is that they have lasted as long as they have. No whale species have been made extinct, and although it looked for a while as if some of the fur seals and the northern elephant seals were going to be shoved over the precipice, only the Caribbean monk seal has been extinguished. First the fur hunters came close to killing them all, then the friends of the sea otter stepped in and saved the remainder, and the population is now on the upswing. The penguins, sleek swimmers of the Southern Hemisphere, are thriving, for the most part because their usual haunts are far from the path beaten by *Homo sapiens,* the most destructive force on Earth since that comet slammed into the Yucatán 65 million years ago.

EVERYBODY BACK INTO THE WATER

THE AQUATIC APE

Some groups of mammals became completely aquatic like the whales and the Sirenia (dugongs and manatees), others like seals almost so, and many others, such as Polar bears, otters, beavers, water voles, etc., became partially aquatic. I then put forward the thesis that perhaps man himself had such a phase of semi-aquatic life.

—Sir Alister Hardy, 1960

In 1695, John Woodward (1665–1728), a teacher of physics at Gresham College in London, published *Essay Toward the Natural History of the Earth,* in which he espoused the idea that Noah's flood had not only changed the face of the Earth but had created fossils as well. Some of Woodward's contemporaries, including the Danish naturalist Nicolaus Steno, believed that fossils showed that momentous changes had taken place over time; mountains thrown up and land conveyed from one place to another, but Woodward believed that the fossils demonstrated that the planet had been completely reconfigured by the flood. "The *Terraqueous Globe,*" he wrote, "is to *this Day* nearly in the same *Condition* that the *Universal Deluge* left it; being also likely to *continue* so till the Time of its final *Ruin* and *Dissolution,* preserved to the *same End* for which 'twas first *formed.*"

Woodward believed that the flood waters had receded with such violence that great mountains and valleys were raised up or gouged out, and what we now know as fossils were actually organisms forced into the stone by the violence of the waters. Among his supporters was the mathematician and philosopher Gottfried Wilhelm Leibniz (1646–1716), who believed in the reality of the flood but recognized that so much water could not have fallen as rain. He proposed that water that was contained in hollows within the Earth was squeezed out and flooded the planet. But a Swiss physician named Johann Jacob Scheuchzer (1672–1733) became the most prominent advocate of the Deluge theory; in 1708, he wrote *Piscium querelae et vindiciae*

(*Complaints and Justifications of the Fishes*), in which he argued, through the voice of a Latin-speaking fish, that fossils are the remains of various sea creatures that had been carried to the mountains by the flood.

When Scheuchzer unearthed a partial skeleton of a large vertebrate in a limestone quarry near Oeningen, Germany, in 1725, he believed he had found a man who had drowned in the flood. The fossil consisted of a vertebral column, a pair of rather shortish forelegs, and a flattened skull. In his *Physica Sacra* (1730), Scheuchzer wrote:

> It is certain that this . . . is the half, or nearly so, of the skeleton of a man: that the substance even of the bones, and, what is more, of the flesh and parts still softer than the flesh, are there incorporated into the stone. We see there the remains of the brain . . . of the roots of the nose . . . and some vestiges of the liver. In a word it is one of the rarest relics which we have of that cursed race which was buried under the waters.

Scheuchzer calculated that the man had drowned in 2306 B.C. and named the fossil *Homo diluvii testis* ("Man, a Witness of the Deluge"). It was placed on exhibit in the Teyler Museum in Haarlem, the Netherlands, where it can still be seen. In 1811, when Baron Georges Cuvier, Napoleon's minister of education, who has been called the founder of comparative anatomy and paleontology, came to the museum and examined the skeleton, he saw immediately that the "hands" were actually the clawed forelimbs of a giant salamander. Some two decades later, the German explorer Philipp Franz von Siebold saw the 5-foot-long giant salamanders of southern Japan, and *Homo diluvii testis* was reclassified as an amphibian and renamed *Andrias scheuchzeri*. Although Scheuchzer's "Witness to the Deluge" turned out to be a salamander, it would not be the last theory about *Homo aquaticus*—"water man."

In 1960, Sir Alister Hardy, a venerable British marine biologist, published an article in the British magazine *New Scientist* entitled "Was Man More Aquatic in the Past?" In the article, he asked whether ichthyosaurs, plesiosaurs, turtles, water-snakes, whales, dolphins, dugongs, manatees, seals, sea lions, polar bears, otters, shrews, and the platypus "were forced back into the water to make a living," why could not the same thing have happened to the primate now known as *Homo sapiens*? He wrote, "A branch of the primitive ape-stock was forced by competition from life in the trees to feed on the sea-shores, and to hunt for food, shellfish, sea-urchins, etc., in the shallow waters off the coast." In this hypothetical scenario, he suggested that these semi-aquatic creatures were soon wading into deeper water, and eventually began to swim and dive for food at even greater depths. His hypothesis is based on certain observations: humans, he says, are excellent swimmers, which "indicates to my mind that there must have been a long period of natural selection improving Man's qualities for such feats." Our enjoyment of the seaside is another factor in Hardy's ruminations—and "does not the vogue of the aqua-lung indicate a latent urge in Man to swim below the surface?" Then there is the business of hairlessness. Since humans have lost almost all of their hair except that on the top of the head, perhaps they need hair only to protect their heads from the sun's rays while swimming. Hardy also commented on the graceful streamlined shape of humans "compared with the clumsy form of the ape," and concluded that "all the curves of the human body have the beauty of a well-designed boat." In the "tentacle-like fingers" of the human hand, Hardy sees great possibilities for exploring the sea bed and capturing crabs and other crustaceans, and for "turning over stones to find worms and other creatures sheltering underneath." And to him, the fat beneath the skin so resembles the blubber layers of whales, seals, and penguins that he could not but think that "perhaps Man had been aquatic too."

Hardy promised that "next week I shall treat of the future," but his subsequent article, "Will Man Become More Aquatic in the Future?" is about utilizing the resources of the sea, not becoming *Homo aquaticus.* "I am not supposing," he wrote, "that we shall have an actual race of aquatic men, but that a large proportion of the population will become sub-aquatic artisans, cultivating the continental shelf, and husbanding the stocks of fish. It is the pioneer aqua-lung men of today who point the way to such a development." He predicted a greater efficiency in bottom trawling, but did not forsee that this greater efficiency would wreak terrible destruction upon the bottom and, in some areas, destroy the fish stocks and often the very bottom itself. Another suggestion of Hardy's for increasing the harvest of the sea's bounty was to eliminate the invertebrate predators, such as starfishes, that feed on other invertebrates that would otherwise be eaten by the fishes that humans want to eat. This ecologically unsound idea, however, is followed by the realization that scientists will have to enter the sea to study it, and "We shall want for more marine biologists who are trained in ecology, but they must be men who are also trained to be sub-aquatic naturalists, and prepared to spend a great deal of their time as frogmen making observations down below."

Elaine Morgan was more than willing to pick up where Hardy left off in his first article, about an early aquatic hominid. Born in Wales and educated at Oxford, she started as a writer for television, working on shows that were broadcast on the BBC between 1962 and 1971. In 1972, enraged by the omission of women in almost all discussions of human evolutionary theory, she published a book called *The Descent of Woman,* which was a revolutionary work about the unrecognized role of women in human evolution. Traditional anthropologists, whom she calls "Tarzanists," emphasized the idea of "man the hunter," which was based on his role as breadwinner—or, more accurately, meatwinner—while the women were relegated to a footnote, passively cooking and raising the babies. She wrote: "Most of the books forget about her for most of the time. They drag her onstage rather suddenly for the obligatory chapter in Sex and Reproduction, and then say, 'All right, love, you can go now,' while they deal with the real meaty stuff about the mighty hunter with his lovely new weapons and his lovely new straight legs racing along the Pleistocene plains."

Inspired by Hardy's article, she began to formulate the notion that the traditional ideas about the descent of man (that is, *Homo sapiens,* not just the guys), were also unsatisfactory, and began writing articles for *New Scientist;* eventually, she wrote three further books, *The Aquatic Ape* (1982), *The Scars of Evolution* (1990),* and *The Aquatic Ape Hypothesis* (1997).

The "aquatic ape hypothesis" is far from a trivial book, although conservative anthropologists may not accept her hypothesis. Nevertheless, attention must be paid, if for no other reason than that many of her propositions make considerably more sense than the accepted ones. She has addressed audiences at Oxford, Cambridge, Harvard, and Tufts universities, and in 1998, she was invited to participate in the conference of the International Association for the Study of Human Paleontology in South Africa. A television film of *The Aquatic Ape* was made by the BBC, and shown in the United States in 1999.

Elaine Morgan questions many aspects of accepted hominid evolutionary theory. For example, if our ancestors inhabited the African plains along with zebras, antelopes, lions, and baboons, why are we the only ones without fur? For that matter, why are we the only *primates* without fur? Why do we have more body fat than any other primates? Why is it that along with the penguins, *Homo sapiens* is the only animal that walks with a perpendicular, bipedal gait? Why would a primate that previously had a hairy coat move to the plains—scorchingly hot by day and uncomfortably cold at night—and then *lose* its protective covering? Then there is our breathing: humans and aquatic mammals such as whales, seals, and manatees breathe *consciously*—that is, they take in the appropriate amount of air for the activity they are about to perform—whereas all other mammals breathe involuntarily. Why were we the ones to develop a big brain, when such a development would have been just as advantageous for a chimpanzee? We also developed the power of speech, which no other mammals have.

Morgan has a simple explanation for these discrepancies: humans are not descended from terrestrial hominids, like *Australopithecus,* but from aquatic apes. In the past 5 million years, Morgan says, humans have lost many of their aquatic traits but the vestiges remain. Hairlessness, she writes, had evolved long before humans moved from the forests to the savannas; they had become hairless when they were semi-aquatic. The only hairless mammals are those that live underground, like naked mole rats, or aquatic ones, like whales, dolphins, manatees, and hippos. The argument that humans became hairless to prevent overheating doesn't seem to work when you realize how many animals that live in the hottest climates—camels, for instance—have a full hairy coat.

Bipedalism—walking upright on two legs—now occurs only in humans and penguins. Walking on two legs—although not necessarily vertically—was also commonplace among dinosaurs. Birds also walk on two legs, but except for the penguins, their spines are inclined toward the horizontal. Certain other mammals, such as

* The "scars," which are the price we have to pay for having achieved various evolutionary enhancements, such as the power of speech and an upright stance, consist of "the propensity to suffer from lower back pains, obesity, enlarged adenoids, acne, varicose veins, cot deaths, sunburn, sleep apnoea, gynaecological and sexual malfunctions, dandruff, inguinal hernia, hemorrhoids."

chimps, gorillas, orangutans, and bears, are capable of standing up on their hind legs and walking for a while, but their natural gait is quadrupedal. The apes "knuckle-walk." Many rodents—chipmunks, squirrels, and prairie dogs, for example—also assume a vertical posture, usually when looking for predators, but when the time comes to escape, they dash off on all fours. Kangaroos are among the only other mammals that move faster on two legs than four, but their two-legged hopping gait is completely different from that of other mammals. Kangaroos cannot move by putting one hind foot in front of the other—they cannot actually *walk*—and swing their hind legs forward while leaning over with their weight on their forelegs. The heavy tails helps them maintain their balance.

Because the idea of a quadruped rising up on its hind legs and deciding to become a biped is so preposterous, Elaine Morgan asks us to imagine what it must have been like for the first apes who did so to stand up on their hind legs and walk:

> Millions of years ago a population of apes on the savanna chose to walk on two limbs, instead of running rapidly and easily on four, like a baboon or a chimpanzee. They stood up. With their unmodified pelves, their inappropriate single-arched spines, their absurdly under-muscled thighs and buttocks, and their heads stuck on at the wrong angle, they doggedly shuffled along on the sides of their long-toed, ill-adapted feet.

The proboscis monkey, a long-nosed creature that lives in and around the mangrove swamps of Borneo, is a semiaquatic primate. Morgan cites many instances of these monkeys walking through the water and says that wild proboscis monkeys, "having acquired their bipedal gait in water, are seen calmly walking on the ground through the trees in single file." She points out that monkeys walking on all fours in the water would likely have their heads submerged, so bipedalism in proboscis monkeys may be a function of water depth. Some other primates, baboons, for example, have a protruding muzzle, but none except proboscis monkeys and humans have protruding noses. Morgan sees this as another aquatic adaptation: "If a gorilla attempted to dive or to swim under water, the water would be forced into her nasal cavities and cause her the most acute discomfort. A seal avoids this by having nostrils which it can open and close at will. The aquatic ape avoided it just as efficiently by modifying the shape of her face so that the water would be deflected by a splendid new streamlined structure and her sinuses would be safe."

Morgan's thesis is that there was a period during which early hominids lived a semiaquatic existence; they were never aquatic like cetaceans, but, rather, went in and out of the water frequently, like pinnipeds. In a 1997 discussion of the aquatic ape theory, she wrote,

> Whales and dolphins have been aquatic for about 70 million years and seals for between 25 and 30 million years. For most of these periods the cetaceans have been fully aquatic, never returning to land; and the seals need to go ashore only to breed. The hypothetical aquatic phases of the ancestral apes

> during the fossil gap would have been brief, a matter of two or three million years. Nobody has suggested that they turned into mermen or mermaids. They would have been water-adapted apes in the same sense that an otter is a water-adapted mustelid.

They lived in flooded forests, swamps, lagoons, or near the shallows in places like the Afar Triangle, formerly the Afar Sea in northeast Africa. Her preferred location is the Danakil Strait, at the southern end of the Red Sea, before the Somalian Plate came crashing into it and closed it up, at the beginning of the Pliocene about 5 million years ago. The Danakil Strait was open to the Gulf of Aden, and Danakil Island sat in the middle of it. The Rift Valley, where so many early hominid fossils have been found, lies just south of this formation, and this suggests to Morgan that the aquatic apes were isolated by tectonic plate movements and developed their watery ways some 5 million years ago. This may have been the place where the upright phase of human evolution actually began, and in *The Descent of Woman*, she describes the hypothetical process:

> She spent so much time in the water that her fur became nothing but a nuisance to her. Oftener than not, mammals who return to the water and stay there long enough, especially in warm climates, lose their hair as a perfectly logical consequence. Wet fur on land is no use to anyone, and fur in the water tends to slow down your swimming. She began to turn into a naked ape for the same reason as the porpoise turned into a naked cetacean, the hippopotamus turned into a naked ungulate, the walrus into a naked pinniped, and the manatee into a naked sirenian. As her fur began to disappear, she felt more and more comfortable in the water, and that is where she spent the Pliocene, patiently waiting for conditions in the interior to improve.

Morgan also introduces us to *Oreopithecus,* a hominid of the Late Miocene (9 million to 7 million years ago), commonly known as the "swamp ape," of whom she says, "Since *Oreopithecus* was expunged from our family tree, interest in the species has waned [and] it is now recognized that any resemblances between the swamp apes and man are not due to a close genetic relationship." Morgan wrote that the principle of covergence might link *Oreopithecus* and *Homo sapiens* because *Oreopithecus* "is characterized by the short iliac bones, which are seen in most aquatic mammals, and in man and his ancestors. The evidence suggests that through living in marshes, *Oreopithecus* had become either a good swimmer or a bipedalist—or both."

Interest in *Oreopithecus* picked up again in 1997, when the paleontologists Köhler and Moyà-Solà (1997) examined a skeleton of *Oreopithecus bambolii* that had been found on the island of Sardinia and determined that it was "functionally intermediate between apes and early hominids." In another study of *Oreopithecus,* Rook et al. (1999) determined that its hip bones showed conclusively that "the postural and locomotor behavior of this late Miocene hominid included habitual bipedality."

Humans are the only nonaquatic mammals that have a thick layer of fat under the skin. We tend to feel that this fat layer is a result of eating too much, but in fact, as Morgan points out, humans have at least 10 times the number of fat cells available for the storage of fat as any other land mammal of our size. "They must have been acquired because at some stage of our evolution they were useful or necessary," wrote Morgan. "The unavoidable conclusion is that at some stage in the past our ancestors were considerably fatter than we are today." In the 1998 BBC documentary based on Morgan's theory, *The Aquatic Ape,* the opening sequence consists of human babies happily swimming underwater with their eyes wide open and holding their breath. Viewing this sequence, it is difficult to imagine that the babies' behavior is anything but instinctive. Because fat is an insulator against cold and adds buoyancy, it stands to reason that the fattest animals—whales, seals, manatees, hippos, and humans—would be aquatic.

In her discussion of breathing, Morgan points out that only humans and certain diving mammals have a "descended larynx," a larynx that has "has slid right down onto the throat, below the back of the tongue, so that . . . the opening of the windpipe is in front of the gullet opening and on the same level. Our soft palate is no longer a continuous sheet of tissue with a neat, closable hole in it to accommodate the larynx. . . . We can therefore breathe with equal facility either through our noses or through our mouths. But we cannot breathe and drink at the same time, and if anything makes us gasp while we are drinking, some of the liquid is inhaled into the lungs." This arrangement not only makes it awkward to gasp (or laugh) and drink, but it also seems to be responsible for a variety of human ills, including sleep apnea, in which the sleeper's throat closes periodically, shutting off breath (apnea means "without breath"), and sudden infant death syndrome.

Several aquatic species, such as penguins and crocodiles, have a movable flap in the throat that closes the throat when the nostrils are employed for breathing; the presence of a similar structure in humans, the velum, suggests to Morgan that we used to do more diving than we do now. Humans, and aquatic mammals, and penguins can "hold their breath," which few other creatures can do. Morgan asserts that "the really indispensable pre-adaptation for speech is the enhanced degree of conscious breath control which we share with all diving mammals and no purely terrestrial ones. The pattern of inhaling deeply and quickly, and exhaling slowly at a controlled rate, is characteristic of aquatic mammals when they dive—and of humans when they speak."

Our senses of smell, hearing, and vision are considerably less efficient than those of many other mammals—Morgan writes that "a dog's sense of smell is not ten times better than his master's, nor a hundred times, nor a thousand times, but much nearer a million times." But we have managed to aquire something that sets us far apart from all other mammals: we can talk. Explanations for this amazing development range from the sublime (God willed it so) to the ridiculous ("*Homo sapiens* needed to increase his ability to communicate attack strategies to his fellows"), but nobody really knows why this naked, bipedal ape began to speak. Morgan speculates that immersion in water made the use of smell, vision, visual signaling, hand gestures, or even sounds like barking or hooting less valuable—how would the recipient of the hoot know what the immersed ape was trying to tell her? She couldn't really see anything in the water. Consequently, says Morgan, "The individuals who were good at producing noises at will were the most genetically successful, so that in the end everybody could do it." Are there any other mammals that rely almost exclusively on "vocal" communications? Of course there are: those subcutaneously fat, hairless, large-brained, completely aquatic mammals, the whales and dolphins.

Researchers have long known that elephants and sirenians shared a common ancestor, and Australian scientists have recently discovered proof that the ancestor was aquatic. Gaeth, Short, and Renfree (1999) examined African elephant fetuses and embryos ranging in age from 58 to 166 days that they obtained from females culled in South Africa's Kruger National Park. They found that all the elephant fetuses contained a physiological curiosity called a nephrostome, a funnel-shaped kidney duct found only in freshwater fish, frogs, and egg-laying reptiles and mammals. No other mammal that produces live offspring has nephrostomes. They appear very early in embryo development and then disappear, suggesting that they are probably ancestral. Further embryonic evidence that elephants once swam is that, unlike other land-living mammals, they have internal testicles and always have had them. Terrestrial animals have testes in the scrotum, but in elephants they are retracted into the abdomen. Seals, whales, and sirenians also have internal testicles, but they acquired them only when their terrestrial ancestors took to the seas 50 million years ago. Being carried internally protects the testes from the chilling and sterilizing effects of water. The trunk also appears extremely early on in fetal development, even in the earliest embryo examined, prompting Gaeth et al. to write, "The trunk might have first evolved as an adaptation to an aquatic environment. For example, it could have been used as a snorkel, as it is today when elephants swim in deep water." They concluded: "Our embryological data strongly suggest that the mesonephric kidney, the testis, the trunk, and the lungs of the elephant all originally were adapted to its aquatic environment and that some of these unusual adaptations have persisted in present day terrestrial elephants."

The news about the aquatic descent of elephants was startling enough to make many newspapers and television news programs when the paper was published in May 1999, but Elaine Morgan was way ahead of Gaeth et al. In her 1985 revision of *The Descent of Woman,* she wrote:

> We know that one of the elephant's nearest kin went into the sea and stayed there permanently, for the elephant's closest cousin is the sea cow. We know that early species of elephant developed weird and pointless-looking dental arrangements quite useless to land dwellers. There were, for instance, the shovel-tuskers, *Ambledon* and *Platybelodon grangeri.* Now shovels and spoons as a natural endowment are invaluable to water-feeders, like ducks and spoonbills and platypuses, but why on Earth would a land-dwelling animal want to scoop up a shovelful of Earth? Another primitive elephant had tusks that pointed down, like a walrus's.

Elephants are good swimmers now, so there is every reason to assume that they were good swimmers in the past, and for swimmers, she wrote, "a snorkel makes sense." Furthermore, elephants are the largest living land mammals. "Why," asked Morgan,

> did he achieve a bulk so ponderous and needing such vast quantities of sustenance? . . . Surely he did his growing, like a whale, in an element where weight is no hindrance and size a positive advantage for heat conservation. . . . Then again, if the elephant went marine, it would be logical to suppose that he would have learned to bring the emission of vocal noises under control of his will. He did, too. If you tell him to trumpet, he will trumpet. . . .
>
> Most of the attempts to refute AAT, consist of the alleged reasons why an aquatic ape could never have survived. . . . It was not stream-lined; it was quite the wrong shape for a swimmer. It was warm-blooded; once its fur was wet, the water would have drained it of its body heat. Its movements would have been slow and inefficient; it could never move underwater fast enough to catch a fish, and it would have starved. In the meantime, its hapless young would have been greedily devoured by crocodiles and sharks, and it would have left no descendants.

But, she says, these are the reasons why it would be impossible for *any* land mammal to survive in water. The "dog-like animal that evolved into a seal . . . the bear-like creature that turned into a walrus, the mole-like creature that turned into a platypus, and the elephant's cousin that turned into a manatee, the flying sea bird that turned into a penguin, and the ancient unknown quadrupeds that turned into whales and dolphins" would have faced the same problems, and yet they made it.

In 1995, Christian de Muizon and Greg McDonald published a description of an aquatic variant of a most unaquatic mammal: the ground sloth. Found in 1977 in the same Peruvian fossil beds as *Odobenocetops* (see pp. 220–21), *Thalassocnus natans* dates from the Pliocene, about 2 million years ago. This "slow swimmer of the sea" is, of all the specimens of fish, crocodiles and other mammals from the Sud-Sacaco Formation of Peru, "the only mammal belonging to an order, Edentata, traditionally regarded as terrestrial." The incomplete skeleton consisted of the skull and mandibles and the cervical, thoracic, lumbar, and caudal vertebrae, and some limb and tail bones. Differences from the large (sometimes gigantic) terrestrial ground sloths clearly indicated that the grizzly bear–sized *Thalassocnus natans* had been at least partially aquatic, and spent its days grazing on seagrass or kelp. The morphology of the femur "indicates great mobility and more powerful movement at the hip joint in *Thalassocnus* than in other nothrotheres. This feature, together with a tibia proportionally longer relative to the femur (a classic aquatic feature) than in other ground sloths, is consistent with a shift in function of the hind limb toward aquatic or semiaquatic life." The authors conclude:

> *Thallasocnus* therefore indicates another pathway by which a terrestrial group has shifted to an aquatic habitat. The anatomical features are different from those in other semi-aquatic mammals (hippos, desmostylians, aquatic rodents, aquatic carnivores) because of the anatomical constraints of being a sloth. In fact, there seem to be many ways in which a mammal can become aquatic or semi-aquatic.

Elaine Morgan doesn't mention ground sloths—the description of *Thallasocnus* was published long after *The Aquatic Ape* first appeared—but the fossil evidence of a land mammal that adapted to an aquatic existence a couple of million years ago means that it is possible to imagine that another land mammal might have done the same thing. (Of course, cetaceans, sirenians, and pinnipeds are also descended from land mammals, but their re-adaptation to an aquatic or semiaquatic existence took many millions of years.)

In *The Naked Ape*, published in 1967, the zoologist Desmond Morris discussed Hardy's aquatic ape theory, writing that it "needles the traditional fossil hunters by pointing out that they have been singularly unsuccessful in unearthing the vital missing links in our ancient past, and gives them the hot tip that if they would only take the trouble to search around the areas that constituted the African coastal sea-shores of a million years ago or so, they might find something that would be to their own advantage." But, he writes, "Despite its most appealing indirect evidence, the aquatic theory lacks solid support." It obviously made some sense to him, for he concluded his discussion by noting, "Even if eventually it does turn out to be true . . . it will simply mean that the ground ape went through a rather salutary christening ceremony."

In 1977, when Niles Eldredge, an invertebrate paleontologist, and Ian Tattersall, an anthropologist, wrote the article "Fact, Theory, and Fanstasy in Human Paleontology," they discussed the utter impossibility of formulating a scenario for the origin of *Homo sapiens,* writing:

> In devising a scenario, one is limited only by the bounds of one's imagination and by the credulity of one's audience—hence scenarios like Sir Alister Hardy's aquatic theory of human origins, recently popularized and elaborated by Elaine Morgan in her book *The Descent of Woman* (1972). Infinitely too involved to do justice here, this scenario, among many other things, ascribes the origin of human uprightness at least partly to the propensity of a remote ancestor to escape predators by fleeing into the water, where, submerged to the neck, he (or rather, she) would stand and wait for the predator, fearful of swimming, to leave in search of easier game.

Hardly any professionals have published serious assessments of Morgan's aquatic ape theory; when they comment on it at all, they mostly poke fun at it. In 1980, Jerold Lowenstein, a medical doctor and frequent writer on oceanic subjects, joined with Adrienne Zihlman, an anthropologist who propounded the theory that pygmy chimpanzees (bonobos) are the closest living prototype of the common ancestor of humans and African apes, to write "The Wading Ape: A Watered-Down Version of Human Evolution." They didn't accept a word of Morgan's theory, saying that it "does not hold water, anatomically, biochemically, or archaeologically." Lowenstein and Zihlman dismiss the idea that some apes took to the water during the Miocene Period to avoid the heat. They say the climate in subsaharan Africa was not hot and dry, but "a mixture of wet and dry, warm and cool, with a patchwork of savannas not radically different from the African savannas of the present day." They ask: "If the common ancestor of humans and apes went to sea for ten million, or even one million, years, the residual aquatic adaptations should show up in all the descendant species." In fact, Morgan does not say that the common ancestor of apes and humans took to the water; rather, that after the branching that led to humans and apes, the *humans* took to the water. They then say that "humans show none of the anatomical modifications common to aquatic animals, such as reduction in the limbs (especially the hind limbs, as in cetaceans); reduction in the thickness of the pelvis, which in water is not a structural weight-bearing truss as it is on land: the pelvis of whales, dolphins and manatees has nearly disappeared." But whales, dolphins, and manatees are *completely* aquatic, and, with no hind legs, have no need for a pelvis, whereas the *semi*aquatic animals, like pinnipeds, sea otters, and aquatic apes, if they existed, still need a fully developed, weight-bearing pelvis. "Humans," they say, "display none of the skeletal adaptations of horizontal streamlining seen in whales and dolphins, walruses and seals, manatees and dugongs, otters, beavers, water shrews, hippopotami or platypi—to name aquatic or partially aquatic members of eight different orders of mammals." Yet, as even Lowenstein and Zihlman point out, "The aquatic theory is . . . intended to explain how we became *different* from the apes," not how we are structurally similar to whales, walruses, beavers, or platypuses. We do share some characteristics with aquatic mammals: hairlessness, sophisticated communications, large brains, and a substantial subcutaneous fat layer. These modifications, with bipedality, still unexplainable by anthropologists, form the core of Elaine Morgan's theory.

Despite Morgan's willingness to subject her hypothesis to any test that anyone might suggest, with a couple of exceptions, mainstream anthropologists have ignored her.

The philosopher Daniel C. Dennett is Distinguished Arts and Sciences Professor and director of the Center for Cognitive Studies at Tufts University in Medford, Massachusetts. The author of *Brainstorms, Elbow Room, Consciousness,* and *Darwin's Dangerous Idea: Evolution and the Meanings of Life,* he invited Elaine Morgan to speak at Tufts. The obstacle may be, as Dennett has written (1995), "that she is not only a woman, but a science writer, an amateur without proper official credentials. . . . But in this case I wonder," says Dennett.

> Many of the counterarguments seem awfully thin and *ad hoc.* During the last few years, when I have found myself in the company of distinguished biologists, evolutionary theorists, paleoanthropolgists, and other experts, I

have often asked them to tell me, please, exactly why Elaine Morgan must be wrong about the aquatic ape theory. I haven't yet had a reply worth mentioning, aside from those who admit, with a twinkle in their eyes, that they have often wondered the same thing. There seems to be nothing *inherently* impossible about the idea; other mammals have made the plunge, after all. Why couldn't our ancestors have started back into the ocean and then retreated, bearing some telltale scars of this history?

In *The Aquatic Ape Hypothesis,* published in 1997, Morgan included numbered references, because, she said, "After disputing for 25 years with professional scientists, I have learned to respect the high standards they set themselves, and expect from others, in identifying their sources." She also wrote:

> The questions posed by the Aquatic Ape Theory are important and valid. The answers it offers are speculative, but no more so than any other available model. It is now generally agreed that the last common ancestor of apes and men lived in Africa in a landscape which was a mosaic, a mixture of trees and grassland. One sub-group of these animals—for some reason—began to change. First they stood up on their hind legs and began to walk bipedally; at some point the hair on their bodies changed direction and ultimately they became functionally naked; the larynx descended and was relocated below the tongue; they became fatter; forgot how to pant; lost their apocrine glands and much of their sense of smell; their sebaceous glands proliferated; their nostrils pointed in a new direction; finally, they evolved larger brains, gave birth to more immature babies, and learned to speak.
>
> There may be sound reasons why the aquatic model, like the savanna one, will in the end after careful scrutiny have to be abandoned. But there is no case for rejecting it out of hand. Over the past ten years it has been adjusted and modified to meet valid objections and to accommodate new data. For those who have assumed that there is something inherently untenable about it, it is time to think again.

Similarly, in a 2000 article published in *BBC Wildlife,* Simon Bearder, a professor of anthropology at Oxford Brookes University in Oxford, England, and a member of the International Union for the Conservation of Nature Primate Specialist Group, wrote a comprehensive summary of the aquatic ape theory that began with the question "How do you upset a gathering of biological anthropologists?" and answered it with "Just mention the words 'aquatic ape hypothesis.' " Bearder believes that Morgan's suggestions are worthy of consideration, and concludes:

> So, is it really true that the ancient causes of our unique way of life are still a mystery? Or is it more of a mystery why the ideas of Elaine Morgan and others have often been systematically ignored by professional anthropologists? Why are the possibilities of an amphibious origin not subjected to rational

> debate, scrutiny and research? These questions are hard to answer without concluding that prejudice and dogma have reigned far too long. . . . It is a challenge for future generations of anthropologists to expose these ideas to research and analysis to see if they survive close scrutiny, rather than continuing the denial of their predecessors and pretending there is no case to answer.

Finally—one is tempted to say *mercifully*—on November 25, 2000, *New Scientist,* the journal that published Hardy's original aquatic ape hypothesis and many of Elaine Morgan's articles, came back around and published a cover story emblazoned with the headline "Eve's Watery Origins: How the Sea Shore Made Us Human," the title of a piece by the writer Kate Douglas. Douglas begins her article by introducing Phillip Tobias of South Africa, one of the world's foremost anthropologists—he is one of the discoverers, with Louis Leakey and John Napier, of *Homo habilis*—who has decided that early man didn't live on the savannas at all, but, rather, in heavily wooded areas, thus dismissing the entire "savanna theory," which held that the upright stance of *Homo sapiens* was a response to its living in the open grasslands and standing up to look off into the distance. In a 1998 article, "Water and Human Evolution," Tobias had written:

> In our bodily functions, chemistry and microscopical anatomy, we should be hopeless as savannah-dwellers. So Marc Verhaegen and Elaine Morgan, in her remarkable book, *The Scars of Evolution,* came to the same conclusion that we had reached from quite different lines of evidence: the old Savannah Hypothesis was not tenable. All former savannah supporters must recant. . . . As Elaine Morgan has chronicled in *The Aquatic Ape Hypothesis,* this view [the savanna hypothesis] was supported, directly or indirectly, by numerous scholars, including Sherwood Washburn, Kenneth Oakley, Richard Leakey, Peter Wheeler, Alan Walker. It was a paradigm that lasted for about 70 years of this century. . . . By 1995, when I gave the Daryll Forde Memorial Lecture at University College, London, I stated of the Savannah Hypothesis, "We were all profoundly and unutterably wrong!"

Marc Verhaegen, a paleontologist at the Center for Anthropological Studies in Putte, Belgium, and an early collaborator of Morgan's, wrote that "humans . . . descended from a hominid population that remained near the coast and which gave rise to efficient waders and divers, and eventually to the various species of the *Homo* genus, some of which later returned to a more terrestrial lifestyle."

Does Tobias's support mean that the aquatic ape hypothesis will now be accepted? Probably not. Anthropologists are concerned that most of Morgan's evidence comes from the physiology and form of modern humans, not from the fossil record. But claiming that water-adapted fossils have not been found is a circular argument, because the theory of water-adaptedness purports to explain the very erectness of those fossil skeletons that have been found. But as more mainstream anthropologists recognize that something has to replace the savanna hypothesis, the aquatic ape theory is gaining more and more support. At the conclusion of a paper given at the Symposium of Water and Human Evolution, in 1999 in Ghent, Belgium, Elaine Morgan said: "Hardy's aquatic hypothesis, although highly speculative, is based on Darwinian assumptions. It outlines a scenario which could conceivably account for a number of hitherto unexplained human characteristics. Attempts to depict it as on a par with pseudo-scientific fringe fantasies are misconceived."

Tobias prefers to call the hypothesis the theory of water and human evolution. In fact, he believes one of the problems is the name. In his 1998 article he wrote,

> It seems . . . that the name Aquatic Ape Theory has become a handicap. For nearly 40 years since Hardy first put the idea forward, AAT has been a bit of a joke to many scientists, conjuring up visions of a creature that spent all—or almost all—of its time in the water. Yet Hardy's original 1960 article was modestly entitled, "Was Man More Aquatic in the Past?" In scientific writing a name can send very misleading messages and the term "Aquatic Ape" does just that. Replace it with something else, I urged Elaine Morgan. Then, I think the implications of those apparently water-adapted features like humans' loss of hair will receive less cynical attention from those who have hitherto smirked at the mere mention of "The Aquatic Ape"!

Tobias's support will certainly add credibility to Morgan's arguments, but her lack of scientific credentials suggests it will be some time before many anthropologists come around. In an essay in his book *Eight Little Piggies,* Stephen Jay Gould wrote (1993): "Professionals have always tried to seal the borders of their trade, and to snipe at an outsider."

Obviously, the fact that she is a woman doesn't automatically validate (or negate) Morgan's theory, and, given the nature of *The Descent of Woman*—and its publication in 1972 at the height of the burgeoning women's movement—Morgan may be considered an outsider by the paleontology and anthropology establishments. But when they refuse to acknowledge and seriously debate her theories they do exactly what she says they have done throughout discussions of human evolutionary history: relegate women to the kitchen and nursery.

THE END OF THE BEGINNING

From the fossil evidence, stratigraphy, and molecular analysis, it is difficult enough to understand the evolution of the past. Even so, a few paleontologists have ventured to predict *future* evolution, and Dougal Dixon is one of them. In *After Man: A Zoology of the Future* (1981), Dixon wrote, "The raw materials for this reparation are the kinds of animals that do well despite, or because of, man's presence, and which will outlive him—those that man regards as pests and vermin. . . . The result is a zoology of the world set, arbitrarily, 50 million years in the future, which I have used to expound some of the basic principles of evolution and ecology. The result is speculation built on fact. What I offer is not a firm prediction—more a exploration of possibilities."*

Although his "reparation" of terrestrial life forms is fascinating (humans are long gone; rabbits have evolved into the predominant large herbivores; rats have grown long legs and become the major predators), it is life in the sea that concerns us here, so let's have a look at Dixon's oceans. Fifty million years from now the Polar Ocean has become almost land-locked, and the "totally defenseless" flightless auk (*Nataralces maritimus*) has reappeared. (Dixon gives his future creatures plausible scientific names; *Nataralces maritimus* can be translated as "swimming auk of the sea.") They are preyed upon by "pytherons," which are "aquatic carnivorous mammals related to predator rats." Dixon's northerly polar oceans are also the home of Distarterops (*Scinderedens solungulatus*), "by far the most massive aquatic relative of the predator rats . . . occupying the same niche that seals occupied in the Age of Mammals, and like them have developed streamlined blubbery bodies and fin-shaped limbs." But "Its most unusual feature is its teeth; the upper incisors form long pointed tusks—the left-hand one projects forward, whereas the right-hand one points straight down and is used as a pick for removing shells from the sea bottom." It is truly remarkable that Dixon's hypothetical Distarterops bears such a close resemblance to a creature that actually existed but had not been discovered when Dixon wrote his book. Although unrelated to rats, the walruslike cetacean *Odobenocetops,* discovered in 1984 in Peru, also had asymmetrical tusks and fed on shellfish.

In Dixon's Southern Ocean, evolution runs wild. Like the whales in the Age of Mammals, sea creatures reach enormous proportions, and the vortex has become the largest creature on Earth. *Baleornis vivipera* ("live-birthing whale-bird") is "descended from the penguins, which although they were birds, had long since lost the power of flight and were totally adapted to an aquatic life except for one thing—they had to come on shore to lay eggs. This remained so until, shortly after the extinction of the whales, one species of penguin developed the ability to retain its single egg internally until it was ready to hatch and gave birth to live young in the open ocean." Relegated to the order Pelagornidae ("sea-birds"), the vortex has developed a plankton sieve in its beak not unlike that of the now extinct baleen whales. Other pelagornids include the porpin (*Stenavis piscivora*), which has a serrated bill "that enables it to catch larger fish than would otherwise be possible," and the skerns (a combination of "skua" and "tern"?), wingless birds that cannot walk and crawl out of the water to lay their eggs

* In *Crucible of Creation,* Simon Conway Morris offers Dixon's *After Man* as an alternative way of interpreting Gould's "rerunning of the evolutionary tape." He says it is "an exercise rich in imagination in its depiction of the riot of species that quickly radiate to fill the vacant ecological niches left after a time of devastation. All the animals, of course, are hypothetical. Certainly they look very strange, sometimes almost alien. When we look more closely at their peculiarities, however, they turn out to be little more than skin-deep. . . . The book *After Man* was published some years before *Wonderful Life.* It is notable that the fundamental message, that even with an effectively clean slate the re-emergence of new forms of life has a basic predictability, is not addressed in *Wonderful Life.*"

in warm volcanic sands. (There is still a controversy about whether the long-extinct *Hesperornis* could walk or even stand upright, because its forelimbs were so reduced as to have been useless for anything—including helping it up, the way the flippers of penguins are used—so it might have had to crawl out of the water like a skern.)

Fifty million years from now, in the swamps of tropical Africa, some unexpected creatures will appear, such as the mud gulper (*Phocapotamus lutuphagus*), a rat that has evolved into a hippopotamus-like animal; a swimming anteater (*Myrmevenarius amphibius*), and, lo and behold, a swimming monkey! Unlike Elaine Morgan's aquatic ape, Dixon's *Natopithecus ranapes* ("frog-footed swimming ape") is furry and has a kind of protruding dorsal ridge. "Descended from the swamp monkey (*Allenopithecus nigroviridis*) this creature has developed a froglike body with webbed hind feet, long clawed fingers for catching fish, and a ridge down its back to give it stability in the water." Although Elaine Morgan discusses the 9-million-year-old swamp ape *Oreopithecus*, she does not mention *Allenopithecus*, which is an African monkey related to the guenons. In Walker's *Mammals of the World* (1991) the following description appears: "They frequent swampy areas, so probably go into water freely. Captives appeared to enjoy wading in shallow water in the U.S. National Zoological Park."

Charles Darwin's theory of evolution, as detailed in *The Origin of Species,* is the very cornerstone of the entire structure of modern biology. The actual mechanisms responsible for the origin of species are still only vaguely understood, but all we have to do is look around us to see the results of the process he identified. The diversity of life can only have happened by earlier forms somehow evolving into later forms, and the later ones ultimately becoming the ones we can see out the window. The very presence of snakes, snails, puppy dogs, redwood trees, centipedes, giant squid, seahorses, hummingbirds, seaweeds, cows, horses, chickens, and wombats is vibrant, irrefutable proof of evolution. As E. O. Wilson wrote in his 1992 *The Diversity of Life,* "Darwinism is both the greatest idea in nineteenth-century science and the simplest. Its power arises from the fact that natural selection is protean in form. In some cases selection is lethal, mediated by predation, disease, and starvation. In others, it is benign, arising from family size, without increasing mortality in the least. Its products range in magnitude from fixing the number of hairs on a fly's wing to the human brain."

We can spot particular trends over time, but those trends do not necessarily provide the answers to the whys and wherefores; they only enable us to identify certain patterns that we can superimpose on the data, which might give us the idea that something other than random selection—or random evolution, or random extinction—has been occurring for these billions of years. For us the origin of species is not the problem, but rather, the origin of the various phyla and genera; we still don't know what causes new forms to spring up—or old ones to die off. Paleontologist Henry Fairfield Osborn published a book in 1917 titled *The Origin and Evolution of Life* in which he wrote:

> In contrast to the unity of opinion on the *law* of evolution is the wide diversity of opinion on the *causes* of evolution. In fact, the causes of the evolution of life are as mysterious as the law of evolution is certain. Some contend that we already know the chief causes of evolution, others contend that we know little or nothing of them.

The evolutionary biologist Richard Dawkins, who holds the Charles Simonyi Chair of Public Understanding of Science at Oxford University, argues in *The Selfish Gene*, *The Blind Watchmaker*, and *Climbing Mount Improbable* that Darwin was absolutely, unequivocally correct about the process of evolution, and that the mechanism causing change is the gene itself. He argues that genes are little bits of software that have only one goal: to make more copies of themselves. All living things are just vessels that have been created by these "copy-me programs" to help them to reproduce. Dawkins is such an avid Darwinian that he believes that Darwinism is the "only known theory that is in principle *capable* of explaining certain aspects of life." He goes on to predict, "If a form of life is ever discovered in another part of the universe, however outlandish and weirdly alien that form of life may be in detail, it will be found to resemble life on Earth in one key respect: it will have evolved by some kind of Darwinian natural selection."

Evolution rarely retraces its own steps. An adaptation once lost is usually not reacquired. There is even a "law" to this effect: "Dollo's Law," formulated by the Belgian evolutionary biologist Louis Dollo (1857–1931), also known as the Law of Irreversible Evolution, basically states that organisms cannot re-evolve along lost pathways, but must find alternate routes. The reason for this is that the same fortuitous train of mutational events, being totally random, will never repeat. Whales will never again walk on land with re-evolved pelvic appendages that derive from the current remnant structures that correspond to legs. They might, however, evolve appendages that derive from other biological provenance—especially if there were some pressure to do so—say, if the oceans began to dry up.

So even if Morgan's aquatic apes did emerge from the water, it would seem that they are not likely to return. Yet that is exactly what various creatures did, and this anomalous readaptation occurred not once, but many times. Scientists from numerous disciplines have convened two conferences called Secondary Adaptations to Life in Water, the first in Poitiers, France, in 1996, and the second in Copenhagen in 1999. The subject of the meetings was the modifications required by cetaceans, pinnipeds, sirenians, marine reptiles, turtles, sea snakes, and aquatic birds to reenter the sea. In the preface to the proceedings of the Poitiers meeting (published in *Historical Biology* in 2000), Jean-Michel Mazin, Patrick Vignaud, and Vivian de Buffrénil wrote:

> Paradoxically, in the course of their evolution on land, some animal groups turned about face and returned to the oceans. Marine ecosystems offered them abundant food resources, and comparatively protective living condi-

> tions. Consequently, a second succession of fundamental adaptive modifications, associated with readaptation or secondary adaptation to life in water, took place. Once again, complex and strictly irreversible evolutionary transformations were involved. Of course, subaerial, terrestrial animals could not simply be transformed into fully aquatic forms such as the fishes, but they had to be able to compete with them in the marine food webs.

The ancestors of whales, having established a beachhead on land, did not "need" to return to the water: the land was not flooded, so it was not necessary for terrestrial creatures to acquire modifications that would enable them to survive. In fact, only cetaceans and sirenians lost their hair: with the exception of the walrus, the pinnipeds kept their pelage, and in some cases, improved it.

We know that land mammals are descended from reptiles or amphibians that emerged from the sea, but what stimulus could have encouraged them to reverse this trend? Not enough food on land? Less competition in the sea? It is almost impossible to imagine the sequence of events that led a land mammal *gradually* to evolve into a sea mammal. Did its ancestors wade into the water ankle-deep to feed? Then knee-deep? Then shoulder-deep? And did they finally take the plunge and submerge while holding their breath? And did this breath-holding lead to enlarged lung capacity or the ability to store oxygen in the muscle tissue? And did their legs disappear and turn into flippers to make it easier to swim?

There seems to be little question that something happened to the mesonychids to encourage them to move toward an aquatic existence, but we are completely ignorant as to what it might have been. Yes, their legs got shorter and eventually disappeared altogether; they swapped their fur for blubber and their forelegs for flippers—but *why*? Was it somehow preordained that certain mammals would return to the sea? It can hardly have been, to use a traditional explanation, "to fill an available niche." That niche—the ocean—was already filled with assorted large predators, such as sharks, fishes, and squids that were there before the whales decided on this inexplicable volte-face, so the missing piece of this puzzle is some idea of why it happened. Carl Zimmer describes the probable course the mammals' transformation took, without identifying the mechanism that propelled these changes. Writing that mesonychids "stuck to this kind of life [scavenging along the shore], raising hundreds of thousands of generations of young mesonychids, until 34 million years ago, when their particular brand of anatomical confusion could survive no longer" suggests that the failure of paleontologists to place the mesonychids in a convenient category is part of the reason they became extinct. Thirty-four million years is a pretty good run (so far, the hominids have had a measly 4 million), and it is ridiculous to identify "anatomical confusion" as the reason for their extinction. The platypus, an animal that may be as "anatomically confused" as any that has ever lived, is still paddling around in Australian streams.

There is no obvious explanation for the disappearance of hind legs—the main propulsive engine for the protowhales—and their replacement by flukes. Natural selection, which dictates that the more fit will survive at the expense of their less well adapted siblings, cannot explain how the hind legs of a terrestrial creature shrank to the point where they were useless for walking or swimming, and eventually disappeared altogether. In the early stages of whale evolution, what possible advantage would reduced hind legs confer? It is worth noting that the marine reptiles (ichthyosaurs, plesiosaurs, mosasaurs, crocodiles, turtles), all of which are believed to be descended from four-legged terrestrial ancestors, *retained* their hind legs, even though they didn't seem to have much use for them, and they had more time—100 million years compared with 50 million for the cetaceans—to lose them. Sea snakes lost their legs, like their terrestrial relatives.

Describing the process as if it were a speeded-up film where we can see legs morphing into fins and nostrils migrating to the top of the head completely ignores the fundamental tenet of evolutionary theory: it has to have taken place slowly and gradually, so the fossil record should be chockful of transitional forms. But it isn't; we have only a few scattered signposts—*Himalayacetus, Pakicetus, Ambulocetus, Rodhocetus, Basilosaurus*—along a long and tortuous road, and some of them point us in the wrong direction. Zimmer describes the chaotic world of early whale evolution as if it were a series of fits and starts; of sometimes failed experiments, that somehow led us to a world where the oceans are populated by a great variety of cetaceans, from the little porpoises and dolphins to the gigantic blue and sperm whales. Are these also way stations, or have we arrived at the place that all this micro- and macroevolution was supposedly leading us?

If we accept the phylogenetic explanation, we "know" that there was a progression from the mesonychids to the protocetaceans, with various branches that led to dead ends. These animals apparently died off without leaving any direct descendants, and while the *Himalayacetus-Pakicetus-Rodhocetus-Ambulocetus* progression may indeed be, as Gould says, "a triumph in the history of paleontology," it is still only a series of unrelated animals that are quite different from one another, that all lived at different times, and that all went extinct. In an article in *Nature* (1994), Michael Novaček wrote, "*Ambulocetus, Rodhocetus,* and other more aquatically specialized archaeocetes cannot be strung in procession from ancestor to descendant in a *scala naturae.* Nonetheless, these fossils are real data on the early evolutionary experiments on whales." These are fossils of different animals that happen to look as if one might possibly have led to another.

When the description of *Ambulocetus* was published, Stephen Jay Gould saw the evolution of whales as a vindication of the efforts of paleontologists because it resoundingly answered critics who said there were no transitional forms. It is curious that Gould chose this group of fossils to demonstrate the existence of the long-sought transitional forms, because it was he who identified the fallacies in the durable and popular, but unfortunately incorrect, arrangement that was originally chosen by W. D. Matthew to show the "linear" nature of the evolution of horses. In an exhibit case in the American Museum of Natural History, where he was a curator of paleontology, Matthew carefully arranged the fossils of horses in an orderly progression from fox-terrier-sized *Eohippus* (now known as *Hyracotherium*) through *Protorohippus, Mesohippus, Pliohippus,* and at the top of the ladder, the modern genus *Equus,* which includes modern horses, zebras, and wild asses. In a booklet published in 1936 by the museum Matthew wrote,

> The history of the evolution of the horse through the Tertiary Period of the Age of Mammals affords the best known illustration of the doctrine of evolution by means of natural selection and the adaptation of a race of animals to its environment. The ancestry of this family has been traced back to near the beginning of the Tertiary without a single important break. During this long period of time, estimated at nearly three millions of years, these animals passed through important changes in all parts of the body, but especially in the teeth and feet, adapting them more and more perfectly to their particular environment, namely the open plains of a great plateau region with their scanty stunted herbage, which is the natural habitat of the horse.

Long before Matthew had proposed this progression, the English biologist Thomas Henry Huxley (1825–1895) had suggested a similar arrangement during an 1873 visit to Yale University. He said, "And if the horse has been thus evolved, and the remains of the different stages of its evolution have been preserved, they ought to present us with a series of forms in which the number of the digits becomes reduced; the bones of the forearm and leg gradually take on the equine condition; and the form and arrangement of the teeth successively approximate to those which obtain in existing horses. Let us turn to the facts and see how far they fulfill these requirements of the doctrine of evolution."

Obviously an admirer of fine horseflesh, Matthew saw the modern race horse as "the finest example of what nature, acting through millions of years, has been able to accomplish in the way of adapting a large quadruped to speed over long distances, and likewise, to the extent to which man, during the few thousand years he has controlled its development, has been able to improve upon nature, in the sense of adapting it to serve more exactly his own purpose." But, as Gould (1987) observed,

> The lineage of *Hyracotherium* to *Equus* represents only one pathway through a very elaborate bush of evolution that waxed and waned in a remarkably complex pattern through the last 55 million years. This particular pathway cannot be interpreted as a summary of the bush; or as an epitome of the larger story; or, in any legitimate sense, as a central tendency in equine evolution.

Matthew's display in the American Museum of Natural History has long since been dismantled, but many of the original fossils have now become part of a new exhibit on equine evolution in the Hall of Advanced Mammals, where the "progression" from small to large has been downplayed. But the inference lives on. Included in the exhibit is a sentence written in 1951 by George Gaylord Simpson, an esteemed evolutionary biologist who was at one time the curator of fossil mammals and birds at the museum: "The history of the horse family is still one of the clearest and most convincing for showing that organisms really have evolved, for demonstrating that, so to speak, an onion can turn into a lily." In the book it appears as part of a section written to show that there can be no argument that "evolution is a fact." Because the sentence appears as part of the exhibit—indeed, it dominates it—it is possible to come away with the idea that the *Hyracotherium* really did evolve into *Equus*.

It is interesting that nobody seems to be able to account for the disappearance of wild horses. Matthew said, "It is also unknown why the various species which inhabited North and South America and Europe during the early part of the Age of Man should have become extinct, while those of Asia (horse and wild ass) and of Africa (wild ass and zebra) still survive. Since their appearance, humans have played an important part in the extermination of the larger animals; but there is nothing to show how far humans are responsible for the disappearance of the native American species of horse." And Simpson wrote, "The extinction of the horse over the whole of North and South America, where they had roamed in vast herds during the Pleistocene, is one of the most mysterious episodes of animal history." The known evolutionary history of equids is therefore similar to that of cetaceans; rather than a smooth sequence of transitions, it consists of isolated fossils that might (or might not) be ancestral to other forms, which then disappeared, leaving no descendants.

Gould, in "Hooking Leviathan by Its Past," celebrating the discovery of the transitional forms of cetaceans, makes the same error that he accuses Matthew of making: he describes the evolutionary stages of whales as if they constituted a linear progression that clearly shows how natural selection could make an aquatic mammal out of a terrestrial one. Gould crows a little too loudly at his critics ("I cannot imagine a better tale for popular presentation of science, or a more satisfying, and intellectually based, political victory over lingering creationist opposition") but he chooses to ignore those—including Darwin himself, who was not exactly a creationist—who are still troubled by the absence of true transitional forms, which would demonstrate the gradual and unmistakable transformation of one species into another. In his refutation of Matthew's too-neat "ladder," Gould wrote, "Evolution rarely proceeds by the transformation of a single population from one stage to the next." In *Full House* (1996) he describes what he believes to be the actual process:

> Evolution rarely proceeds by the transformation of a single population from one stage to the next. Such an evolutionary style, technically called *anagenesis,* would permit a ladder, a chain, or some similar metaphor of lineality to serve as a proper icon of change. Instead, evolution proceeds by an elaborate and complex series of branching events, or episodes of speciation (technically called *cladogensis,* or branch-making). A trend is not a march along a path, but a complex series of transfers, or side-steps, from one event of speciation to another.

We should therefore view the protocetid fossils as branches of a complex bush on which there were many branches that died out, and not as a vision of land mammals becoming better and better adapted to life in the water as if this were some preordained eventuality.

In his book *Fossils*, Niles Eldredge (coauthor with Gould of the punctuated equilibium theory) discusses George Gaylord Simpson's ideas about the evolution of whales:

> Rather than blaming gaps in the fossil record for the absence of intermediate, transitional forms between the terrestrial ancestor and the Eocene whales, Simpson saw quite another message: Simpson realized that the fossil record of whales had something important to say about the very nature of the evolutionary process. He calculated that, if one took the conventional Darwinian position—i.e., that evolution is generally even-handed, steady, slow, gradual and progressive—and applied it to a case like the whales, the results were absurd. One can measure the average rate of evolution for various anatomical features, in the 50 million years it took to modify Eocene whales into fully modern forms. Let us then take the measured rate of evolution *within* whales and calculate how long it would have taken for Eocene whales to evolve from their terrestrial ancestors. Extrapolating back, it would have taken at least 100 million years (possibly even considerably more) for the transition from terrestrial ancestor to aquatic, primitive whale descendant to have occurred—assuming that whales evolved from terrestrial ancestors at the same rate of evolution we see in the 50 million years that elapsed between Eocene and modern whales. . . .
>
> Patently absurd . . . The only sensible conclusion, Simpson realized, is that evolution, especially episodes of large scale, truly *macro*evolution, must occur much more rapidly than the gentler pace typical of subsequent evolutionary transformation that takes place after a group becomes well established.

Darwin's "imperfection of the fossil record" raises not only the question of transitional forms, but also the question of numbers of individuals: Why are so many more fossils found of some species than of others? Many fossils of *Dorudon* and *Basilosaurus* have been found, but only one of *Himalayacetus,* one of *Ambulocetus,* and two of *Pakicetus.* What happened to all the others? Is it conceivable that of an entire population of pakecetids, only two could be found? Are paleontologists looking in the right places? How could an entire population vanish? Of course the paleontologists are looking for bony needles in stony haystacks, but when they find one, they immediately postulate a phylogenetic family tree and the details of its lifestyle.

It is difficult to picture *Pakicetus* in action, since we cannot tell from the fossils whether it lived on land or in water. We do not know whether living animals are undergoing a process of alteration and therefore are not very good at what they are supposed to do. Most likely speciation is taking place right now, but we can't see it because it is only visible on hindsight. Of course, if evolution consists of "punctuated equilibria," as Eldredge and Gould have suggested, we might have to wait for an ecological catastrophe—caused by *Homo sapiens,* for example—to drive rapid speciation after a long period of stasis. As far as we can see, however, the major effect that *Homo sapiens* has had on evolution is to eliminate species at an alarming rate, with few replacement species visible. To find living examples of biological ineptitude, we need look no further than some of the living marine mammals. The "hunching" locomotion of the phocids as they drag themselves with the claws on the foreflippers might appear to be inefficient (is this the way *Pakicetus* moved?), and the placid awkwardness of the living sirenians—not to mention the bizarre forelegs of *Hydrodamalis*—might lead to the assumption that in the sirenians we are looking at creatures that have reached an evolutionary impasse, or perhaps we are watching the process of extinction take place before our eyes.

It has often been said that every living animal is successful because it is capable of earning a living with its regulation equipment, regardless of how inefficient it might appear to us. The true seals, dragging themselves over ice and rocks with their foreflippers, might be seen as transitional, only because they are quadrupeds with hind legs that are almost useless for quadrupedal locomotion on land. Because these hind legs have a full complement of bones, paleontologists looking at a fossil would not necessarily arrive at the conclusion that the animal dragged itself. Ancestral whales like *Ambulocetus* and *Rodhocetus,* which are believed to have led an amphibian existence, might have dragged themselves over land just as the phocids do today, and therefore, we might view the phocids as "transitional" to a more aquatic existence, with the eventual loss of the hind legs to be replaced by flukes, but they are probably not on the way to becoming whales.

Every living creature must successfully adapt to the conditions of its environment, or it will disappear. Those creatures that "succeed" live on; those that "fail" become extinct. Some define evolutionary success as diversity, the proliferation of different species to occupy available niches. By this definition, the most successful animals in the history of the Earth other than microbes are the insects living today. Many species, like the protopinnipeds, vanished without leaving any recognizable descendants, whereas others, through a process we don't fully understand, were modified into today's living species. Human beings have occasionally helped to define "failure" by speeding up or even initiating the extinction process. There is no question that humans were directly responsible for the elimination of Steller's sea cow, the passenger pigeon, the Carolina parakeet, the great auk, the dodo, the Tasmanian wolf, the Hawaiian honeycreepers, and hundreds of other less charismatic animals.

Among marine mammals, the pinnipeds raise the most questions about their evolution. First of all, we can only guess as to their ancestors; everything that might be a transitional ancestral pinniped already looks like a seal or a sea lion. Furthermore, they are not creatures of either the land or of the sea, but are at home in both environments. They feed in the water but mate and deliver their pups on land. The pinnipeds' terrestrial locomotion may appear awkward, but their only rivals for fluid grace in the water are the dolphins. Where most other marine mammals have lost their hairy coats, the pinnipeds have improved upon theirs, and have ended up with some of the densest pelage in the animal kingdom, contradicting Elaine Morgan's statement that "wet fur on land is no use to anyone, and fur in the water tends to slow down your swimming." With the exception of the anomalous walrus, all pinnipeds have thick fur coats, and they do not appear to be handicapped by their fur, either in or out of the water. As a group they have evolved two types of terrestrial locomotion: the otariids waddle quadrupedally, and the phocids hunch themselves along with their clawed foreflippers. Their four legs clearly indicate a terrestrial ancestor, but what this animal might have looked like before it hunkered down onto its flippers is not evident.

What prompted the pinnipeds' ancestors to leave dry land and reenter the water might never be known, but Charles Repenning of the U.S. Geological Survey, who has written extensively on the evolution of pinnipeds, thinks he may know *where* it happened. Repenning believes there were two points of entry into the sea for the ancestors of today's seals and sea lions, one in the North Atlantic and the other in the North Pacific, both occurring in the Late Oligocene, about 30 million years ago. They did not arrive in the Southern Hemisphere until 20 million years later. He believes that the Atlantic stock became the phocid seals, those that use their hind legs only for swimming because they cannot rotate them forward for walking.

The physiological changes that had to occur before a terrestrial mammal could become a fully aquatic one like a whale are indeed profound, but for a terrestrial mammal to become amphibious required almost as much retrofitting. Adaptation to even a semiaquatic existence requires major modifications to eating, seeing, hearing, and conservation of body heat, not to mention the obvious changes that have to occur to allow a terrestrial mammal to swim and dive efficiently. (Think of the changes that had to take place in a land mammal before it could evolve into a Weddell seal, capable of diving to depths of 2,000 feet under the Antarctic ice.) One of the more remarkable adaptations to deep diving is the development of the "diving reflex," which slows the heartbeat and dramatically reduces blood circulation to peripheral organs like the flippers, shunting more of it to the brain, where a loss of oxygen would be most damaging. This ability is enhanced by enlarged passageways in the skull for conducting an increased blood supply to the brain. Unlike the whales, which developed a blubber coating to insulate them from the ocean's heat-draining propensities, the pinnipeds retained their fur coats. (Only the elephant seals have a significant blubber layer; they are the largest of the pinnipeds, and also the deepest diving.)

Evolution is always going on, but because it is so slow—a million years is but an eyeblink in the process—we cannot see it happening. Because we know that the pinnipeds are descended from terrestrial mammals, and because they must be evolving, as is every animal, including us, we can assume that they are somehow on the way to becoming something else or to becoming extinct. The cetaceans, which evidently began their return to the sea before the pinnipeds, present a parallel example with which to compare them. Certainly the pinnipeds are well adapted to their amphibious existence, but the fully aquatic whales and dolphins are better at living in the water. Whither the pinnipeds? Will they become fully aquatic like the cetaceans and lose their flippers and fur, or will they become transmogrified into some other, unpredictable marine creatures? (In his 1960 article, "Was Man More Aquatic in the Past?," Sir Alister Hardy included a photograph of sea lions with this caption: "The seals are well on their way to an almost completely aquatic future.") We cannot run a program to determine what they will evolve into, but we can predict with some assurance that in 10 million or 20 million years, there will be no seals or sea lions as we know them, and it is a sad but safe bet that the rest of the sirenians will soon follow their cousin *Hydrodamalis* into eternity.

Darwin, of course, was absolutely correct about natural selection. The marine animals that returned to the sea evolved from terrestrial ancestors and then continued their aquatic development by a process whereby minor modifications, encouraged by weather variations, climate change, predation, food shortages, change of diet, bacterial infections, asteroid impacts, or any combination of factors, selected for the better-adapted variations, which in time became new species. Along with the trilobites and the ammonites, every ancestor of the living whales, seals, manatees, and penguins is extinct. The fossil record provides only tantalizing hints of the process, but the abundance and variety of returnees is living proof that it works, and works wonderfully well.

Of all evolutionary events, the return to the sea by the various mammals, reptiles, and birds is among the most mysterious and the most spectacular. It seems to directly contradict the theory that creatures cannot regain attributes and physical characteristics that their ancestors lost. Such a turnaround is not completely anomalous in evolution; flightless birds, such as the ratites, are probably descended from birds that flew. And the dodo, great auk, and Steller's spectacled cormorant, whose living relatives fly perfectly well, are believed to have been affected by a sort of evolutionary atrophy: with no need to fly, the power of flight was lost. (All the aforementioned birds, however, are extinct, which may suggest that loss of flight may be somewhat maladaptive.) For some mammals to return to the sea and then acquire the modifications that would allow them to become fully aquatic is a miracle. But then, everything about this wonderful story is a miracle, isn't it?

REFERENCES

AHLBERG, P. E. 1991. Tetrapod or near-tetrapod fossils from the Upper Devonian of Scotland. *Nature* 354:298–301.

———. 1995. *Elginerpeton pancheni* and the earliest tetrapod clade. *Nature* 373:420–25.

———. 1999. Something fishy in the family tree. *Nature* 397:564–65.

AHLBERG, P. E., and R. A. MILNER. 1994. The origin and early diversification of tetrapods. *Nature* 338:507–14.

ALBINO, A. M. 1993. Snakes from the Paleocene and Eocene of Patagonia (Argentina): paleoecology and coevolution with mammals. *Historical Biology 7:51–69.*

ALBRIGHT, L. B. 1996. A protocetid cetacean from the Eocene of South Carolina. *Jour. Paleo.* 70(3):519–23.

ALDRICH, F. A. 1991. Some aspects of the systematics and biology of squid of the genus *Architeuthis* based on a study of specimens from Newfoundland waters. *Bull. Mar. Sci.* 49(1–2):457–81.

ALDRIDGE, R. J., and D. E. G. BRIGGS. 1989. A soft body of evidence. *Natural History* 5/89:6–11.

ALDRIDGE, R. J., D. E. G. BRIGGS, M. P. SMITH, E. N. K. CLARKSON, and N. D. L. CLARK. 1993. The anatomy of conodonts. *Phil. Trans. Roy. Soc. London* 340:405–21.

ALEXANDER, R. McN. 1990. Size, speed and buoyancy adaptations in aquatic mammals. *American Zoologist* 30:189–96.

ALLEN, J. A. 1880. *History of North American Pinnipeds.* U.S. Geological Survey, Misc. Pub. 12. Reprint, Arno Press, 1974.

AMEND, J. P., and E. L. SHOCK. 1998. Energetics of amino acid synthesis in hydrothermal ecosystems. *Science* 281:1659–62.

ANDRÉ, M., and C. KAMMINGA. 2000. Rhythmic dimension in the echolocation click trains of sperm whales: a possible function of identification and communication. *Jour. Mar. Biol. Assoc. U. K.* 80:163–69.

ANTONELIS, G. A., M. S. LOWRY, D. P. DeMaster, and C. H. FISCUS. 1987. Assessing northern elephant seal feeding habits by stomach lavage. *Mar. Mam. Sci.* 3(4):308–32.

APPLEGATE, S. P. 1967. A survey of shark hard parts. In P. W. Gilbert, R. F. Mathewson, and D. P. Rall, eds., *Sharks, Skates, and Rays,* pp. 37–64. Johns Hopkins University Press.

APPLEGATE, S. P., and L. ESPINOSA-ARRUBARRENA. 1996. The fossil history of *Carcharodon* and its possible ancestor, *Cretolamna:* a study in tooth identification. In A. P. Klimley and D. G. Ainley, eds., *Great White Sharks,* pp. 37–47. Academic Press.

ARANDA-MANTECA, F. J., D. P. DOMNING, and L. G. BARNES. 1994. A new Middle Miocene sirenian of the genus *Metaxytherium* from Baja California and California: relationships and paleobiogeographic implications. *Proc. San Diego Hist. Soc.* 29:191–204.

ARNOLD, J. M. 1987. Reproduction and embryology of *Nautilus.* In W. B. Saunders and N. H. Landman, eds., *Nautilus: The Biology and Paleobiology of a Living Fossil,* pp. 353–72. Plenum.

ARNOLD, J. M., and B. A. CARLSON. 1986. Living *Nautilus* embryos: preliminary observations. *Science* 232:73–76.

ARNOLD, J. M., N. H. LANDMAN, and H. MUTVEI. 1987. Development of the embryonic shell of *Nautilus.* In W. B. Saunders and N. H. Landman, eds., *Nautilus: The Biology and Paleobiology of a Living Fossil,* pp. 373–400. Plenum.

AZMI, R. J. 1998. Fossil discoveries in India. *Science* 282:627.

BACKUS, R. H., and SCHEVILL, W. E. 1966. *Physeter* clicks. In K. S. Norris, ed., *Whales, Dolphins and Porpoises,* pp. 510–528. University of California Press.

Bagla, P. 2000. Team rejects claim of early Indian fossils. *Science* 289:1273.
Bajpai, S., and P. D. Gingerich. 1998. A new Eocene archaeocete (Mammalia, Cetacea) from India and the time of origin of whales. *Proc. Nat. Acad. Sci.* 95:15464–68.
Bajpai, S., and J. M. G. Thewissen. 1998. Middle Eocene cetaceans from the Harudi and Subathu Formations of India. In J. M. G. Thewissen, ed., *The Emergence of Whales,* pp. 213–34. Plenum.
Bajpai, S., J. M. G. Thewissen, and A. Sahni. 1996. *Indocetus* (Cetacea, Mammalia) endocasts from Kachchh (India). *Jour. Vert. Paleo.* 16(3):582–84.
Ball, P. 1999. *H_2O: A Biography of Water*. Weidenfeld & Nicolson.
Balter, M. 1998. Did life begin in hot water? *Science* 280:31.
———. 2000. Evolution on life's fringes. *Science* 289:1866–67.
Barnes, L. G. 1976. Outline of Eastern North Pacific fossil cetacean assemblages. *Syst. Zool.* 25(4):321–43.
———. 1978. A review of *Lophocetus* and *Liolithax* and their relationships to the delphinoid family Kentriodointidae (Cetacea: Odontoceti). *Nat. Hist. Mus. L. A. County Science Bull.* 28:1–35.
———. 1984. Search for the first whale: retracing the ancestry of cetaceans. *Oceans* 17(2):20–23.
———. 1985. Evolution, taxonomy, and antitropical distribution of the porpoises. *Marine Mammal Science* 1(2):149–165
Barnes, L. G., and S. A. MacLeod. 1984. The fossil record and phyletic relationships of gray whales. In M. L. Jones. S. L. Swartz, and S. Leatherwood, eds., *The Gray Whale,* Eschrichtius robustus, pp. 3–32. Academic Press.
Barnes, L. G., and E. D. Mitchell. 1975. Late Cenozoic Northeast Phocidae. *Rapp. P. -v. Réun. Cons. int. Explor. Mer.* 169:34–42.
———. 1978. Cetacea. In V. J. Maglio and H. B. S. Cooke, eds., *Evolution of African Mammals,* pp. 582–602. Harvard University Press.
———. 1984. *Kentriodon obscurus* (Kellogg, 1931), a fossil dolphin (Mammalia: Kentriodontidae) from the Miocene Sharktooth Hill bonebed in California. *Contrib. Sci. Nat. Hist. Mus. L. A. County* 353:1–23.
Barnes, L. G., D. P. Domning, and C. E. Ray. 1985. Status of studies on fossil marine mammals. *Mar. Mam. Sci.* 1(1):15–53.
Baross, J. A. 1998. Living on the edge. *Science* 395:136.
Baross, J. A., and J. W. Deming. 1995. Growth at high temperatures: isolation, taxonomy, physiology and ecology. In D. M. Karl, ed., *The Microbiology of Hydrothermal Vent Habitats,* pp. 168–77. CRC Press.
Baross, J. A., and S. E. Hoffman. 1985. Submarine hydrothermal vents and associated gradient environments as sites for the origin and evolution of life. *Origins of Life* 15:327–45.
Basden, A. M., G. C. Young, M. I. Coates, and A. Ritchie, 2000. The most primitive osteichthyan braincase? *Nature* 403:185–88.
Basil, J. A., R. T. Hanlon, S. I. Sheikh, and J. Atema. 2000. Three-dimensional odor tracking by *Nautilus pompilius. Jour. Exp. Biol.* 203:1409–14.
Batt, R. J. 1989. Ammonite shell morphotype distributions in the Western Interior Greenhorn Sea and some paleocological implications. *Palaios* 4:32–42
Batten, R. L. 1984. *Neopilina, Neomphalus,* and *Neritopsis,* living fossil molluscs. In N. Eldredge and S. M. Stanley, eds., *Living Fossils,* pp. 196–213. Springer-Verlag.
Bearder, S. 2000. Flood brothers. *BBC Wildlife* 18(6):64–68.
Beebe, W. 1934. *Half Mile Down.* Harcourt, Brace.
Bel'kovich, V. M., and A. V. Yablokov. 1963. The whale—an ultrasonic projector. *Yuchnyi Teknik* 3:76–77.
Bendix-Almgreen, S. 1966. New investigations on *Helicoprion* from the Phosphoria Formation of south-east Idaho, U. S. A. *Biol. Skr. Dan. Vid. Selskab.* 14(5):1–54.
———. 1970. The anatomy of *Menaspis armata* and the phyletic affinities of the menaspid bradyodonts. *Lethaia* 4:21–49.
Bengston, S. 1991. Oddballs from the Cambrian start to get even. *Nature* 351:184.
———. 1994. The advent of animal skeletons. In S. Bengston, ed., *Early Life on Earth,* pp. 412–25. Columbia University Press.
Benton, M. J. 1985. Classification and phylogeny of the diapsid reptiles. *Zool. Jour. Linn. Soc. London* 84:97–164.
———. 1990. *Vertebrate Palaeontology.* Chapman & Hall.
Benton, M. J., M. A. Wills, and R. Hitchin. 2000. Quality of the fossil record through time. *Nature* 403:534–36.
Bergquist, D. C., F. M. Williams, and C. R. Fisher. 2000. Longevity record for deep-sea invertebrate. *Nature* 403:499–500.
Bergström, J. 1994. Ideas on early animal evolution. In S. Bengston, ed., *Early Life on Earth,* pp. 460–66. Columbia University Press.

BERGSTRÖM, J., W. W. NAUMANN, J. VIEHWEG, and M. MARTÍ-MUS. 1998. Conodonts, calcichordates and the origin of vertebrates. *Geowissenschaftliche Reihe* 1:93–102.

BERTA, A. 1994. What is a whale? *Science* 263:180–81.

BERTA, A., and T. D. DEMÉRÉ, eds., 1994. Contributions in marine mammal paleontology honoring Frank C. Whitmore, Jr. *Proc. San Diego Hist. Soc.* 29:1–268.

BERTA, A., and G. S. MORGAN. 1985. A new sea otter (Carnivora, Mustelidae) from the Late Pliocene of North America. *Jour. Paleo.* 59:809–19.

BERTA, A., and C. E. RAY. 1990. Skeletal morphology and locomotor capabilities of the archaic pinniped *Enaliarctos mealsi. Jour. Vert. Paleo.* 10(2):141–57.

BERTA, A., and J. L. SUMICH. 1999. *Marine Mammals: Evolutionary Biology.* Academic Press.

BERTA, A., and A. WYSS. 1994. Pinniped phylogeny. *Proc. San Diego Hist. Soc.* 29:33–56.

BERTA, A., C. E. RAY, and A. R. WYSS. 1989. Skeleton of the earliest known pinniped, *Enaliarctos mealsi. Science* 244:60–62.

BERZIN, A. A. 1972. *The Sperm Whale.* Translated from the Russian by Israel Program for Scientific Translation, Jerusalem.

BIGELOW, H. B., and W. C. SCHROEDER. 1948. *Lancelets, Cyclostomes, Sharks. Part I, Fishes of the Western North Atlantic.* Memoirs of the Sears Foundation for Marine Research. Yale University.

BOCKELIE, T., and R. A. FORTEY. 1978. An early Ordovician vertebrate. *Nature* 260:36–38.

BODEN, B. P., and E. M. KAMPA. 1957. Records of bioluminescence in the ocean. *Pacific Science* 11:229–35.

———. 1967. The influence of natural light on the vertical migrations of an animal community in the sea. *Symp. Zool. Soc. London* 19:15–26.

BOLIN, R. L. 1961. The function of the luminous organs of deep-sea fishes. *Proc. 9th Pacific Sci. Congress* 10:37–39.

BONNER, J. T. 1988. *The Evolution of Complexity by Means of Natural Selection.* Princeton University Press.

BONNER, W. N. 1980. *Whales.* Blandford.

BOWEN, W. D., and D. B. SINIFF. 1999. Distribution, population biology, and feeding ecology of marine mammals. In J. E. Reynolds and S. A. Rommel, eds., *Biology of Marine Mammals,* pp. 423–84. Smithsonian Institution Press.

BOWLER, P. J. 1998. Cambrian conflict: *Crucible* an assault on Gould's Burgess Shale interpretation. Review of *Wonderful Life,* by S. J. Gould. *American Scientist* 86(5):472–75.

BRANDES, J. A. et al. 1998. Abiotic nitrogen reduction in the early Earth. *Nature* 395:365–67.

BRASIER, M. 1998. From deep time to late arrivals. *Nature* 395:547–48.

BRIGGS, D. E. G., and E. N. K. CLARKSON. 1982. The Lower Carboniferous Granton "Shrimp-bed," Edinburgh. *Spec. Papers Paleont.* 30:161–77.

BRIGGS, D. E. G., D. H. ERWIN, and F. J. COLLIER. 1994. *The Fossils of the Burgess Shale.* Smithsonian Institution Press.

BROAD, W. J. 1985. Authenticity of fossil is challenged. *New York Times,* May 7, C1–3

BROWN, G. 2000. Symbionts and assassins. *Natural History* 109(6):66–71

BROWNLEE, S. 1988. Did nature give dolphins a nose job? *Discover* 9(2):81–83.

BUCHHOLTZ, E. A. 1998. Implications of vertebral morphology for locomotor evolution in early cetacea. In J. G. M. Thewissen, ed., *The Emergence of Whales,* pp. 325–52. Plenum.

BUCKLEY, G. A., C. A. BROCHU, D. W. KRAUSE, and D. POL. 2000. A pug-nosed crocodyliform from the Late Cretaceous of Madagascar. *Nature* 405:941–44.

BUFFETAUT, E. 1989. Evolution. In C. A. Ross, ed., *Crocodiles and Alligators,* pp. 14–25. Facts On File.

BUFFRÉNIL, V. DE, L. ZYLBERBERG, W. TRAUB, and A. CASINOS. 2000. Structural and mechanical characteristics of the hyperdense bone of the rostrum of *Mesoplodon densirostris* (Cetacea, Ziphiidae): summary of recent obesrvations. *Historical Biology* 14(1–2):57–65.

BULLARD, E. C. 1969. The origin of the oceans. *Scientific American* 221(3):68–75.

BURKENROAD, M. D. 1943. A possible function of bioluminescence. *Jour. Mar. Res.* 5: 161–64.

BUSCHBAUM, R., and L. J. MILNE. 1966. *Living Invertebrates of the World.* Doubleday.

BUSCHBAUM, R., M. BUSCHBAUM, J. PEARSE, and V. PEARSE. 1987. *Animals Without Backbones.* University of Chicago Press.

BUSHBECK, E., B. EHMER, and R. HOY. 1999. Chunk versus point sampling: visual imaging in a small insect. *Science* 286:1178–80.

BUTTERFIELD, N. J., A. H. KNOLL. and K. SWETT. 1988. Exceptional preservation of fossils in an Upper Proterozoic shale. *Nature* 334:424–27.

CALDWELL, D. K., and M. C. CALDWELL. 1972. Senses and communication. In S. H. Ridgway, ed., *Mammals of the Sea: Biology and Medicine,* pp. 466–502. Charles C. Thomas.

———. 1985. Manatees—*Trichecus manatus, Trichecus senegalensis,* and *Trichecus inunguis.* In S. H. Ridgway and R. J. Harrison, eds., *Handbook of Marine Mammals.* Vol. 3, *The Sirenians and Baleen Whales,* pp. 33–36. Academic Press.

CALDWELL, D. K., M. C. CALDWELL, and D. W. RICE. 1966. Behavior of the sperm whale, *Physeter catodon* L. In K. S. Norris, ed., *Whales, Dolphins and Porpoises*, pp. 677–717. University of California Press.

CALDWELL, M. W. 1997. Modified perichondrial ossification and the evolution of paddlelike limbs in ichthyosaurs and plesiosaurs. *Jour. Vert. Paleo.* 17(32):534–47.

———. 2000. On the phylogenetic relationships of *Pachyrachis* within snakes: a response to Zaher. *Jour. Vert. Paleo.* 20(1):187–190.

CAMPAGNO, L. J. V. 1984. *FAO Species Catalogue.* Vol 4, *Sharks of the World. Part 1—Hexanchiformes to Lamniformes.* U. N. Development Programme, Rome.

CAMPBELL, R. R. 1988. Status of the sea mink *Mustela macrodon* in Canada. *Canadian Field Naturalist* 102(2):304–6.

CANFIELD, D. E. 1998. A new model for Proterozoic ocean chemistry. *Nature* 396:450–53.

———. 1999. A breath of fresh air. *Nature* 400:503–5.

CANFIELD, D. E., and A. TESKE. 1996. Late Ptoterozoic rise in atmospheric oxygen concentration inferred from phylogenetic and sulphur-isotope studies. *Nature* 382:127–32.

CARLSON, B. A. 1987. Collection and aquarium maintenance of *Nautilus.* In W. B. Saunders and N. H. Landman, eds., *Nautilus: The Biology and Paleobiology of a Living Fossil*, pp. 563–78. Plenum.

CARLSON, B. A., M. AWAI, and J. ARNOLD. 1992. Waikiki Aquarium's chambered nautilus reach their first "hatch-day" anniversary. *Hawaiian Shell News* 40(1):1–3.

CARR, A. 1995. Notes on the behavioral ecology of sea turtles. In K. Bjorndal, ed., *Biology and Conservation of Sea Turtles.* First published in 1982. Smithsonian Institution Press.

CARROLL, R. L. 1988. *Vertebrate Paleontology and Evolution.* Freeman.

———. 1995. Between fish and amphibian. *Nature* 373:398–90.

CASE, G. R. 1973. *Fossil Sharks: A Pictorial Review.* Pioneer Litho.

CASE, J. F., J. WARNER, A. T. BARNES, and M. LOWENSTINE. 1977. Bioluminescence of lanternfish (Myctophidae) in response to changes in light intensity. *Nature* 265:179–81.

CHANG, M. 2000. Fossil fish up for election. *Nature* 403:152–53.

CHATTERJEE, S. 1997. *The Rise of Birds.* Johns Hopkins University Press.

CHEN, J.-Y., J. DZIK, G. D. EDGECOMBE, L. RAMSKÖLD, and G.-Q. ZHOU. 1995. A possible Early Cambrian chordate. *Nature* 377:720–22.

CHEN, J.-Y., HOU, X., and L. LU. 1985. Early Cambrian netted scale-bearing worm-like sea animal. *Acta Paleontologica Sinica* 28:1–16.

CHEN, J.-Y., D.-Y. HUANG, and C.-W. LI. 1999. An early Cambrian craniate-like chordate. *Nature* 402:518–22.

CHEN, J.-Y., J. DZIK, G. D. EDGECOMBE, L. RAMSKÖLD, and G.-Q. ZHOU. 1995. A possible Early Cambrian chordate. *Nature* 377:720–22.

CHIAPPE, L. M. 1995. The first 85 million years of avian evolution. *Nature* 378:349–55.

CHILDRESS, J. J., H. FELBECK, and G. N. SOMERO. 1987. Symbiosis in the deep sea. *Scientific American* 256(5):115–20.

CHYBA, C. F. 1990. Impact delivery and erosion of planetary oceans in the early inner Solar System. *Nature* 343:129–33.

———. 1998. Origins of life: buried beginnings. *Nature* 395:329–30.

CLACK, J. A. 1997. Devonian tetrapod trackways and trackmakers: a review of the fossils and footprints. *Palaeogeography, Palaeoclimatology, Palaeoecology* 130:227–50.

———. 1998. A new Early Carboniferous tetrapod with a mélange of crown-group characters. *Nature* 394:66–69.

CLARKE, A. H. 1962a. On the composition, zoogeography, origin and age of the deep-sea mollusk fauna. *Deep-Sea Research* 9:291–306.

CLARKE, J. M. 1889. The structure and development of the visual area in the trilobite *Phacops rana,* Green. *Journal of Morphology* 2:253–267.

CLARKE, M. R. 1970. Growth and development of *Spirula spirula. Jour. Mar. Biol. Assoc. U.K.* 50:53–64.

CLARKE, W. D. 1963. Function of bioluminescence in mesopelagic organisms. *Nature* 198:1244–46.

CLINES, F. X. 2000. U.S. acts to protect embattled horseshoe crab. *New York Times*, August 9, A12.

COATES, M. I. 1994. The origin of vertebrate limbs. *Development* 120(1):169–80.

COATES, M. I., and J. A. CLACK. 1990 Polydactyly in the earliest known tetrapod limbs. *Nature* 347:66–69.

CODY, G. D., N. Z. BOCTOR, T. R. FILLEY, R. M. HAZEN, J. H. SCOTT, A. SHARMA, and H. S. YODER. 2000. Primordial carbonylated iron-sulfur compounds and the synthesis of pyruvate. *Science* 289:1337–40.

COLBERT, E. H. 1955. *Evolution of the Vertebrates.* John Wiley.

———. 1963. *Dinosaurs: Their Discovery and Their World.* Hutchinson.

———. 1965. *The Age of Reptiles.* Dover.

———. 1973. *Wandering Lands and Animals.* Dutton.

COLLINS, D. 1996. The "evolution" of *Anomalocaris* and its classification in the arthropod class Dinocardia (nov.) and order Radiodonta (nov.) *Jour. Paleo.* 70(2):280–93.

CONE, J. 1991. *Fire Under the Sea: Volcanic Hot Springs on the Ocean Floor.* Quill.

CONWAY MORRIS, S. 1993. The fossil record and the early evolution of the Metazoa. *Nature* 361:219–25.

———. 1994. Early metazoan evolution: first steps to an integration of molecular and morphological data. In S. Bengston, ed., *Early Life on Earth,* pp. 450–59. Columbia University Press.

———. 1997. *The Crucible of Creation: The Burgess Shale and the Rise of Animals.* Oxford University Press.

———. 1998–99. Showdown on the Burgess Shale: the challenge. *Natural History* 107(10):48–51.

CONWAY MORRIS, S., and J. S. PEEL. 1990. Articulated halkieriids from the Lower Cambrian of north Greenland. *Nature* 345:802–5.

CONWAY MORRIS, S., S. JENSEN, and N. J. BUTTERFIELD. 1998. Fossil discoveries in India: continued. *Science* 282:1265.

CORLISS, J. B., and R. D. BALLARD. 1977. Oases of life in a cold abyss. *National Geographic* 152(4):441–53.

CORLISS, J. B., J. DYMOND, L. I. GORDON, J. M. EDMOND, R. P. VON HERZEN, R. D. BALLARD, K. GREEN, D. WILLIAMS, A. BAINBRIDGE, K. CRANE, and T. H. VAN ANDEL. 1979. Submarine thermal springs on the Galápagos Rift. *Science* 203:1073–83.

COWEN, R. 1996. Locomotion and respiration in aquaitc air-breathing vertebrates. In D. Jablonski, D. H. Erwin, and J. H. Lipps, eds., *Evolutionary Paleobiology,* pp. 337–53. University of Chicago Press.

———. 2000. *History of Life.* Blackwell Science.

CRANFORD, T. W. 1999. The sperm whale's nose: sexual selection on a grand scale? *Marine Mammal Science.* 15(4):1133–57.

DAESCHLER, E. B. 2000. Early tetrapod jaws from the Late Devonian of Pennsylvania, USA. *Jour. Paleo.* 74:301–8.

DAESCHLER, E. B., and N. SHUBIN. 1995. Tetrapod origins. *Paleobiology* 21(4):404–9.

———. 1998. Fish with fingers? *Nature* 391:133.

DAHLGREN, U. 1916. The production of light by animals; light production in cephalopods. *Jour. Franklin Inst.* 81:525–56.

———. 1917. The production of light by animals; luminosity in fishes. *Jour. Franklin Inst.* 183(6):735–54.

———. 1928. The bacterial light of *Ceratias. Science* 68:65–66.

DANIEL, T. L., B. S. HELMUTH, W. B. SAUNDERS, and P. D. WARD. 1997. Septal complexity in ammonoid cephalopods increased mechanical risk and limited depth. *Paleobiology* 23(4):470–81.

DARWIN, C. 1855. *The Voyage of the Beagle.* London.

———. 1859. *The Origin of Species by Means of Natural Selection.* London.

DAS, H. S., and S. C. DEY. 1999. Observations on the dugong, *Dugong dugon* (Muller), in the Andaman and Nicobar Islands, India. *Jour. Bombay Nat. Hist. Soc.* 96(2):195–98.

DAVIDSON, J. P. 1999. Dating the timeline of Earth's history. In J. W. Schopf, ed., *Evolution: Facts and Fallacies,* pp. 15–36. Academic Press.

DAVIES, P. 1999. *The Fifth Miracle (The Search for the Origin and Meaning of Life).* Simon & Schuster.

DAWKINS, R. 1996. *Climbing Mount Improbable.* W. W. Norton.

———. 2000. Branching out. [Review of *Extinct Humans,* by Tattersall and Schwartz.] *New York Times Book Review,* August 6, 18–19.

DEAN, B. 1909. The giant of ancient sharks. *Amer. Mus. Jour.* 9(8):232–34.

DE LA BECHE, H. T. and W. D. CONYBEARE. 1821. Notice of the discovery of a new fossil animal, forming a link between the *Ichthyosaurus* and the crocodile, together with general remarks on the osteology of the *Ichthyosaurus. Trans. Geol. Soc. London* 5:559–94

DEMÉRÉ, T. A. 1986. The fossil whale *Balaenoptera davidsonii* (Cope 1872), with a review of other Neogene species of *Balaenoptera* (Cetacea: Mysticeti). *Marine Mammal Science* 2(4):277–98.

———. 1994a. Two new species of fossil walruses (Pinnipedia: Odobenidae) from the Upper Pliocene San Diego Formation, California. *Proc. San Diego Hist. Soc.* 29:77–98.

———. 1994b. The family Odobenidae: a phylogenetic analysis of fossil and living taxa. *Proc. San Diego Hist. Soc.* 29:99–124.

DENISON, R. H. 1941. The soft anatomy of *Bothriolepis. Jour. Paleo.* 15:533–61.

———. 1978. Placodermi. In H.-P. Schultze, ed., *Handbook of Paleoichthyology,* vol. 2. Gustav Fischer Verlag.

———. 1979. Acanthodii. In H.-P. Schultze, ed., *Handbook of Paleoichthyology,* vol. 5. Gustav Fischer Verlag.

———. 1984. Further consideration of the phylogeny and classification of the order Arthrodira (Pisces; Placodermi). *Jour. Vert. Paleo.* 4(3):396–412.

DENNETT, D. C. 1995. *Darwin's Dangerous Idea.* Simon & Schuster.

DENTON, R. K., J. L. DOBIE, and D. C. PARRIS. 1997. The marine crocodilian *Hyposaurus* in North America. In J. M. Callaway and E. L. Nicholls, eds., *Ancient Marine Reptiles*, pp. 375–97. Academic Press.

DE SYLVA, D. P. 1966. Mystery of the silver coelacanth. *Sea Frontiers* 12(3):172–75.

DIETZ, T. 1992. *The Call of the Siren: Manatees and Dugongs*. Fulcrum.

DIXON, D. 1981. *After Man: A Zoology of the Future*. St. Martin's.

DIXON, D., B. COX, R. J. G. SAVAGE, and B. GARDINER. 1988. *The Macmillan Illustrated Encyclopedia of Dinosaurs and Prehistoric Animals*. Macmillan.

DOINO, J. A. and M. J. MCFALL-NGAI. 1995. Transient exposure to competent bacteria initiates symbiosis-specific squid light organ morphogenesis. *Biol. Bull.* 189:347–55.

DOMNING, D. P. 1976. An ecological model for Late Tertiary sirenian evolution in the North Pacific Ocean. *Systematic Zoology* 25(4):352–61.

———. 1981. Manatees of the Amazon. *Sea Frontiers* 27(1):18–23.

———. 1982. Evolution of manatees: a speculative history. *Jour. Paleo.* 56:599-619.

———. 1994. A phylogenetic analysis of the Sirenia. *Proc. San Diego Hist. Soc.* 29:177–90.

———. 1999a. Fossils explained 24: sirenians (seacows). *Geology Today*, March–April, 75–79.

———. 1999b. The earliest sirenians: what we know and what we would like to know. In E. Hoch and A. K. Brantsen, eds., *Secondary Adaptation to Life in Water*, pp. 12–13. Papers presented at the University of Copenhagen Geological Museum, 13–17 September 1999.

———. 2000. The readaptation of Eocene sirenians to life in water. *Historical Biology* 14(1–2):115–19.

DOMNING, D. P., and V. DE BUFFRÉNIL. 1991. Hydrostasis in the Sirenia: quantitative data and functional interpretation. *Marine Mammal Science* 7(4):331–68.

DOMNING, D. P., and P. D. GINGERICH. 1994. *Protosiren smithae,* new species (Mammalia, Sirenia), from the late middle Eocene of Wadi Hitan, Egypt. *Contrib, Mus. Paleo. Univ. Mich.* 29(3):69–87.

DOMNING, D. P., and C. E. RAY. 1986. The earliest sirenian (Mammalia: Dugongidae) from the Eastern Pacific Ocean. *Marine Mammal Science* 2(4):263–76.

DOMNING, D. P., G. S. MORGAN, and C. E. RAY. 1982. North American Eocene sea cows (Mammalia: Sirenia). *Smithsonian Contrib. Paleobiol.* 52:1–69.

DOMNING, D. P., C. E. RAY, and M. C. MCKENNA. 1986. Two new Oligocene desmostylians and a discussion of Tethytherian systematics. *Smithsonian Contrib. Biol.* 59:1–55.

DONOVAN, D. T. 1977. Evolution of the dibranchiate cephalopoda. *Symp. Zool. Soc. London* 38:15–48

DOUGLAS, K. 2000. Eve's watery origins: how the sea shore made us human. *New Scientist* 168(226):28–33.

DUDLEY, P. 1725. An essay upon the natural history of whales. *Phil. Trans. Roy. Soc. London* 33:256–69.

DU TOIT, A. I. 1927. A geological comparison of South America with South Africa. *Carnegie Inst. Wash. Publ.* 381:1–157.

———. 1937. *Our Wandering Continents: An Hypothesis of Continental Drifting*. Oliver & Boyd.

EASTMAN, C. R. 1900. Karpinsky's genus *Helicoprion*. *American Naturalist* 34:579–82.

———. 1902. Some carboniferous cestraciont and acanthodian sharks. *Bull. Mus. Comp. Zool.* 39:55–99.

———. 1903. Carboniferous fishes from the central western States. *Bull. Mus. Comp. Zool.* 39:163–226.

EBEL, K. 1990. Swimming abilities of ammonites and limitations. *Paläontologische Zeitschrift* 62:25–37.

———. 1992. Mode of life and soft body shape of heteromorph ammonites. *Lethaia* 25:179–93.

ELDREDGE, N. 1975. Survivors from the good old, old, old days. *Natural History* 81(10):52–59.

———. 1976. Collecting trilobites in North America. Part one: the East. *Fossils* 1(1):58–67.

———. 1984. Simpson's Inverse: bradytely and the phenomenon of living fossils. In N. Eldredge and S. M. Stanley, eds., *Living fossils*, pp. 272–77. Springer-Verlag.

———. 1991. *Fossils: The Evolution and Extinction of Species*. Princeton University Press.

———. 2000. *The Triumph of Evolution and the Failure of Creationism*. Freeman.

ELDREDGE, N., and I. TATTERSALL. 1977. Fact, theory, and fantasy in human paleontology. *American Scientist* 65:204–211.

ELLIOT, D. K., G. V. IRBY, and J. H. HUTCHISON. 1997. *Desmatochelys lowi,* a marine turtle from the Upper Cretaceous. In J. M. Callaway and E. L. Nicholls, eds., *Ancient Marine Reptiles*, pp. 243–258. Academic Press.

ELLIS, R. 1975. *The Book of Sharks*. Grosset & Dunlap.

———. 1998. *The Search for the Giant Squid*. Lyons.

ELLIS, R., and J. E. MCCOSKER. 1991. *Great White Shark*. Stanford University Press.

EMLONG, D. R. 1966. A new archaic cetacean from the Oligocene of northwest Oregon. *Univ. Ore. Nat. Hist. Bull.* 3:1–51.

ENGESSER, T. 1998. The fossil Coleoidea page. http://userpage.fu-berlin.de

ENSERINK, M. 1999. Fossil opens window on early animal history. *Science* 286:1829.

ERDMANN, M. V. 1998. Sulawesi coelacanths. *Ocean Realm*, Winter 1998–99, 26–28.

ERDMANN, M. V., and R. L. CALDWELL. 2000. How new technology put a coelacanth among the heirs of Piltdown Man. *Nature* 406:343.

ERDMANN, M. V., R. L. CALDWELL, and M. KASIM MOOSA. 1998. Indonesian "king of the sea" discovered. *Nature* 395:335.

ERICKSON, B. R. 1990. Paleoecology of crocodile and whale-bearing strata of Oligocene age in North America. *Historical Biology* 4:1–14.

ERWIN, D. H. 1994. The Permo-Triassic extinction. *Nature* 367:231–36.

———. 1996. The mother of mass extinctions. *Scientific American*, 273(1):72–78.

ERWIN, D. H., J. W. VALENTINE, and D. JABLONSKI. 1997. The origin of animal body plans. *American Scientist* 85(2):126–37.

ESTES, J. A. 1980. Enhydra lutris. *Mammalian Species* (American Society of Mammalogists) 133:1–8.

EVERHART, M. 1999. *Cretoxyrhina mantelli*—the Ginsu shark. http://www.oceansofkansas.com/ginsu.html

FEDONKIN, M. A. 1994. Vendian body fossils and trace fossils. In S. Bengston, ed., *Early life on Earth*, pp. 370–88. Columbia University Press.

FEDONKIN, M. A., and B. M. WAGGONER. 1997. The Late Precambrian fossil *Kimberella* is a mollusc-like bilaterian organism. *Nature* 388:868–71.

FELDMAN, H. R., R. LUND, C. G. MAPLES. and A. W. ARCHER. 1994. Origin of the Bear Gulch beds (Namurian, Montana, USA). *Geobios* 16:283–91.

FISH, F. E. 1998. Biomechanical perspective on the origin of cetacean flukes. In J. G. M. Thewissen, ed., *The Emergence of Whales*, pp. 303–24. Plenum.

FISHER, D. C. 1984. The Xiphosurida: archetypes of bradytely? In N. Eldredge and S. M. Stanley, eds., *Living Fossils*, pp. 196–213. Springer-Verlag.

FITCH, J. E., and R. J. LAVENBERG. 1968. *Deep-Water Fishes of California*. University of California Press.

FLOWER, W. H. 1872. On a subfossil whale (*Eschrichtius robustus*) discovered in Cornwall. *Ann. Mag. Nat. Hist.* 4(9):440–42.

FOLKENS, P. A., and L. G. BARNES. 1984. Reconstruction of an archaeocete. *Oceans* 17(2):22–23.

FÖRCH, E. C. 1998. The marine fauna of New Zealand: Cephalopoda: Oegopsida: Architeuthidae (giant squid). *NIWA Biodiversity Memoir* 110:1–113.

FORDYCE, R. E. 1977. The development of the Circum-Antarctic Current and the evolution of the Mysticeti (Mammalia: Cetacea). *Palaeogeography, Palaeoclimatology, Palaeoecology* 21:256–71.

———. 1980. Whale evolution and Oligocene Southern Ocean environments. *Palaeogeography, Palaeoclimatology, Palaeoecology* 31:319–36.

———. 1982. A review of Australian fossil cetacea. *Mem. Nat. Mus. Victoria* 43:43–58.

———. 1994. *Waipatia maerewhenua,* new genus and new species (Waipatiidae, new family), an archaic Late Oligocene dolphin (Cetacea: Odontoceti: Platanistoidea) from New Zealand. *Proc. San Diego Hist. Soc.* 29:147–76.

FORDYCE, R. E., and L. G. BARNES. 1994. The evolutionary history of whales and dolphins. *Ann. Rev. Earth. Planet. Sci.* 22:419–55.

FORDYCE, R. E., and C. M. JONES. 1990. Penguin history and new fossil material from New Zealand. In L. S. Davis and J. T. Darby, eds., *Penguin Biology*, pp. 419–46. Academic Press.

FORDYCE, R. E., C. M. JONES, and B. D. FIELD. 1986. The world's oldest penguin? *Geol. Soc. N.Z. Newsletter* 74:56–57.

FOREY, P. 1984. The coelacanth as a living fossil. In N. Eldredge and S. M. Stanley, eds., *Living Fossils*, pp. 166–69. Springer-Verlag.

———. 1998a. *History of the Coelacanth Fishes*. Chapman & Hall.

———. 1998b. A home from home for the coelacanths. *Nature* 395:319–20.

FOREY, P., and P. JANVIER. 1993. Aganthans and the origin of jawed vertebrates. *Nature* 361:129–34.

———. 1994. Evolution of the early vertebrates. *American Scientist* 82(6):554–65.

FORSTEN, A., and P. M. YOUNGMAN. 1982. Hydrodamalis gigas. *Mammalian Species* 165:1–3.

FORTEY, R. 1991. *Fossils: The Key to the Past*. Natural History Museum Publications (London).

———. 1998. *Life: A Natural History of the First Four Billion Years of Life on Earth*. Knopf.

———. 2000a. *Trilobite!* HarperCollins.

———. 2000b. Olenid trilobites: the oldest known chemoautotrophic symbionts? *Proc. Nat. Acad. Sci.* 97(12):6574–78.

FOSTER, J. S., and M. J. MCFALL-NGAI. 1998. Induction of apoptosis by cooperative bacteria in the morphogenesis of host epithelial tissues. *Dev. Genes Evol.* 208:295–303.

FRICKE, H., K. HISSMANN, J. SCHAUER, M. ERDMANN, M. K. MOOSA, and R. PLANTE. 2000. Biogeography of Indonesian coelacanths. *Nature* 403:38–39.

FUSON, R. H. 1987. *The Log of Christopher Columbus*. International Marine Publishing.

GAETH, A. P., R. V. SHORT, and M. B. RENFREE. 1999. The developing renal, reproductive, and respiratory systems of the African elephant suggest an aquatic ancestry. *Proc. Natl. Acad. Sci.* 96:5555–58.

GALT, C. P. 1978. Bioluminescence: dual mechanism in a planktonic tunicate produces brilliant surface display. *Science* 200:70–71.

GATESY, J. 1998. Molecular evidence for the phylogenetic affinities of Cetacea. In J. G. M. Thewissen, ed., *The Emergence of Whales*, pp. 63–112. Plenum.

GEHLING, J. G. 1999. Microbial mats in terminal Proterozoic siliciclastics: Ediacaran death masks. *Palaios* 14:40–57.

GEISLER, J. H., and Z. LUO. 1996. The petrosal and inner ear of *Herpetocetus* spp. (Mammalia: Cetacea) and their implications for the phylogeny and hearing of archaic mysticetes. *Jour. Paleo.* 70(6):1045–66.

———. 1998. Relationships of cetacea to terrestrial ungulates and the evolution of cranial vasculature in Cete. In J. G. M. Thewissen, ed., *The Emergence of Whales*, pp. 163–212. Plenum.

GIDLEY, J. W. 1975. A recently mounted Zeuglodon skeleton in the United States National Museum. *Proc. U.S. Natl. Mus.* 44:649–54.

GINGERICH, P. D. 1991. Partial skelton of a new archaeocete from the earliest middle Eocene Habib Rahi Limestone, Pakistan. *Jour. Vert. Paleo.* 11:31A.

———. 1998. Paleobiological perspectives on Mesonychia, Archaeoceti, and the origin of whales. In J. G. M. Thewissen, ed., *The Emergence of Whales*, pp. 423–50. Plenum.

GINGERICH, P. D., and D. E. RUSSELL. 1981. *Pakicetus inachus,* a new archaeocete (Mammalia, Cetacea) from the early-middle Eocene Kuldana Formation of Kohat (Pakistan). *Contrib. Mus. Paleo. Univ. Mich.* 25(11):235–246.

———. 1991. Dentition of Early Eocene *Pakicetus* (Mammalia, Cetacea). *Contrib. Mus. Paleo. Univ. Mich.* 28(1):1–20.

———. 1995. Unusual mammalian limb bones (Cetacea? Archaeoceti?) from the early-to-middle Eocene Subathu Formation of Kashmir (Pakistan). *Contrib. Mus. Paleo. Univ. Mich.* 29:109–17.

GINGERICH, P. D., and M. D. UHEN. 1996. *Analectus simonsi,* a new dorudontine archaeocete (Mammalia, Cetacea) from the early late Eocene of Wadi Hitan, Egypt. *Contrib. Mus. Paleo. Univ. Mich.* 29:359–401.

———. 1997. The evolution of whales. *LSAmagazine* (University of Michigan) 20(2):4–10.

GINGERICH, P. D., M. ARIF, and W. C. CLYDE, 1995. New archaeocetes (Mammalia, Cetacea) from the middle Eocene Domanda Formation of the Sulaiman Range, Punjab (Pakistan). *Contrib. Mus. Paleo. Univ. Mich.* 29:291–330.

GINGERICH, P. D., B. H. SMITH, and E. L. SIMONS. 1990. Hind limbs of Eocene *Basilosaurus:* evidence of feet in whales. *Science* 246:154–57.

GINGERICH, P. D., M. ARIF, M. A. BHATTI, and W. C. CLYDE. 1998. Middle Eocene stratigraphy and marine mammals (Mammalia: Cetacea and Sirenia) of the Sulaiman Range, Pakistan. *Bull. Carnegie Mus. Nat. Hist.* 34:239–59.

GINGERICH, P. D., M. A. BHATTI, H. A. RAZA, and S. M. RAZA. 1995. *Protosiren* and *Babiacetus* (Sirenia and Cetacea) from the Middle Eocene Drazinda Formation, Sulaiman Range, Punjab (Pakistan). *Contrib. Mus. Paleo. Univ. Mich.* 29(12):331–57.

GINGERICH, P. D., D. P. DOMNING, C. E. BLAINE, and M. D. UHEN. 1994. Cranial morphology of *Protosiren fraasi* (Mammalia, Sirenia) from the Middle Eocene of Egypt: a new study using computed tomography. *Contrib. Mus. Paleo. Univ. Mich.* 29(2):41–67.

GINGERICH, P. D., S. M. RAZA, M. ARIF, M. ANWAR, and X. ZHOU. 1993. Partial skeletons of *Indocetus ramani* (Mammalia, Cetacea) from the lower middle Eocene Domanda Shale in the Sulaiman Range of Punjab, Pakistan. *Contrib. Mus. Paleo. Univ. Mich.* 16:393–416.

GINGERICH, P. D., N. A. WELLS, D. E. RUSSELL, and S. M. I. SHAH. 1983. Origin of whales in epicontinental remnant seas: new evidence from the Early Eocene of Pakistan. *Science* 220:403–6.

GLAESSNER, M. F. 1958. New fossils from the base of the Cambrian in South Australia. *Trans. Roy. Soc. S. Aust.* 81:185–88.

———. 1961. Pre-Cambrian animals. *Scientific American* 204(3):72–78.

GLAUSNIUZ, J. 1999. The old fish of the sea. *Discover* 20(1):49.

GORE, R. 1993. The Cambrian Period: explosion of life. *National Geographic* 184(4):120–36.

GOTTFRIED, M. D., L. J. V. COMPAGNO, and S. C. BOWMAN. 1996. Size and skeletal anatomy of the giant "Megatooth" shark *Carcharodon megalodon.* In A. P. Klimley and D. G. Ainley, eds., *Great White Sharks*, pp. 55–66. Academic Press.

GOULD, S. J. 1980a. An early start. In *The Panda's Thumb: More Reflections on Natural History* (pp. 217–26). Norton.

———. 1980b. Introduction. In B. Kurtén, *Dance of the Tiger.* Pantheon.

———. 1983. Nature's great era of experiments. *Natural History* 7/83:12–20.

———. 1987. The fossil fraud that never was. *New Scientist* 113(1551):32–36.

———. 1989. *Wonderful Life*. W. W. Norton.

———. 1993. *Eight Little Piggies*. W. W. Norton.

———. 1994a. The evolution of life on the earth. *Scientific American* 271(4):85–91.

———. 1994b. Hooking Leviathan by its past. *Natural History* 103(5):8116.
———. 1998–99. Showdown on the Burgess Shale: the reply. *Natural History* 107(10):49–55.
———. 1999. The evolution of life. In J. W. Schopf. ed., *Evolution: Facts and Fallacies*, pp. 1–13. Academic Press.
———. 2000a. Will we figure out how life began? *Time* 155(14):92–93.
———. 2000b. Linnaeus's luck? *Natural History* 109(7):18–25, 66–76.
GOULD, S. J., and N. ELDREDGE. 1977. Punctuated equilibria: the tempo and mode of evolution reconsidered. *Paleobiology* 3:115–51.
GRAHAM, A., and P. BEARD. 1973. *Eyelids of Morning: The Mingled Destinies of Crocodiles and Men*. New York Graphic Society.
GREEN, R. G. 1974. Teuthids of the Late Cretaceous Niobrara Formation of Kansas and some ecological implications. *The Compass* 51(3):53–60.
———. 1977. *Niobrarateuthis walkeri*, a new species of teuthid from the Upper Cretaceous Niobrara Formation of Kansas. *Jour. Paleo.* 51(5):992–95
GREENE, H. W. 1997. *Snakes: The Evolution of Mystery in Nature*. University of California Press.
GREENE, H. W., and D. CUDNALL. 2000. Limbless tetrapods and snakes with legs. *Science* 287:1939–41.
GREER, A. E., J. D. LAZELL, and R. M. WRIGHT. 1973. Anatomical evidence for a counter-current heat exchanger in the leatherback turtle *(Demochelys coriacea). Nature* 244:181.
GREGORY, J. T. 1951. Convergent evolution: the jaws of *Hesperornis* and the mosasaurs. *Evolution* 5:345–54.
GROGAN, E. D., and R. LUND. 1997. Soft tissue pigments of the Upper Mississippian chondrenchelyid *Harpagofututor volsellorhinus* (Chondrichthyes, Holocephali) from the Bear Gulch Limestone of Montana, USA. *Jour. Paleo.* 71(2):337–42.
———. 2000. *Debeerius ellefseni* (fam. nov., gen. nov., spec. nov.), an autodiastylic chondrichthyan from the Mississippian Bear Gulch Limestone of Montana (USA), the relationships of the chondrichthyes, and comments on gnathostome evolution. *Jour. Morphology* 243(3):219–45.
GURA, T. 2000. Bones, molecules . . . or both? *Nature* 406:230–33.
HAGADORN, J. W., and B. WAGGONER. 2000. Ediacaran fossils from the southwestern Great Basin, United States. *Jour. Paleo.* 74(2):349–59.
HAGADORN, J. W., C. M. FEDO, and B. WAGGONER. 2000. Early Cambrian Ediacaran-type fossils from California. *Jour. Paleo.* 74(4):731–40.
HALEY, D. 1978. Saga of Steller's sea cow. *Natural History* 87(9):9–17.
HALL, A. J. 1984. Man and manatee: can we live together? *National Geographic* 166(3):400–413.
HALL, R. L. 1985. *Paraplesioteuthis hastata* (Münster), the first teuthid squid recorded from the Jurassic of North America. *Jour. Paleo.* 59(4):870–74.
HALLAM, A. 1972. Continental drift and the fossil record. *Scientific American* 227(11):56–66.
HALLAM, A., and P. B. WIGNALL. 1997. *Mass Extinctions and Their Aftermath*. Oxford University Press.
HALSTEAD, L. B. 1968. *The Pattern of Vertebrate Evolution*. Freeman.
———. 1985. On the posture of desmostylians: a discussion of Inuzuka's "Herpetiform mammals." *Mem. Fac. Sci. Kyoto Univ. Ser. Biol.* 10:137–44.
HANLON, R. T., and J. B. MESSENGER. 1996. *Cephalopod Behaviour*. Cambridge University Press.
HARDY, A. 1960a. Was man more aquatic in the past? *New Scientist* 7(174):642–45.
———. 1960b. Will man be more aquatic in the future? *New Scientist* 7(175):730–33.
HARLAN, R. 1828. Note on the examination of the large bones disinterred at the mouth of the Mississippi River, and exhibited in the city of Baltimore, January 22, 1828. *Amer. Jour. Sci. and Art* 14:186–87.
HARRISON, R. J., and J. E. KING. 1980. *Marine Mammals*. Hutchinson.
HARTMAN, D. S. 1969. Florida's manatees: mermaids in peril. *National Geographic* 36(3):342–53.
———. 1979. Ecology and behavior of the manatee (*Trichechus manatus*) in Florida. Spec. Pub. No. 5. American Society of Mammalogists.
HARVEY, E. N. 1921. A fish with a luminous organ, designed for the growth of luminous bacteria. *Science* 53:314–15.
HASTINGS, J. W. 1971. Light to hide by: ventral luminescence to camouflage the silhouette. *Science* 173:1016–17.
———. 1983. Biological diversity, chemical mechanisms, and the evolutionary origins of bioluminescent systems. *Jour. Molec. Evol.* 19:309–21.
———. 1998. Bioluminescence. In N. Sperelakis, ed., *Cell Physiology Source Book* 2nd ed., pp. 984–1000. Academic Press.
HAY, O. P. 1907. A new genus and species of fossil shark related to *Edestus* Leidy. *Science* 26:22–24.
———. 1909. On the nature of *Edestus* and related genera, with descriptions of one new genus and 3 new species. *Proc. U. S. Nat. Mus.* 37:43–61.
HAYGOOD, M. G., and D. L. DISTEL. 1993. Bioluminescent symbionts of flashlight fishes and deep-sea anglerfishes form unique lineages related to the genus *Vibrio*. *Nature* 363:155–57.

HEATWOLE, H. 1999. *Sea Snakes.* Krieger.
HECHT, J. 2000. Prehistoric pins. *New Scientist* 165:15.
HEEZEN, B. C. 1957. Whales entangled in deep-sea cables. *Deep-Sea Research* 4:105–15.
HEIDELBERG, J. F., et al. 2000. DNA sequence of both chromosomes of the cholera pathogen *Vibrio cholerae. Nature* 406:477–83.
HEIRTZLER, J. R., and W. B. BRYAN. 1975. The floor of the Mid-Atlantic Ridge. *Scientific American* 233(2):78–90.
HENDEY, Q. B., and C. A. REPENNING. 1972. A Pliocene phocid from South Africa. *Ann. S. Afr. Mus.* 59(4):71–98.
HENDRICKSON, J. R. 1980. The ecological strategies of sea turtles. *American Zoologist* 20:597–603.
HERRING, P. J. 1974. New observations on the bioluminescence of echinoderms. *Jour. Zool. London.* 172:401–18.
———. 1977. Bioluminescence of marine organisms. *Nature* 267:788–93.
———. 1993. Light genes will out. *Nature* 363:110–11.
HERRING, P. J., M. R. CLARKE, S. VON BOLETZKY and K. P. RYAN. 1981. The light organs of *Sepiola atlantica* and *Spirula spirula* (Mollusca: Cephalopoda): bacterial and intrinsic systems in the order Sepioidea. *Jour. Mar. Biol. Ass. U.K.* 61:901–16.
HERRING, P. J., A. K. CAMPBELL, M. WHITFIELD, and L. MADDOCK, eds. 1990. *Light and Life in the Sea.* Cambridge University Press.
HEYNING, J. E. 1997. Sperm whale phylogeny revisited: analysis of the morphological evidence. *Marine Mammal Science* 13(4):596–13.
———. 1999. Whale origins—conquering the seas. Review of *The Emergence of Whales,* J. G. M. Thewissen, ed. *Science* 283:943.
HINDELL, M. A., D. J. SLIP, H. R. BURTON, and M. M. BRYDEN. 1992. Physiological implications of continuous, prolonged, and deep dives of the southern elephant seal (*Mirounga leonina*). *Canadian Jour. Zool.* 70:370–79.
HIRAYAMA, R. 1997. Distribution and diversity of Cretaceous chelonioids. In J. M. Callaway and E. L. Nicholls, eds., *Ancient Marine Reptiles,* pp. 225–41. Academic Press.
HISSMANN, K., H. FRICKE, and J. SCHAUER. 1997. Population monitoring of the coelacanth (*Latimeria chalumnae*). *Conservation Biology* 12(4):759–65.
HOCH, E. 2000. Olfaction in whales: Evidence from a young odontocete of the Late Oligocene North Sea. *Historical Biology* 14(1–2):67–89.
HOOKER, S. K., and R. W. BAIRD. 1999. Deep-diving behaviour of the northern bottlenose whale, *Hyperoodon ampullatus* (Cetacea: Ziphiidae). *Proc. Royal. Soc. London* 266:671–76.
HORIKAWA, H. 1994. A primitive odobenine walrus of early Pliocene age from Japan. *The Island Arc* 3:390–428.
HORVATH, G., E. N. K. CLARKSON, and W. PIX. 1997. Survey of modern counterparts of schizochroal trilobite eyes: Structural and functional similarities and differences. *Historical Biology* 12:229–63.
HOUSE, M. R. 1989. Ammonoid extinction events. *Phil. Trans. Roy. Soc. London* B325:307–26.
HOYLE, F. 1950. *The Nature of the Universe.* Basil Blackwell.
———. 1983. *The Intelligent Universe.* Holt, Rinehart & Winston.
HOYLE, F., and C. WICKRAMASINGHE. 1978. *Lifecloud: The Origin of Life in the Universe.* Harper & Row.
———. 1982. *Evolution from Space.* Simon & Schuster.
HSÜ, K. J. 1986. *The Great Dying: Cosmic Catastrophe.* Harcourt, Brace, Jovanovich.
HUA, S., and E. BUFFETAUT. 1997. Crocodylia. In J. M. Callaway and E. L. Nicholls, eds., *Ancient Marine Reptiles,* pp. 357–74. Academic Press.
HUBBELL. G. 1996. Using tooth structure to determine the evolutionary history of the white shark. In A. P. Klimley and D. G. Ainley, eds., *Great White Sharks,* pp. 9–18. Academic Press.
HUBER, C., and G. WÄCHTERSHÄUSER. 1997. Activated acetic acid by carbon fixation on (Fe,Ni)S under primordial conditions. *Science* 276:245–47.
———. 1998. Peptides by activation of amino acids with CO on (Ni,Fe)S surfaces: implications for the origin of life. *Science* 281:670–72.
HULBERT, R. C. 1998. Postcranial osteology of the North American Middle Eocene protocetid *Georgiacetus.* In J. G. M. Thewissen, ed., *The Emergence of Whales,* pp. 235–68. Plenum.
HUNT, R. M., and L. G. BARNES. 1994. Basicranial evidence for the ursid affinity of the oldest pinnipeds. *Proc. San Diego Hist. Soc.* 29:57–68.
HUSAR, S. L. 1977. *Trichechus inunguis* [Amazonian manatee]. *Mammalian Species* (American Society of Mammalogists) 72:1–4.
———. 1978a. *Dugong dugon* [Dugong]. *Mammalian Species* (American Society of Mammalogists) 88:1–7.
———. 1978b. *Trichechus senegalensis* [African manatee]. *Mammalian Species* (American Society of Mammalogists) 89:1–3.
HUXLEY, J. 1964. *Evolution: The Modern Synthesis.* John Wiley.

IDYLL, C. P. 1964. *Abyss: The Deep Sea and the Creatures That Live in It.* Crowell.

———. 1965. Living fossils of the deep sea. *Sea Frontiers* 11(3):178–87.

INUZUKA, N. 1984. Skeletal restoration of the Desmostylians: Herpetiform mammals. *Mem. Faculty Sci. Kyoto Univ.* 9:157–253.

———. 1985. Are "herpetiform mammals" really impossible? A reply to Halstead's discussion. *Mem. Fac. Sci. Kyoto Univ. Ser. Biol.* 10(2):145–50.

———. 1988. The skeletons of *Desmostylus* from Utanobori, Hakkaido. In Japanese; English abstract. *Bull. Geol. Surv. Japan* 39(3):139–90.

———. 1989. Reconsideration of tooth class identification in *Desmostylus,* with special reference to the holotype of *D. japonicus.* In Japanese; English abstract. *Jour. Geol. Soc. Japan* 95(1):17–31.

———. 1996. Body size and mass estimates of desmostylians (Mammalia). *Jour. Geol. Soc. Japan* 102(9):816–19.

———. 1997. Fossil footprints of desmostylians predicted from a restored skeleton. *Ichnos* 5:163–66.

———. 1999. Evolution of aquatic adaptation in Desmostylia. Abstract. In E. Hoch and A. K. Brantsen, eds., *Secondary Adaptation to Life in Water,* pp. 21–22. Papers presented at University of Copenhagen, Geological Museum, 13–17 September 1999.

———. 2000. Aquatic adaptations in desmostylians. *Historical Biology* 14(1–2):97–113.

INUZUKA, N., D. P. DOMNING, and C. E. RAY. 1994. Summary of taxa and morphological adaptations of the desmostylia. *The Island Arc* 3:522–37.

IRVINE, A. B., and H. W. CAMPBELL. 1978. Aerial census of the West Indian manatee *Trichechus manatus* in the southeastern United States. *Jour. Mammal* 59(3):613–17.

JACOBS, D. K., and N. H. LANDMAN. 1993. *Nautilus*—a poor model for function and behavior of ammonoids? *Lethaia* 26:101–11.

JANVIER, P. 1996. *Early Vertebrates.* Clarendon Press.

———. 1999. Catching the first fish. *Nature* 402:21–22.

JANNASCH, H. W., and M. J. MOTTL. 1985. Geomicrobiology of deep-sea hydrothermal vents. *Science* 229:717–25.

JARVIK, E. 1952. On the fish-like tail in the ichthyostegid stegocephalians. *Medd. Grønl.* 114:1–90.

JELETZKY, J. A. 1966. Comparative morphology, phylogeny and classification of fossil Coleoidea. *Paleont. Contrib. Univ. Kansas* 7:1–62.

JERISON, H. 1973. *Evolution of the Brain and Intelligence.* Academic Press.

JOHNSEN, S., E. J. BALSER, and E. A. WIDDER. 1999. Light-emitting suckers in an octopus. *Nature* 398:113–14.

JOHNSON, F. H., and O. SHIMOMURA. 1975. Bacterial and other "luciferins." *BioScience* 25(11):718–22.

JOHNSON, K. 1999. Night of the giant ammonites. *Natural History* 108(6):14–17.

JONES, C. M., and A. MANNERING. 1997. New Paleocene fossil bird material from the Waipara Greensand, North Canterbury, New Zealand. Geological Society of New Zealand, Miscellaneous Publication 95A, 88.

KARPINSKY, A. 1899. Uber die Reste von *Edistien* und die neue Gattung *Helicoprion. Verhandl. Russ. Mineral. Ges.* 26(2):361–475.

KASATKINA, A. P. 2000. Conodont (Eucondontophytes), a living fossil. *Doklady Biological Sciences* 373:419–22.

KAUFFMAN, E. G., and R. V. KESLING. 1960. An Upper Cretaceous ammonite bitten by a mosasaur. *Univ. Mich. Mus. Paleont. Contrib.* 15:193–248.

KELLER, M., E. Blochl, G. WÄCHTERSHÄUSER, and K. O. STETTER. 1994. Formation of amide bonds without a condensation agent and implications for origin of life. *Nature* 368:836–38.

KELLEY, K. V. 1971. *Kelley's Guide to Fossil Sharks.* M & M Printing.

KELLOGG, R. 1922. Pinnipeds from the Miocene and Pleistocene deposits of California . . . and a résumé of curent theories regarding the origin of Pinnipedia. *Bull. Dept. Geol. Univ. Calif.* 23:13–132.

———. 1925. New pinnipeds from the Miocene diatomaceous earth near Lompoc, California. In Additions to the Tertiary history of the pelagic mammals on the Pacific coast of North America. *Contrib. Paleont. Carnegie Inst.* part 4, pp. 71–95.

———. 1931. Pelagic mammals from the Temblor Formation of the Kern River region, California. *Proc. Calif. Acad. Sci.* 19:217–397.

———. 1936. *A Review of the Archaeoceti.* Carnegie Institution of Washington.

———. 1944. Fossil cetaceans from the Florida Tertiary. *Bull. Mus. Comp. Zool.* 44(9):432–71.

KEMP, B., and H. A. A. OELSCHLÄGER. 2000. Evolutionary strategies of odontocete brain development. *Historical Biology* 14(1–2):41–45.

KENYON, K. W. 1969. *The Sea Otter in the Eastern Pacific Ocean.* Bureau of Sport Fisheries and Wildlife. Reprint, Dover, 1975.

KERR, R. A. 1998. Tracks of billion-year-old animals? *Science* 282:19–21.

———. 1999. Early life thrived despite travails. *Science* 284:2111–13.

———. 2000. Stretching the reign of early animals. *Science* 288:789.

KIRK, T. W. 1888. Brief description of a new species of large decapod (*Architeuthis longimanus*). *Trans. N.Z. Inst.* 20:34–39.

KITCHELL, J. 1978. Buoyancy architecture: evolution and invention of *Nautilus. Oceans* 11(6):44–49.

KLIMA, M. 1995. Cetacean phylogeny and systematics based on the morphogenesis of the nasal skull. *Aquatic Mammals* 21(2):79–89.

KLIMLEY, A. P., and D. G. AINLEY, eds. 1996. *Great White Sharks.* Academic Press.

KNOLL, A. H., and S. B. CARROLL. 1999. Early animal evolution: emerging views from comparative biology and geology. *Science* 248:2129–37.

KNOLL, A. H., R. K. BAMBACH, D. E. CANFIELD, and J. P. GROTZINGER. 1996. Comparative earth history and the Late Permian extinction. *Science* 273:452–57.

KÖHLER, M., and S. MOYÀ-SOLÀ. 1997. Ape-like or hominid-like? The positional behavior of *Oreopithecus bambolii* reconsidered. *Proc. Natl. Acad. Sci.* 94:11747–50.

KÖHLER, R., and R. E. FORDYCE. 1997. An archaeocete whale (Cetacea: Archaeoceti) from the Eocene Waihao Greensand, New Zealand. *Jour. Vert. Paleo.* 17(3):574–83.

KOOYMAN, G. L. 1985. Physiology without restraint in diving mammals. *Marine Mammal Science* 1(2):166–78.

KORETSKY, I. 1999. Ecomorphotypes in modern and fossil Phocinae seals. In E. Hoch and A. K. Brantsen, eds., *Secondary Adaptation to Life in Water* (p. 28). Papers presented at University of Copenhagen Geological Museum, 13–17 September 1999, p. 28.

KORETSKY, I., and P. HOLEC. In press. A primitive seal (Mammalia: Phocidae) from the Early Middle Miocene of Central Paratethys.

KORETSKY, I., and A. E. SANDERS. In press. The oldest known seal (Mammalia: Carnivora: Phocidae) from the Late Oligocene of South Carolina, U.S.A.

KRAMP, P. L. 1956. Pelagic fauna. In A. F. Bruun, S. Greve, H. Melche, and R. Spärck, eds., *The* Galathea *Deep Sea Expedition 1950–52,* pp. 65–86. George Allen & Unwin.

KRISTOFFERSEN, A. V. 1999. Adaptive specialization to life in water through the evolutionary history of birds. Abstract. In E. Hoch and A. K. Brantsen, eds., *Secondary Adaptation to Life in Water,* pp. 29–30. Papers presented at University of Copenhagen Geological Museum, 13–17 September 1999.

KUBOTA, K., S. SHIBANAI, J. KUBOTA, and S. TOGAWA. 2000. Developmental transition to monophylodonty in adaptation to marine life by the northern fur seal, *Callorhinus ursinus* (Otariidae). *Historical Biology* 14(1–2):91–95.

KULU, D. D. 1972. Evolution and cytogenetics. In S. H. Ridgway, ed., *Mammals of the Sea: Biology and Medicine,* pp. 503–27. Charles C. Thomas.

KUMAR, K., and A. SAHNI. 1986. *Remingtonocetus harudiensis,* new combination, a middle Eocene archaeocete (Mammalia, Cetacea) from western Kutch, India. *Jour. Vert. Paleo.* 6(4):326–49.

KURTÉN, B. 1991. *The Innocent Assassins.* Columbia University Press.

KUWABARA, S. 1954. Occurrence of luminous organs on the tongue of two scopelid fishes, *Neoscopelus macrolepidotis* and *M. michrochir. Jour. Shimonoseki Coll. Fish.* 3(3):83–87.

LAMARCQ, L. H. and M. J. MCFALL-NGAI. 1998. Induction of a gradual, reversible morphogenesis of its host's epithelial brush border by *Vibrio fischeri. Infect. Immun.* 66:777–85.

LAMBERT, W. D. 1997. The osteology and paleoecology of the giant otter, *Enhydritherium terraenovae. Jour. Vert. Paleo.* 17(4):738–49.

LANCASTER, W. C. 1990. The middle ear of the Archaeoceti. *Jour. Vert. Paleo.* 10:117–27.

LANDMAN, N. H. 1984. Not to be or to be? *Natural History* 8/84:34–40.

———. 1988. Early ontogeny of Mesozoic ammonites and nautilids. In J. Weidmann and J. Kullman, eds., *Cephalopods—Present and Past,* pp. 215–28. Schweizerbart'sche Verlagsbuchhandlung.

LARSEN, E. R., and J. B. SCOTT. 1955. *Helicoprion* from Elko County, Nevada. *Jour. Paleo.* 29:918–19.

LARSEN, N. L., S. D. JORGENSEN, R. A. FARRAR, and P. L. LARSEN. 1997. *Ammonites and Other Cephalopods of the Pierre Seaway.* Geoscience Press.

LEE, M. S. Y., and M. W. CALDWELL. 1998. Anatomy and relationships of *Pachyrhachis problematicus,* a primitive snake with hindlimbs. *Phil. Trans. Roy. Soc. London* 353:1521–22.

———. 2000. *Adriosaurus* and the affinities of mosasaurs. dolichosaurs, and snakes. *Jour. Paleo.* 74(5):915–37.

LEE, M. S. Y., G. L. BELL, and M. W. CALDWELL. 1999. The origin of snake feeding. *Nature* 400:655–59.

LEE, M. S. Y., M. W. CALDWELL, and J. D. SCANLON. 1999. A second primitive marine snake: *Pachyophis woodwardi* from the Cretaceous of Bosnia-Herzgovina. *Jour. Zool. London* 248: 509–20.

LEMCHE, H. 1957. A new living deep-sea mollusc of the Cambro-Devonian (Class Monoplacophora). *Nature* 179:413–16.

———. 1972. The discovery of *Neopilina.* In M. Jenkins, *The Curious Mollusks* (pp. 195–201). Holiday House.

LEMCHE, H., and K. G. WINGSTRAND. 1959. The anatomy of *Neopilina galatheae* Lemche 1957 (Mollusca, Tryblidiacea). *Galathea Report* 3:9–72.
LEPAGE, M. 2001. There's a lot of them around. *New Scientist* 169(2272):12.
LEVINTON, J. 1992. The big bang of animal evolution. *Scientific American* 267(5):84–91.
LEY, W. 1948. *The Lungfish, the Dodo and the Unicorn.* Viking.
LIEM, K. F. 1990. Aquatic versus terrestrial feeding modes: possible impacts on the trophic ecology of vertebrates. *Amer. Zool.* 30:209–21.
LINGHAM-SOLIAR, T. 1994. Going out with a bang: the Cretaceous-Tertiary extinction. *Biologist* 41:215–18.
LONG, D. G., and B. M. WAGGONER. 1996. Evolutionary relationships of the white shark: a phylogeny of lamniform sharks based on dental morphology. In A. P. Klimley and D. G. Ainley, eds., *Great White Sharks*, pp. 37–47. Academic Press.
LONG, J. A. 1995. *The Rise of Fishes: 500 Million Years of Evolution.* Johns Hopkins University Press.
———. 1998. *Dinosaurs of Australia and New Zealand, and Other Animals of the Mesozoic Era.* Harvard University Press.
LOWENSTEIN, J. M. 1983. Very like a whale. *Oceans* 16(5):65.
———. 1986. The pinniped family tree puzzle. *Oceans* 19(2):72.
LOWENSTEIN, J. M., and A. L. ZIHLMAN. 1980. The wading ape. *Oceans* 13(3):3–6.
LUCAS, F. A. 1916. Sea cows, past and present. *Amer. Mus. Jour.* 16(5):315–18.
LUND, R. 1974a. *Squatinactis caudispinatus*, a new elasmobranch from the Upper Mississippian of Montana. *Ann. Carnegie Mus. Nat. Hist.* 45(4):43–55.
———. 1974b. *Stethacanthus altonensis* (Elasmobranchii) from the Bear Gulch Limestone of Montana. *Ann. Carnegie Mus. Nat. Hist.* 45(8):161–78.
———. 1977a. A new petalodont (Chondrichthes, Bradyodonti) from the Upper Mississippian of Montana. *Ann. Carnegie Mus. Nat. Hist.* 46(10):129–55.
———. 1977b. *Echinomera meltoni*, new genus and species (Chimaeriformes) from the Upper Mississippian of Montana. *Ann. Carnegie Mus. Nat. Hist.* 46(13):195–219.
———. 1980. Viviparity and intrauterine feeding in a new Holocephalan fish from the Lower Carboniferous of Montana. *Science* 209:697–99.
———. 1982. *Harpagofututor volsellorhinus* new genus and species (Chondrichthyes, Condrenchelyiformes) from the Namurian Bear Gulch Limestone, *Chondrenchelys problematica* Traquair (Visean), and their sexual dimorphism. *Jour. Paleo.* 56(4):938–58.
———. 1983. On a dentition of *Polyrhizodus* (Chondrichthyes, Petalodontiformes) from the Namurian Bear Gulch Limestone of Montana. *Jour. Vert. Paleo.* 3(1):1–6.
———. 1984. On the spines of the Stethacanthidae (Chondrichthyes), with a description of a new genus from the Mississippian Bear Gulch Limestone. *Geobios* 17(3):281–95.
———. 1985. The morphology of *Falcatus falcatus* (St. John and Worthen), a Mississippian stethacanthid chondrichthyan from the Bear Gulch Limestone of Montana. *Jour. Vert. Paleo.* 5(1):1–19.
———. 1986a. New Mississippian Holocephali (Chondrichthyes) and the evolution of the Holocephali. In D. E. Russell, J.-P. Santoro, and D. Sigogneau-Russell, eds., *Teeth Revisited: Proceedings of the Seventh International Symposium on Dental Morphology, Paris. Mém. Mus. Nat. Hist. Paris* 53:195–205.
———. 1986b. The diversity and relationships of the Holocephali. In T. Uyeno, R. Arai, T. Taniuchi, and K. Matsuura, eds., *Indo-Pacific Fish Biology: Proceedings of the Second International Conference on Indo-Pacific Fishes.* Ichthyological Society of Japan.
———. 1986c. On *Damocles serratus*, nov. gen et sp. (Elasmobranchii: Cladodontida) from the Upper Mississippian Bear Gulch Limestone of Montana. *Jour. Vert. Paleo.* 6(1):12–19.
———. 1988. New information on *Squatinactis caudispinatus* (Chondrichthyes, Cladodontida) from the Chesterian Bear Gulch Limestone of Montana. *Jour. Vert. Paleo.* 8(3):340–42.
———. 1990. Chondrichthyan life history styles as revealed by the 320 million year old Mississippian of Montana. *Env. Biol. Fishes* 27:1–19.
———. 1998. The Bear Gulch. http://www.adelphi.edu/~lund/
LUND, R., and E. D. GROGAN. 1997a. Relationship of the Chimaeraformes and the basal radiation of the Chondrichthyes. *Rev. Fish Biol. Fisheries* 7:65–123.
———. 1997b. Cochliodonts from the Mississippian Bear Gulch Limestone (Heath Formation; Big Snowy Group; Chesterian) of Montana and the relationships of the Holocephali. *Proc. Dinofest Intl.* 477–92.
LUND, R., and P. JANVIER. 1986. A second lamprey from the Lower Carboniferous (Namurian) of Bear Gulch, Montana, U.S.A. *Geobios* 19(5):647–52.
LUND, R., and W. L. LUND. 1985. Coelacanths from the Bear Gulch Limestone (Namurian) of Montana and the evolution of the coelacanths. *Bull. Carnegie Mus. Nat. Hit.* 25:1–74.
LUND, R., and R. H. MAPES. 1984. *Carcharopsis wortheni* from the Fayetteville Formation (Mississippian) of Arkansas. *Jour. Paleo.* 58(3):709–17.

LUND, R., and W. G. MELTON. 1982. A new actinopterygian fish from the Mississippian Bear Gulch Limestone of Montana. *Palaeontology* 25(3):485–98.
LUND, R., and C. POPLIN. 1999. Fish diversity of the Bear Gulch Limestone, Namurian, Lower Carboniferous of Montana, USA. *Geobios* 32(2):285–95.
LUND, R., H. FELDMAN, W. L. LUND, and C. G. MAPLES. 1993. The depositional environment of the Bear Gulch Limestone, Fergus County, Montana. In *Montana Geological Society Guidebook*, pp. 87–96. Energy and Mineral Resources of Central Montana.
LUND, W. L., R. LUND, and G. A. KLEIN. 1985. Coelacanth feeding mechanisms and ecology of the Bear Gulch coelacanths. *Comptes Rendus, IX Int. Cong. Carboniferous Stratigraphy and Geology* 5:492–500.
LUO, Z. 1998. Homology and transformation of cetacean ectotympanic structures. In J. G. M. Thewissen, ed., *The Emergence of Whales*, pp. 269–302. Plenum.
———. 2000. In search of the whales' sisters. *Nature* 404:235–38.
LUO, Z., and K. MARSH. 1996. Petrosal (periotic) and inner ear of a pliocene kogiid whale (Kogiinae, Odontoceti): implications on relationships and hearing evolution of toothed whales. *Jour. Vert. Paleo.* 16(2):328–48.
LUTCAVAGE, M. E., P. PLOTKIN, B. WITHERINGTON, and P. L. LUTZ. 1997. Human impacts on sea turtle survival. In P. L. Lutz and J. A. Musick, eds., *The Biology of Sea Turtles*, pp. 1–28. CRC Press.
MADAR, S. I. 1998. Structural adaptations of early archaeocete long bones. In J. G. M. Thewissen, ed., *The Emergence of Whales*, pp. 353–78. Plenum.
MAISEY, J. G. 1996. *Discovering Fossil Fishes*. Henry Holt.
———. 1998. Voracious evolution. *Natural History* 107(5):38–41.
MAISEY, J. G., and K. E. WOLFRAM. 1984. "*Notidanus*." In N. Eldredge and S. M. Stanley, eds., *Living Fossils* (pp. 170–80). Springer-Verlag.
MARCHANT, J. 2000. First light. *New Scientist* 167(2248)34–35.
MARGULIS, L. 1998. *Symbiotic Planet: A New Look at Evolution*. Basic Books.
MARGULIS, L., and D. Sagan. 1995. *What Is Life?* Simon & Schuster.
MARSH, O. C. 1888. Notice of a new fossil sirenian, from California. *Amer. Jour. Sci.* 35:94–96.
MARSHALL, C. R. 1999. Missing links in the history of life. In J. W. Schopf. ed., *Evolution: Facts and Fallacies*, pp. 37–69. Academic Press.
MARSHALL, C. R., and P. D. WARD. 1996. Sudden and gradual molluscan extinctions in the Latest Cretaceous of western European Tethys. 1996. *Science* 274:1360–63.
MARSHALL, N. B. 1954. *Aspects of Deep Sea Biology*. Hutchinson.
MARTILL, D. M. 1991. Marine reptiles. In D. M. Martill and J. D. Hudson, eds., *Fossils of the Oxford Clay*, pp. 226–43. Paleontological Association.
MARTILL, D. M., and J. D. HUDSON, eds. 1991. *Fossils of the Oxford Clay*. Paleontological Association.
MARTIN, A. P. 1996. Systematics of the Lamnidae and the origination time of *Carcharodon carcharias* inferred from the comparative analysis of mitochondrial DNA sequences. In A. P. Klimley and D. G. Ainley, eds., *Great White Sharks*, pp. 49–53. Academic Press.
MARTIN, L. D. 1984. A new hesperornithid and the relationships of Mesozoic birds. *Trans. Kansas Acad. Sci.* 87(3–4):141–50.
MARTIN, L. D., and J. D. STEWART. 1982. An ichthyornithiform bird from the Campanian of Canada. *Can. Jour. Earth Sci.* 19:324–27.
MARTIN, M. W., D. V. GRAZHDANKIN, S. A. BOWRING, D. A. D. EVANS, M. A. FEDONKIN, and J. L. KIRSCHVINK. 2000. Age of Neoproterozoic bilatarian body and trace fossils, White Sea, Russia: Implications for metazoan evolution. *Science* 288:841–45.
MATSEN, B., and R. TROLL. 1994. *Planet Ocean: Dancing to the Fossil Record*. Ten Speed Press.
MATTHEW, W. D., and S. H. CHUBB. 1936. *Evolution of the Horse*. American Museum of Natural History.
MAXWELL, G. 1967. *Seals of the World*. Houghton Mifflin.
MAYR, E. 1963. *Animal Species and Evolution*. Harvard University Press.
———. 2000. Darwin's influence on modern thought. *Scientific American* 283(1):79–83.
MAZIN, J.-M., P. VIGNAUD, and V. DE BUFFRÉNIL. 2000. Preface. *Historical Biology* 14(1–2):i–ii.
MAZZOTTI, F. J. 1986. Structure and function. In C. A. Ross, ed. *Crocodiles and Alligators*, pp. 42–57. Facts On File.
MCALLISTER, D. E. 1967. The significance of ventral bioluminescence in fishes. *Jour. Fish. Res. Bd. Canada* 24(3): 537–54.
MCAULIFFE, K. 1998. When whales had feet. *Sea Frontiers* 40(1):20–33.
MCCABE, H., and J. WRIGHT. 2000. Tangled tale of a lost, stolen, and disputed coelacanth. *Nature* 406:114.
MCCOSKER, J. E., and M. D. LAGIOS. 1975. *Les petits peugeots* of Grande Comore. *Pacific Discovery* 28:1–6.

McFall-Ngai, M. J. 1998. The development of cooperative associations between animals and bacteria: establishing détente between domains. *Am. Zool.* 38:3–18.

———. 1999. Consequences of evolving with bacterial symbionts: insights from the squid-vibrio associations. *Ann. Rev. Ecol. Syst.* 30:235–56.

McFall-Ngai, M. J., and E. G. Ruby. 1991. Symbiont recognition and subsequent morphogenesis as early events in animal-bacterial mutualism. *Science* 254:1491–94.

Mchedlidze, G. A. 1984. *General Features of the Paleobiological Evolution of Cetacea.* Smithsonian Inst. Lib. and Nat. Sci. Found. New Delhi: Amerind.

McLeod, M. 2000. One small step for fish, one giant leap for us. *New Scientist* 167:28–32.

McMenamin, M. A. S. 1986. The Garden of Ediacara. *Palaios* 1:178–82.

———. 1987. The emergence of animals. *Scientific American* 256(4):94–102.

———. 1996. Ediacaran biota from Sonora, Mexico. *Proc. Nat. Acad. Sci.* 93:4990–93.

———. 1998. *The Garden of Ediacara: Discovering the First Complex Life.* Columbia University Press.

McMenamin, M. A. S., and D. L. S. McMenamin. 1994. *Hypersea: Life on Land.* Columbia University Press.

McNamara, K., and J. Long. 1998. *The Evolution Revolution.* John Wiley.

McVay, S. 1966. The last of the great whales. *Scientific American* 215(2):13–21.

Mead, J. G. 1975. A fossil beaked whale (Cetacea: Ziphiidae) from the Miocene of Kenya. *Jour. Paleo.* 49(4):745–51.

———. 1986. Twentieth-century records of right whales (*Eubalaena glacialis*) in the northwestern Atlantic Ocean. Special Issue. *Rep. Intl. Whal. Commn* 10:83–105.

Melton, W. G., and H. C. Scott. 1973. Conodont-bearing animals from the Bear Gulch Limestone, Montana. *Spec. Pap. Geol. Soc. Amer.* 31–65.

Messenger, S. L. 1994. Phyologenetic relationships of platanistoid river dolphins: assessing the significance of fossil taxa. *Proc. San Diego Hist. Soc.* 29:125–34.

Meyer, E. R. 1984. Crocodilians as living fossils. In N. Eldredge and S. M. Stanley, eds., *Living Fossils,* pp. 105–31. Springer-Verlag.

Milinkovitch, M. C., M. Bérubé, and P. J. Palsbøll. 1998. Cetaceans are highly derived artiodactyls. In J. G. M. Thewissen, ed., *The Emergence of Whales,* pp. 113–32. Plenum.

Milinkovitch, M. C., G. Orti, and A. Meyer. 1993. Revised phylogeny of whales suggested by mitochondrial ribosomal DNA sequences. *Nature* 361:346–48.

Milius, S. 1998. Second group of living fossils reported. *Science News* 154(13):196.

———. 2000a. Famine reveals incredible shrinking iguanas. *Science News* 157(2):20.

———. 2000b. Pregnant and still macho. *Science News* 157(11):168–70.

Miller, H. 1841. *The Old Red Sandstone or New Walks in an Old Field.* Johnstone.

Miller, H. W., and M. V. Walker. 1968. *Enchoteuthis melanae* and *Kansasteuthis lindneri,* new genera and species of Teuthids, and a sepiid from the Niobrara Formation of Kansas. *Trans. Kansas Acad. Sci.* 71(2):176–83.

Miller, S. L. 1953. A production of amino acids under possible primitive Earth conditions. *Science* 117:528–29.

Miller, S. L., and J. L. Bada. 1988. Submarine hot springs and the origin of life. *Nature* 334:609–11.

Minton, S. A., and H. Heatwole. 1978. Snakes and the sea. *Oceans* 11(2):53–56.

Mitchell, E. D. 1966. Faunal succession of extinct North Pacific marine mammals. *Norsk Hvalfangst-tidende* 55(3):47–60.

———. 1975. Parallelism and convergence in the evolution of Otariidae and Phocidae. *Rapp, P.-v. Réun. Cons. Int. Explor. Mer.* 169: 12–26.

Mitchell, E. D., and R. H. Tedford. 1973. The Enaliarctinae: A new group of extinct Carnivora and a consideration of the origin of the Otariidae. *Bull. Amer. Mus. Nat. Hist.* 151(3):205–84.

Mivart. St. G. 1871. *On the Genesis of Species.* Macmillan.

Mojzsis, S. J., and T. M. Harrison. 2000. Vestiges of a beginning: clues to the emergent biosphere recorded in the oldest known sedimentary rocks. *GSA Today* 10(4):1–7.

Monastersky, R. 1996. Living large on the Precambrian planet. *Science News* 149:308.

———. 1998a. The rise of life on Earth. *National Geographic* 193(3):54–81.

———. 1998b. The rise of life on Earth: life grows up. *National Geographic* 193(4):100–115.

———. 1999a. Out of the swamps. *Science News* 155(21):328–30.

———. 1999b. Earliest evidence of complex life. *Science News* 156(9):141.

———. 1999c. Waking up to the dawn of vertebrates. *Science News* 156:292.

———. 1999d. The whale's tale: Searching for the landlubbing ancestors of marine mammals. *Science News* 156:296–98.

———. 2000. All mixed up over birds and vertebrates. *Science News* 157:38.

Monks, N. 1999. Half a billion years of floating slugs and racing snails: Fossil cephalopods FAQ. http://is.dal.ca/~ceph/TCP/FosCephs.html.

———. 2000a. Functional morphology and evolution of the scaphitaceae gill, 1871 (Cephalopoda). *Jour. Mollusc. Studies* 66:205–16.

———. 2000b. Mid-Cretaceous heteromorph ammonite shell damage. *Jour. Mollusc. Studies* 66:283–85.

Monks, N., and J. R. Young. 1998. Body position and the functional morphology of Cretaceous heteromorph ammonites. *Palaeontologica Electronica*, http://luna.geol/niu.edu/1998_1/monks/main.htm

Monks, N., J. D. Hardwick, and A. S. Gale. 1996. The function of the belemnite guard. *Paläontologie Zeitschrift* 70(3/4):425–31.

Montgomery, M. K., and M. J. McFall-Ngai. 1994. Bacterial symbionts induce host organ morphogenesis during early postembryonic development of the squid *Euprymna scolopes. Development* 120:1719–29.

———. 1998. Late postembryonic development of the symbiotic light organ of *Euprymna scolopes. Biol. Bull.* 195:326–36.

Moody, R. T. J. 1997. The paleogeography of marine and coastal turtles of the North Atlantic and Trans-Saharan regions. In J. M. Callaway and E. L. Nicholls, eds., *Ancient Marine Reptiles*, pp. 259–78. Academic Press.

Morgan, E. 1972. *The Descent of Woman*. Stein & Day.

———. 1982. *The Aquatic Ape*. Stein & Day.

———. 1984. The aquatic hypothesis. *New Scientist* 102(1405):11–13.

———. 1985. Sweaty old man and the sea. *New Scientist* 105(1428):27–28.

———. 1987. Lucy's child. *New Scientist* 112(1540–41):13–15.

———. 1990. *The Scars of Evolution*. Oxford University Press.

———. 1997. *The Aquatic Ape Hypothesis*. Independent Voices.

———. 1999. Human evolution: the water theory. *Proc. Symp. Water and Human Evolution University of Ghent, April 30, 1999*, 1–10.

Morgan, E., and M. Verhaegen. 1986. In the beginning was the water. *New Scientist* 109(1498):62–63.

Morris, D. 1967. *The Naked Ape*. Dell.

Morton, J. E. 1958. *Molluscs*. Hutchinson University Library.

Moy-Thomas, J. A. 1936. The structure and affinities of the fossil elasmobranch fishes from the Lower Carboniferous rocks of Glencartholm, Eksdale. *Proc. Zool. Soc. London* 1936:761–87.

———. 1939. The early evolution and relationships of the elasmobranchs. *Biol. Rev.* 14:1–26.

Muir, H. 2000. Smash hits: did asteroids have a hand in evolution of life on Earth? *New Scientist* 165(2230):14.

Muizon, C. de. 1993a. *Odobenocetops peruvianus*: una remarcable convergencia de adaptación alimentaria entre morse y delfín. *Bull. Inst. Fr. Études Andines* 22(3):671–83.

———. 1993b. Walrus-like feeding adaptation in a new cetacean from the Pliocene of Peru. *Nature* 365:745–48.

———. 1994. Are the squalodonts related to the platanistoids? *Proc. San Diego Hist. Soc.* 29: 135–46.

Muizon, C. de., and T. J. DeVries. 1985. Geology and paleontology of the Late Cenozoic marine deposits in the Sacaco Area (Peru). *Geologische Rundschau* 74(3):547–63.

Muizon, C. de., and H. G. MacDonald. 1995. An aquatic sloth from the Pliocene of Peru. *Nature* 375:224–27.

Muizon, C. de., D. P. Domning, and M. Parish. 1999. Dimorphic tusks and adaptive strategies in a new species of walrus-like dolphin (Odobenocetopsidae) from the Pliocene of Peru. *Comptes Rendus Acad. Sci. Paris, Sci. Terre Planètes* 329:449–55.

Naef, A. 1921. *Die Fossilen Tintenfische*. G. Fischer.

Naish, D. 2000. Placodonts—bizarre "walrus turtles" of the Triassic, http://www.oceansofkansas.com/placodont.html

Narbonne, G. M., B. Z. Saylor, and J. P. Grotzinger. 1997. The youngest Ediacaran fossils from southern Africa. *Jour. Paleo.* 71(6):933–67.

Nash, J. M. 1995. When life exploded. *Time* 146(23):50–58.

Neill, W. T. 1971. *The Last of the Ruling Reptiles: Alligators, Crocodiles, and Their Kin*. Columbia University Press.

Nesis, K. N. 1987. *Cephalopods of the World*. Translated from the Russian. T. F. H. Publications.

Nicholls, E. L. 1997. Testudines. In J. M. Callaway and E. L. Nicholls, eds., *Ancient Marine Reptiles*, pp. 219–23. Academic Press.

Nicholls, E. L., and H. Isaak. 1987. Stratigraphic and taxonomic significance of *Tusoteuthis longi* Logan (Coleoidea, Teuthida) from the Pembina Member, Pierre Shale (Campanian), of Manitoba. *Jour. Paleo.* 61(4):727–37.

Nicol, J. A. C. 1961. Luminescence in marine organisms. *Smithsonian Rep.* 1960:447–56.

Nishiwaki, M., and H. Marsh. 1990. Dugong *Dugong dugon* (Muller 1976). In S. H. Ridgway and R. Harrison, eds., *Handbook of Marine Mammals*, vol. 3: *The Sirenians and Baleen Whales*, pp. 1–31. Academic Press.

NISHIWAKI, M., T. KASUYA, T. TOBAYAMA, N. MIYAZAKI, and T. KATAOKA. 1981. Distribution of the dugong in the world. *Sci. Rep. Whales Res. Inst.* 31:131–41.

NISHIWAKI, M., M. YAMAGUCHI, S. SHOKITA, S. UCHIDA, and T. KATAOKA. 1982. Recent survey on the distribution of the African manatee. *Sci. Rep. Whales Res. Inst.* 34:137–47.

NORMAN, D. 1994. *Prehistoric Life*. Macmillan.

NORRIS, K. S., and G. W. HARVEY. 1972. A theory for the function of the spermaceti organ in the sperm whale (*Physeter catadon* L.). In S. R. Galler, K. Schmidt-Koenig, G. J. Jacobs, and R. E. Belleville, eds., *Animal Orientation and Navigation*, pp. 397–419. NASA.

NORRIS, K. S., and B. MØHL. 1983. Can odontocetes debilitate prey with sound? *American Naturalist* 122(1):85–104.

NOVACEK, M. J. 1993. Genes tell a new whale tale. *Nature* 361:298–99.

———. 1994. Whales leave the beach. *Nature* 368:807.

NOWAK, R. M. 1999. *Walker's Primates of the World*. Johns Hopkins University Press.

NYHOLM, S. V., and M. J. MCFALL-NGAI. 1998. Sampling the microenvironment of the *Euprymna scolopes* light organ: description of a population of host cells with the bacterial symbiont *Vibrio fischeri*. *Biol. Bull.* 195:89–97.

NYHOLM, S. V., E. V. STABB, E. G. RUBY, and M. J. MCFALL-NGAI. 2000. Establishment of an animal-bacterial association: recruiting symbiotic vibrios from the environment. *Proc. Natl. Acad. Sci.* 97(18):10231–35.

OBRUCHEV, D. 1952. The origin and significance of the spiral in *Helicoprion*. *Contrib. U.S.S.R. Sci. Acad.* 87(2):277–80. Translated from the Russian.

———. 1953. Studies on Edistids and the works of A. R. Karpinski. *Contrib. U.S.S.R. Sci. Acad. Sci. Paleo. Inst.* 45:1–85. Translated from the Russian.

O'DAY, W. T. 1973. Luminescent silhouetting in stomiatoid fishes. *Contrib. Sci. Nat. Hist. Mus. Los Angeles County* 246:1–8.

———. 1974. Bacterial luminescence in the deep-sea anglerfish *Oneirodes acanthias* (Gilbert 1915). *Contrib. Sci. L.A. County Mus.* 255:1–12.

O'DOR, R. K., J. WELLS, and M. WELLS. 1990. Speed, jet pressure and oxygen consumption relationships in free-swimming *Nautilus*. *Jour. Exp. Biol.* 154:383–96.

OELSCHLÄGER, H. H. A. 2000. Morphological and functional adaptations of the toothed whale head to aquatic life. *Historical Biology* 14(1–2):33–39.

O'LEARY, M. A. 1998. Phylogenetic and morphometric reassessment of the dental evidence for a mesonychian and cetacean clade. In J. G. M. Thewissen, ed., *The Emergence of Whales*, pp. 133–61. Plenum.

———. 1999a. Parsimony analysis of total evidence from extinct and extant taxa and the Cetacean-Artiodactyl question (Mammalia, Ungulata). *Cladistics* 15:315–330.

———. 1999b. Whale origins. *Science* 283:1641–42.

O'LEARY, M. A., and J. H. GEISLER. 1999. The position of Cetacea within Mammalia: phylogenetic analysis of morphological data from extinct and extant taxa. *Syst. Biol.* 48(3):455–490.

O'LEARY, M. A., and K. D. ROSE. 1995. Postcranial skeleton of the early Eocene mesonychid *Pachyaena* (Mammalia, Mesonychia). *Jour. Vert. Paleo.* 15:401–30.

O'LEARY, M. A., and M. D. UHEN. 1999. The time of the origin of whales and the role of behavioral changes in the terrestrial-aquatic transition. *Paleobiology* 24(4):534–56.

OLIWENSTEIN, L. 1996. Life's grand explosions. *Discover* 17(1):42–43.

OLSON, S. L. 1980. A new genus of penguin-like pelecaniform bird from the Oligocene of Washington (Pelecaniformes: Plotopteridae). In K. E. Campbell, Jr., ed., *Papers in Avian Paleontology Honoring Hildegarde Howard*. Spec. issue. *Contrib. Sci. Nat. Hist. Mus. L.A. County* 330:51–57.

OLSON, S. L., and Y. HASEGAWA. 1979. Fossil counterparts of giant penguins from the North Pacific. *Science* 286:688–89.

———. 1996. A new genus and two new species of gigantic Plopteridae from the Oligocene of Japan (Aves: Pelecaniformes.) *Jour. Vert. Paleo.* 16(4):742–51.

OSBORN, H. F. 1905. *Tyrannosaurus* and other Cretaceous carnivorous dinosaurs. *Bull. Amer. Mus. Nat. Hist.* 32:91–92.

———. 1924. *Andrewsarchus*, giant mesonychid of Mongolia. *Amer. Museum Novitates* 146:1–5.

OWEN, R., 1867. On the dental characters of genera and species, chiefly of fishes from the Lower Main seam and shales of coal, Northumberland. *Trans. Odont. Soc. Gt. Br.* 5:323–76.

PABST, D. A., S. A. ROMMEL, and W. A. MCLELLAN. 1998. Evolution of thermoregulatory function in cetacean reproductive systems. In J. G. M. Thewissen, ed., *The Emergence of Whales*, pp. 379–98. Plenum.

PADIAN, K., and L. CHIAPPE. 1998. The origin of birds and their flight. *Scientific American* 278(2):38–47.

PAGE, K. N., and P. DOYLE. 1991. Other cephalopods. In D. M. Martill and J. D. Hudson, eds., *Fossils of the Oxford Clay*, pp.144–66. The Paleontological Association.

PANDER. C. H. 1856. *Monographie der fossilen Fische des Silurischen Systems der Russisch-Baltischen Gouvernements*. Königliche Akademie Wissenschaften St. Petersburg.
PARRY, D. A. 1948. The anatomical basis of swimming in whales. *Proc. Zool. Soc. London* 119:49–60
PATON, R. L., T. R. SMITHSON, and J. A. CLACK. 1999. An amniote-like skeleton from the Early Carboniferous of Scotland. *Nature* 398:508–13.
PEAT, N. 1992. Penguins from the past. *Forest & Bird* (New Zealand), February, 32–34.
PENNISI, E. 2000. In nature, animals that start and stop win the race. *Science* 288:83–85.
PICKFORD, G. E. 1949. *Vampyroteuthis infernalis* Chun, an archaic dibranchiate cephalopod. II. External anatomy. *Dana-Report* 32:1–32.
PITMAN, R. L., D. M. PALACIOS, P. L. R. BRENNAN, B. J. BRENNAN, K. C. BALCOMB, and T. MIYASHITA. 1999. Sightings and possible identity of a bottlenose whale in the tropical Indo-Pacific. *Indopacetus pacificus? Mar. Mam. Sci.* 15(2):531–49.
POUYAUD, L., S. WIRJOATMODJO, I. RACHMATIKA, A. TJAKRAWIDJAJA, R. HADIATY, and W. HADIE. 1999. Une nouvelle espèce de coelacanthe. Preuves génétiques et morphologiques. *Comp. Rend. Acad. Sci.* 322:261–67.
POWER, J. H. T., I. R. DOYLE, K. DAVIDSON, and T. E. NICHOLAS, 1999. Ultrastructural and protein analysis of surfactant in the Australian lungfish *Neoceratodus forsteri:* Evidence for conservation of composition for 300 million years. *Jour. Exp. Biol.* 202(18):2543–50.
PRITCHARD, P. C. H. 1997. Evolution, phylogeny and current status. In P. L. Lutz and J. A. Musick, eds., *The Biology of Sea Turtles*, pp. 1–28. CRC Press.
PURDY, R. W. 1996. Paleoecology of fossil white sharks. In A. P. Klimley and D. G. Ainley, eds., *Great White Sharks*, pp. 67–78. Academic Press.
PURNELL, M. A., P. C. J. DONOGHUE, and R. J. ALDRIDGE. 2000. Orientation and anatomical notation in conodonts. *Jour. Paleo.* 74(1):113–22.
RAMSKÖLD, L., and X. HOU. 1991. New Early Cambrian animal and onychophoran affinities of enigmatic metazoans. *Nature* 351:225–28.
RANDALL, J. E. 1973. The size of the great white shark *(Carcharodon). Science* 181:169–70.
RASMUSSEN, A. R. 1997. Systematics of sea snakes: a critical review. *Symp. Zool. Soc. London* 70:15–30.
RAUP, D. M., and S. M. STANLEY. 1971. *Principles of Paleontology*. Freeman.
RAVEN, H. C., and W. K. GREGORY. 1933. The spermaceti organ and nasal passages of the sperm whale *(Physeter catadon)* and other odontocetes. *Amer. Mus. Novitates* 677:1–18.
RAY, C. E. 1976a. Geography of Phocid evolution. *Systematic Zoology* 25(4):391–407.
———. 1976b. Fossil marine mammals of Oregon. *Systematic Zoology* 25(4):420–36.
RAY, C. E., and D. P. DOMNING. 1986. Manatees and genocide. *Mar. Mam. Sci.* 2(1):77–79.
RAY, C. E., D. P. DOMNING, and M. C. MCKENNA. 1994. A new specimen of *Behemotops proteus* (order Desmostylia) from the marine Oligocene of Washington. *Proc. San Diego Hist. Soc.* 29:205–22.
REES, J.-F., B. DE WERGIFOSSE, O. NOISET, M. DUBUISSON, B. JANSSENS, and E. M. THOMPSON. 1998. The origins of marine bioluminescence: turning oxygen defence mechanisms into deep-sea communication tools. *Jour. Exp. Biol.* 20:1211–21.
REINHART, R. H. 1953. Diagnosis of the new mammalian order Desmostylia. *Jour. Geol.* 61(2):187.
———. 1959. A review of the Sirenia and Desmostylia. *Univ. Calif. Publ. Geol. Sci.* 36:1–146.
RENOUS, S., V. BELS, and J. DAVENPORT. 2000. Locomotion in marine Chelonia: adaptation to the marine habitat. *Historical Biology* 14(1-2):1–13.
REPENNING, C. A., 1968. Underwater hearing in seals. *Año Nuevo Reports* (University of California, Santa Cruz) 2:60–61.
———. 1972a. Underwater hearing in seals: functional morphology. In R. J. Harrison, ed., *Functional Anatomy of Marine Mammals*, pp. 307–31. Academic Press.
———. 1972b. Otarioid evolution. *Rapp. P.-v. Rèun. Cons. Int. Explor. Mer.* 169:27–33.
———. 1976a. Introduction. C. A. Repenning, ed., Symposium: Advances in systematics of marine mammals. *Systematic Zoology* 25(4):301–3.
———. 1976b. Adaptive evolution of sea lions and walruses. *Systematic Zoology.* 25(4):375–90.
———. 1976c. *Enhydra* and *Enhydriodon* from the Pacific coast of North America. *Jour. Res. U.S. Geol. Surv.* 4(3):305–15.
———. 1980. Warm-blooded life in cold currents: following the evolution of the seal. *Oceans* 13(3):18–24.
———. 1983. New evidence for the age of the Gubik Formation, Alaskan North Slope. *Quaternary Research* 19:356–72.
———. 1990. Oldest pinniped. *Science* 248:499.
REPENNING, C. A., and E. L. PACKARD. 1990. Locomotion of a desmostylian and evidence of ancient shark predation. In A. J. Boucot, ed., *Evolutionary Paleobiology of Behavior and Coevolution*, pp. 199–203. Elsevier.
REPENNING, C. A., and C. E. RAY. 1977. The origin of the Hawaiian monk seal. *Proc. Biol. Soc. Washington* 89(58):667–88.

Repenning, C. A., and R. H. Tedford. 1977. Otarioid seals of the Neogene. *U.S. Geol. Surv. Prof. Paper* 992:1–93.

Repenning, C. A., R. S. Peterson, and C. L. Hubbs. 1970. Classification of fur seals. In *Proceedings of the Sixth Annual Conference on Biological Sonar and Diving Mammals, 1969*, pp. 29–32. Stanford Research Institute, Menlo Park, California.

———. 1971. Contributions to the systematics of the southern fur seals, with particular reference to the Juan Fernández and Guadalupe species. In W. H. Burt, ed., *Antarctic Pinnipedia*, pp. 1–34. American Geophysical Union, Antarctic Research Series, no. 18.

Repetski, J. E. 1978. A fish from the Upper Cambrian of North America. *Science* 200:529–31.

Retallack, G. J. 1994. Were the Ediacaran fossils lichens? *Paleobiology* 20:523–544.

Reynolds, J. E. 1979. The semisocial manatee. *Natural History* 88(2):44–52.

———. 1992. Distribution and abundance of Florida manatees *(Trichechus manatus latirostris)* around selected power plants following winter cold fronts: 1991–1992. Report. Florida Power and Light.

Reynolds, J. E., and D. K. Odell. 1991. *Manatees and Dugongs*. Facts On File.

Reynolds, J. E., and S. A. Rommel, eds. 1999. *Biology of Marine Mammals*. Smithsonian Institution Press.

Rice, D. W. 1998. *Marine Mammals of the World: Systematics and Distribution*. Society for Marine Mammalogy.

Riedman, M. 1990. *The Pinnipeds: Seals, Sea Lions, and Walruses*. University of California Press.

Rieppel, O. 1997. Sauropterygians. In J. M. Callaway and E. L. Nicholls, eds., *Ancient Marine Reptiles*, pp. 107–19. Academic Press.

Rieppel, O., and H. Hagedorn. 1997. Paleogeography of Middle Triassic sauropterygia in central and western Europe. In J. M. Callaway and E. L. Nicholls, eds., *Ancient Marine Reptiles*, pp. 121–144. Academic Press.

Riordan, J. 1999. Hell's teeth [Megalodon]. *New Scientist* 162:32–35.

Robison, B. H., and R. E. Young. 1981. Bioluminescence in pelagic octopods. *Pacific Science* 35(1):39–44.

Roe, L. J., J. G. M. Thewissen. J. Quade, J. R. O'Neill, S. Bajpai, A. Sahni, and S. T. Hussain. 1998. Isotropic approaches to understanding the terrestrial-to-marine transition of the earliest cetaceans. In J. G. M. Thewissen, ed., *The Emergence of Whales*, pp. 399–422. Plenum.

Roman, J. 2000. Is the right whale going down? *Wildlife Conservation* 103(3):26–35.

Romer, A. S. 1933. Eurypterid influence on vertebrate history. *Science* 78:114–17.

———. 1966. *Vertebrate Paleontology*. University of Chicago Press.

———. 1968. *The Procession of Life*. World.

Rook, L., L. Bondioli, M. Köhler, S. Moyà-Solà, and R. Macchiarelli. 1999. *Oreopithecus* was a bipedal ape after all: evidence from the iliac cancellous architecture. *Proc. Natl. Acad. Sci.* 96:8795–99.

Ross, C. A., ed. 1989. *Crocodiles and Alligators*. Facts On File.

Roth, G. 1999. Small brains—large brains. Evolutionary aspects and functional consequences. *Naturwissenschaftliche Rundschau* 52:213–19.

Rothschild, B. M., and L. D. Martin. 1988. Avascular necrosis: occurrence in diving Cretaceous mosasaurs. *Science* 236:75–77.

Rowe, T., C. A. Brochu, and K. Kishi. 1999. Cranial morphology of *Alligator mississippiensis* and phylogeny of Alligatoridea. *Jour. Vert. Paleo.* 19(2) Supp. 1–100.

Ruby, E. G., and K. H. Nealson. 1976. Symbiotic association of *Photobacterium fisheri* with the marine luminous fish *Monocentrus japonica:* model of symbiosis based on bacterial studies. *Biol. Bull.* 151:574–586.

Rudwick, M. J. S. 1997. *Georges Cuvier, Fossil Bones and Geological Catastrophes*. University of Chicago Press.

Saunders, W. B., and N. H. Landman, eds. 1987. *Nautilus: The Biology and Paleobiology of a Living Fossil*. Plenum.

Saunders, W. B., and P. D. Ward. 1987. Ecology, distribution and population characteristics of *Nautilus*. In W. B. Saunders and N. H. Landman, eds., *Nautilus: The Biology and Paleobiology of a Living Fossil*, pp. 137–62. Plenum.

Saunders, W. B., and D. M. Work. 1996. Shell morphology and suture complexity in Upper Carboniferous ammonoids. *Paleobiology* 22(2):189–218.

Saunders, W. B., D. M. Work, and S. V. Nikolaeva. 1999. Evolution of complexity in Paleozoic ammonoid sutures. *Science* 286:760–63.

Savage, R. J. G. 1976. Review of early Sirenia. *Systematic Zoology* 25(4):344–51.

Savage, R. J. G., D. P. Domning, and J. G. M. Thewissen. 1994. Fossil Sirenia of the West Atlantic and Caribbean region. V. The most primitive known sirenian, *Prorastomus sirenoides* Owen 1855. *Jour. Vert. Paleo.* 14(3):427–49.

Scanlon, J. D., and M. S. Y. Lee. 2000. The Pleistocene serpent *Wonambi* and the early evolution of snakes. *Nature* 403:416–420.

Scanlon, J. D., M. S. Y. Lee, M. W. Caldwell, and R. Shine. 1999. The paleoecology of the primitive snake *Pachyrhachis*. *Historical Biology* 13:127–52.
Schaeffer, B. 1967. Comments on elasmobranch evolution. In P. W. Gilbert, R. F. Mathewson, and D. P. Rall, eds., *Sharks, Skates, and Rays*, pp. 3–35. Johns Hopkins University Press.
Scheffer, V. B. 1958. *Seals, Sea Lions, and Walruses*. Stanford University Press.
———. 1976. *A Natural History of Marine Mammals*. Scribner's.
Scheuchzer. J. J. 1708. *Piscium Querelae et Vindiciae*. Zurich.
Schipp. R., Y. S. Chung, and J. M. Arnold. 1990. Symbiotic bacteria in the coelum of *Nautilus* (Cephalopoda, Tetrabanchiata). *Jour. Ceph. Biol.* 1(2):59–74.
Schmidt, J. 1922. Live specimens of *Spirula*. *Nature* 110:788–90.
Schock. E. L. 1992. Chemical environments of submarine hydrothermal systems. *Origin of Life and Evolution of the Biosphere* 22:67–107.
Schopf, J. W. 1993. Microfossils of the Apex Chert: new evidence of the antiquity of life. *Science* 260:640–46
———. 1999. Breakthrough discoveries. In J. W. Schopf. ed., *Evolution: Facts and Fallacies*, pp. 91–117. Academic Press.
Schopf, T. J. M. 1980. *Paleoceanography*. Harvard University Press.
Seilacher, A. 1989. Vendozoa: organismic construction in the Proterozoic biosphere. *Lethaia* 22:229–39.
———. 1994. Early multicellular life: late Proterozoic fossils and the Cambrian explosion. In S. Bengston, ed., *Early Life on Earth*, pp. 389–400. Columbia University Press.
———. 1998. Mosasaurs, limpets or diagenesis: how *Placenticeras* shells got punctured. *Geowissenschaftliche Reihe* 1:93–102.
———. 1999. Biomat-related lifestyles in the Precambrian. *Palaios* 14:86–93.
Seilacher, A., P. K. Bose, and F. Pflüger. 1998. Triploblastic animals more than 1 billion years ago: trace fossil evidence from India. *Science* 282:80–83.
Shimada, K. 1997a. Gigantic lamnoid shark vertebra from the Lower Cretaceous Kiowa Shale of Kansas. *Jour. Paleo.* 71(3):522–24.
———. 1997b. Dentition of the Late Cretaceous lamniform shark *Cretoxyrhina mantelli*, from the Niobrara Chalk of Kansas. *Jour. Vert. Paleo.* 17(2):269–79.
———. 1997c. Skeletal anatomy of Late Cretaceous lamniform shark *Cretoxyrhina mantelli*, from the Niobrara Chalk of Kansas. *Jour. Vert. Paleo.* 17(4):642–52.
Shimada, K., and N. Inuzuka. 1994. Desmostylian tooth remains from the Miocene Tokigawa Group at Kuzubukuro, Saitama, Japan. *Trans. Proc. Paleont. Soc. Japan* 175:553–77.
Shine, R. 1991. *Australian Snakes: A Natural History*. Cornell University Press.
Shu, D.-G., S. Conway Morris, X.-L. Zhang. 1996. A *Pikaia*-like chordate from the Lower Cambrian of China. *Nature* 384:156–57.
Shu, D.-G., H.-L. Luo, S. Conway Morris, X.-L. Zhang, S.-X. Hu, L. Chen, J. Han. M. Zhu, Y. Li, and L.-Z. Chen. 1999. Lower Cambrian vertebrates from south China. *Nature* 402:42–46.
Simpson, G. G. 1944. *Tempo and Mode in Evolution*. Columbia University Press.
———. 1945. The principles of classification and a classification of mammals. *Bull. Amer. Mus. Nat. Hist.* 85:1–350.
———. 1951. *Horses*. Oxford University Press.
———. 1953. *Life of the Past*. Yale University Press.
———. 1975. Fossil penguins. In B. Stonehouse, ed., *The Biology of Penguins*, pp. 19–41. Macmillan.
———. 1976. *Penguins Past and Present, Here and There*. Yale University Press.
———. 1981. Notes on some fossil penguins, including a new genus from Patagonia. *Ameghiniana* 18:266–72.
———. 1996. *The Dechronization of Sam Magruder*. St. Martin's Press, Griffin.
Simpson, S. 1999. Life's first scalding steps. *Science News* 155(2):24–26.
Singley, C. T. 1983. *Euprymna scolopes*. In P. R. Boyle, ed., *Cephalopod Life Cycles*, pp. 69–74. Academic Press.
Siverson, M. 1992. Biology, dental morphology and taxonomy of lamniform sharks from the Campanian of the Kristianstad Basin, Sweden. *Paleontology* 35(3):519–54.
———. 1996. Lamniform sharks of the Mid-Cretaceous Alinga Formation and Beedagong Claystone, Western Australia. *Paleontology* 39(4):813–49.
Slijper, E. J. 1962. *Whales*. Cornell University Press.
Small, A. L., and M. J. McFall-Ngai. 1999. Halide peroxidase in tissues that interact with bacteria in the host squid *Euprymna scolopes*. *Jour. Cell. Biochem.* 72:445–57.
Smith, J. L. B. 1956. *Old Fourlegs: The Story of the Coelacanth*. Longmans Green.
Smith, M. P., I. J. Sansom, and J. E. Repetski. 1996. Histology of the first fish. *Nature* 380:702–4.
Smolker, R. 2001. *To Touch a Wild Dolphin*. Doubleday.

SPOTILA, J. R., R. D. REINA, A. C. STEYERMARK, P. T. PLOTKIN, and F. V. PALADINO. 2000. Pacific leatherback turtles face extinction. *Nature* 405:529–30.

SPRIGG, R. C. 1947. Early Cambrian (?) jellyfishes from the Flinders Ranges, South Australia. *Trans. Roy. Soc. S. Aust.* 71:212–24.

———. 1949. Early Cambrian "jellyfishes" of the Ediacara, South Australia, and Mount John, Kimberly District, Western Austrlia. *Trans. Roy. Soc. S. Aust.* 73:72–99.

STANLEY, S. M. 1984. Does bradytely exist? In N. Eldredge and S. M. Stanley, eds., *Living Fossils*, pp. 278–80. Springer-Verlag.

STEEL, R. 1985. *Sharks of the World*. Facts On File.

STEJNEGER, L. 1884. Contributions to the history of the Commander Islands. No. 2. Investigations relating to the date of the extermination of Steller's sea-cow. *Proc. U. S. Natl. Mus.* 8:181–189.

———. 1887. How the great northern sea-cow (*Rytina*) became exterminated. *Amer. Nat.* 21:1047–54.

———. 1936. *Georg Wilhelm Steller*. Harvard University Press.

STELLER, G. W. 1781. *Journal of a Voyage with Bering, 1741–1742*. Translated by O. W. Frost. Reprint, Stanford University Press, 1988.

STEWART, B. S., and R. L. DELONG. 1991. Diving patterns of northern elephant seal bulls. *Mar. Mam. Sci.* 7(4):369–84.

STEWART, J. D. 1977. Teuthids of the North American Upper Cretaceous. *Trans. Kansas Acad. Sci.* 79:94.

STEWART, J. D., and K. CARPENTER. 1990. Examples of vertebrate predation on cephalopods in the Late Cretaceous of the Western Interior. In J. Boucot, ed., *Evolutionary Paleobiology of Behavior and Coevolution*, pp.203–8. Elsevier.

STIRTON, R. A. 1960. A marine carnivore from the Clallam Formation, Washington. Its correlation with nonmarine faunas. *Univ. Calif. Publ. Geol. Sci.* 36:345–68.

STONEHOUSE, B., ed. 1975. *The Biology of Penguins*. Macmillan.

STRATHMANN, R. R. 1990. Why life histories evolve differently in the sea. *Amer. Zool.* 30:197–207.

SUES, H. -D. 1987. On the skull of *Placodus gigas* and the relationships of Placodontia. *Jour. Vert. Paleo.* 7:138–44.

———. 1989. The place of crocodiles in the living world. In C. A. Ross, ed., *Crocodiles and Alligators*, pp. 14–25. Facts On File.

SWINTON, W. E. 1973. *Fossil Reptiles and Amphibians*. British Museum.

TARASOFF, F. J., A. BISAILLON, J. PIÉRARD, and A. P. WHITT. 1972. Locomotory patterns and external morphology of the river otter, sea otter, and harp seal (Mammalia). *Can. Jour. Zool.* 50:915–29.

TARDUNO, J. A., D. B. BRINKMAN, P. R. RENNE, R. D. COTTRELL, H. SCHER, and P. CASTILLO. 1998. Evidence for extreme climatic warmth from Late Cretaceous Arctic vertebrates. *Science* 282:2241–43.

TAYLOR, K., and T. ADAMEC. 1977. Tooth histology and ultrastructure of a Paleozoic shark, *Edestus heinrichii*. *Fieldiana Geology* 33(24):441–70.

TAYLOR, M. A. 2000. Functional significance of bone ballast in the evolution of buoyancy control strategies by aquatic tetrapods. *Historical Biology* 14(1–2):15–31.

TCHERNOV, E., O. RIEPPEL, H. ZAHER, M. J. POLCYN, and L. J. JACOBS. 2000. A fossil snake with limbs. *Science* 287:2010–12.

TEDFORD, R. H. 1976. Relationships of pinnipeds to other carnivores (Mammalia). *Syst. Zool.* 25(4):363–74.

TEDFORD, R. H., L. G. BARNES, and C. E. RAY. 1994. The early Miocene littoral ursid carnivoran *Kolonomos*: systematics and mode of life. *Proc. San Diego Hist. Soc.* 29:11–32.

TEICHERT, C. 1940. *Helicoprion* in the Permian of Western Australia. *Jour. Paleo.* 14(2):140–49.

TETT, P. B., and M. G. KELLY. 1973. Marine bioluminescence. *Oceanogr. Mar. Biol. Ann. Rev.* 11:89–173.

THEWISSEN, J. G. M. 1994. Phylogenetic aspects of cetacean origins: a morphological perspective. *Jour. Mammal. Evol.* 2:157–84.

———. 1998. Cetacean origins: evolutionary turmoil during the invasion of the oceans. In J. G. M. Thewissen, ed., *The Emergence of Whales*, pp. 451–64. Plenum.

THEWISSEN, J. G. M., and D. P. DOMNING. 1992. The role of phenacodontids in the origin of the modern orders of ungulate mammals. *Jour. Vert. Paleo.* 12(4):494–504.

THEWISSEN, J. G. M., and F. E. FISH. 1997. Locomotor evolution in the earliest cetaceans: functional model, modern analogues, and paleontological evidence. *Paleobiology* 23:482–90.

THEWISSEN, J. G. M., and S. T. HUSSAIN. 1998. Systematic review of the Pakicetidae, early and middle Eocene cetacea (Mammalia) from Pakistan and India. *Bull. Carnegie Mus. Nat. Hist.* 34:220–38.

THEWISSEN, J. G. M., S. T. HUSSAIN, and M. ARIF. 1994. Fossil evidence for the origin of aquatic locomotion in archaeocete whales. *Science* 263:210–12.

THEWISSEN, J. G. M., S. I. MADAR, and S. T. HUSSAIN. 1996. *Ambulocetus natans*, an Eocene cetacean (Mammalia) from Pakistan. *Cour. Forsch.-Inst. Seckenberg* 191:1–86.

THOMPSON, E. M., and J. -F. REES. 1995. Origins of luciferins: ecology of bioluminescence in marine fishes. In P. W. Hochachka and T. P. Mommsen, eds., *Biochemistry and Molecular Biology of Fishes*, vol. 4, pp. 435–66. Elsevier.

THOMSON, K. S. 1976. On the heterocercal tail in sharks. *Paleobiology* 2:19–38.

———. 1991. *Living Fossil: The Story of the Coelacanth*. W. W. Norton.

———. 1997. They must have come from somewhere! *Paleobiology* 23(4):491–93.

THOMSON, K. S., and D. E. SIMANEK. 1977. Body form and locomotion in sharks. *American Zoologist* 17:343–54.

TOBIAS, P. V. 1998. Water and human evolution. *Out There* 35:38–44.

TODAR, K. 1997. Major groups of prokaryotes. *Bacteriology at UW–Madison*. http://www/bact.wisc.edu/bact303

TORRENS, H. S. 1995. Mary Anning (1799–1847) of Lyme: "The greatest fossilist the world ever knew." *Br. Jour. Hist. Sci.* 28:257–84.

TUDGE, C. 2000. *The Variety of Life*. Oxford University Press.

TURNEY, J. 2000. "What fossils don't tell you." review of *In Search of Deep Time*, by Henry Gee. *New Scientist* 165(2231):46–47.

UHEN, M. D. 1997. *Dorudon atrox:* a first for the Exhibit Museum of Natural History. *LSAmagazine* (University of Michigan) 20(2):9.

———. 1998. Middle to Late Eocene basilosuarines and dorudontines. In J. G. M. Thewissen, ed., *The Emergence of Whales*, pp. 29–62. Plenum.

VALENTINE, J. W. 1994. The Cambrian explosion. In S. Bengston, ed., *Early Life on Earth*, pp. 401–11. Columbia University Press.

VALENTINE, J. W., D. JABLONSKI, and D. H. ERWIN. 1999. Fossils, molecules and embryos: new perspectives on the Cambrian explosion. *Development* 126(5):851–59.

VAN DOVER, C. L. 2000. *The Ecology of Deep-Sea Hydrothermal Vents*. Princeton University Press.

VAN DOVER, C. L., E. Z. SZUTS, B. C. CHAMBERLAIN, and J. R. CANN. 1989. A novel "eye" in "eyeless" shrimp from hydrothermal vents of the Mid-Atlantic Ridge. *Nature* 337:458–60.

VAN VALEN, L. 1968. Monophyly or diphyly in the origin of whales. *Evolution* 22:37–41.

VENTER, P., P. TIMM, G. GUNN, E. LE ROUX, C. SERFONTEIN, P. SMITH, E. SMITH, M. BENSCH, D. HARDING and P. HEEMSTRA. 2000. Discovery of a viable population of coelacanths (*Latimeria chalumnae* Smith, 1939) at Sodwana Bay, South Africa. *S. Afr. Jour. Sci.* 96:567–68.

VERHAEGEN, M., and S. MUNRO. 1999. Australopiths wading? *Homo* diving? *Proc. Symp. Water and Human Evolution, University of Ghent, April 30, 1999*, 11–23.

VICKERS-RICH, P., and T. H. RICH. 1999. *Wildlife of Gondwana: Dinosaurs and Other Vertebrates of the Ancient Supercontinent*. Indiana University Press.

VOGEL, G. 1999. Going beyond appearances to find life's history. *Science* 284:2112–13.

VOLSØE, H. 1939. The sea snakes of the Iranian Gulf and the Gulf of Oman, with a summary of the biology of the sea snakes. *Danish Investigations in Iran* 1:1–45.

———. 1956. Sea snakes. In A. F. Bruun, S. Greve, H. Melche, and R. Spärck, eds., *The* Galathea *Deep Sea Expedition 1950–52*, pp. 87–95. George Allen & Unwin.

WÄCHTERSHÄUSER, G. 1994. Vitalysts and virulists: a theory of self-expanding reproduction. In S. Bengston, ed., *Early Life on Earth*, pp. 124–32. Columbia University Press.

———. 2000. Life as we don't know it. *Science* 289:1307–88.

WAGGONER, B. M. 1996. Phylogenetic hypotheses of the relationships of arthropods to Precambrian and Cambrian problematic fossil taxa. *Systematic Biology* 45:190–222.

———. 1998. Interpreting the earliest metazoan fossils: What can we learn? *American Zoologist* 38:975–82.

———. 1999. Biogeographic analyses of the Ediacara biota: a conflict with paleotectonic reconstructions. *Paleobiology* 25(4):440–58.

WALKER, A. D. 1972. New light on the origin of birds and crocodiles. *Nature* 237:257–63.

WALKER, E. P. 1991. *Mammals of the World*. John Hopkins University Press.

WALKER, M. 1999. Waiting to exhale. *New Scientist* 163(2203):25.

WARD, P. D. 1979. Functional morphology of Cretaceous helically-coiled ammonite shells. *Paleobiology* 5:415–22.

———. 1983. *Nautilus pompilius*. In P. R. Boyle, ed., *Cephalopod Life Cycles*. Vol. 1, *Species accounts*, pp. 11–28. Academic Press.

———. 1984. Is *Nautilus* a living fossil? In N. Eldredge and S. M. Stanley, eds., *Living Fossils*, pp. 247–56. Springer-Verlag.

———. 1988. *In Search of Nautilus*. Simon & Schuster.

———. 1992. *On Methuselah's Trail*. Freeman.

———. 1997. *The Call of Distant Mammoths*. Copernicus.

———. 1998. Coils of time. *Discover* 19(3):100–106.

———. 1999. *Time Machine: Scientific Explorations in Deep Time*. Copernicus.

———. 2000. No place like home. *New Scientist* 165(2230):46–47.

WARD, P. D., and D. BROWNLEE. 2000. *Rare Earth*. Copernicus.
WARD, P. D., and W. B. SAUNDERS. 1997. *Allonautilus*: a new genus of living nautiloid cephalopod and its bearing on phylogeny of the Nautilida. *Jour. Paleo.* 71(6):1054–64.
WATKINS, W. A., and W. E. SCHEVILL. 1977. Sperm whale codas. *Jour. Acoust. Soc. America* 62:1485–90. Also phonograph record.
WATKINS, W. A., and D. WARTZOK. 1985. Sensory biophysics of marine mammals. *Marine Mammal Science* 1(3):219–60.
WEBB, S. D. 1995. Ten million years of mammal extinctions in North America. In P. S. Martin and R. G. Klein, eds., *Quaternary Extinctions: A Prehistoric Revolution*, pp. 189–210. University of Arizona Press.
WEED, W. S. 2000. What did the dinosaurs really look like . . . and will we ever know? *Discover* 21(9):74–84.
WEGENER, A. 1929. *The Origin of Continents and Oceans*. Reprint, Dover, 1966.
WEINBERG, S. 1999. *A Fish Caught in Time: The Search for the Coelacanth*. Fourth Estate.
WEINER, J. 1994. *The Beak of the Finch*. Knopf.
WELTON, B. J., and R. F. FARISH. 1993. *The Collector's Guide to Fossil Sharks and Rays from the Cretaceous of Texas*. Before Time.
WENDT, H. 1959. *Out of Noah's Ark*. Houghton Mifflin.
———. 1968. *Before the Deluge*. Doubleday.
WESSON, R. 1993. *Beyond Natural Selection*. MIT Press.
WESTENBERG, K. 1999. The rise of life on Earth: from fins to feet. *National Geographic* 195(5):114–26.
WESTON, R. J., C. A. REPENNING, and C. A. FLEMING, 1973. Modern age of supposed Pliocene seal, *Arctocephalus caninus* Berry (=*Phocarctos hookeri* Gray) from New Zealand. *New Zealand Jour. Sci.* 16(3):591–98.
WHEELER, H. E. 1939. *Helicoprion* in the Anthracolithic (Late Paleozoic) of Nevada and California, and its stratigraphic significance. *Jour. Paleo.* 13(1):103–14.
WHITE, J. R. 1984. Man can save the manatee. *National Geographic* 166(3):414–18.
WHITMORE, F. C., and L. M. GARD. 1977. Steller's sea cow (*Hydrodamalis gigas*) of late Pleistocene age from Amchitka, Aleutian Islands, Alaska. *U.S. Geologic Survey Professional Paper* 1036, 1–19.
WHITMORE, F. C., and A. E. SANDERS. 1976. Review of the Oligocene Cetacea. *Systematic Zoology* 25(4):304–20.
WHITTINGTON, H. B. 1978. The lobopod animal *Aysheaia pedunculata* Walcott, Middle Cambrian, Burgess Shale, British Columbia. *Phil. Trans. Roy. Soc. London* B 284:165–97.
WIDDER, E. A. 1999. Bioluminescence. In S. N. Archer, M. B. A. Djamgoz, E. Loew, J. C. Partridge, and S. Vallerga, eds., *Adaptive Mechanisms in the Ecology of Vision*, pp. 555–81. Kluwer Academic Publishers.
WIKELSKI, M. 1999. Diving dragons of the Galápagos. http://life.uiuc.edu.wikelski/iguana2.html
WIKELSKI, M., and C. THOM. 2000. Marine iguanas shrink to survive El Niño. *Nature* 403:37.
WILLIAMS, E. M. 1998. Synopsis of the earliest cetaceans: Pakicetidae, Ambulocetidae, Remingtonocetidae, and Protocetidae. In J. G. M. Thewissen, ed., *The Emergence of Whales*, pp. 1–28. Plenum.
WILLIAMS, G. C. 1997. *The Pony Fish's Glow* (*and Other Clues to Plan and Purpose in Nature*). Basic Books.
WILLIAMS, J. S., and D. H. DUNKLE. 1948. *Helicoprion*-like fossils in the Phosphoria Formation. *Bull. Geol. Soc. America* 59:1362.
WILLIAMS, M., and K. ELBAUM. 1973. Bendix-Almgreen's recent investigations on *Helicoprion* and the biomechanical significance of Karpinsky's reconstruction. *Journal of Insignificant Research* 5(31):2–6
WILLISTON, S. W. 1914. *Water Reptiles of the Past and Present*. University of Chicago Press.
WILSON, E. O. 1992. *The Diversity of Life*. Harvard University Press.
———. 1994 *Naturalist*. Shearwater.
WILSON, T., and J. W. HASTINGS. 1998. Bioluminescence. *Ann. Rev. Cell Dev. Biol.* 14:197–230.
WINGE, B. 1921. A review of the interrelationships of the cetacea. *Smithsonian Misc. Coll.* 72(8):1–97.
WITZELL, W. N. 1994. The origin, evolution, and demise of the U.S. sea turtle fisheries. *Mar. Fish. Rev.* 56(4):8–23.
WOOD, R. C. 1976. *Stupendemys geographicus*, the world's largest turtle. *Breviora* 436:1–31.
WOODWARD, A. S. 1900. *Helicoprion*—spine or tooth? *Geol. Mag.* 1:33–36.
WORTHINGTON, L. V., and W. E. SCHEVILL. 1957. Underwater sounds heard from sperm whales. *Nature* 180:291.
WRIGHT, K. 1977. When life was odd. *Discover* 18(3):52–61.
WYSS, A. R. 1988. Evidence from flipper structure for a single origin of pinnipeds. *Nature* 334:427–28.
———. 1994. The evolution of body size in phocids: some ontogenetic and phylogenetic observations. *Proc. San Diego Hist. Soc.* 29:69–76.
YOUNG, R. E. 1983. Oceanic bioluminescence: an overview of general functions. *Bull. Mar. Sci.* 33:829–45.

ZANGERL, R. 1966. A new shark of the family Edestidae, *Ornithoprion hertwigi* from the Pennsylvanian Mecca and Logan Quarry shales of Indiana. *Fieldiana Geol.* 16(1):1–43.

———. 1980. Patterns of phylogenetic differentiation in the toxochelyd and cheloniid sea turtles. *American Zoologist* 2(3):585–96.

———. 1981. *Chondrichthyes I. Paleozoic Elasmobranchii.* Gustav Fisher Verlag.

———. 1984. On the microscopic anatomy and possible function of the spine-"brush" complex of *Stethacanthus* (Elasmobranchii: Symmoriida). *Jour. Vert. Paleo.* 4:372–78.

ZANGERL, R., and G. R. CASE. 1973. Iniopterygia, a new order of chondrichthyan fishes from the Pennsylvanian of North America. *Fieldiana Geol. Mem.* 6:1–67.

———. 1976. *Cobelodus aculeatus* (Cope), an anacanthous shark from the Pennsylvania black shales of North America. *Palaeontographica* 154:107–57.

ZANGERL, R., and E. S. RICHARDSON. 1963. The paleoecological history of two Pennsylvania black shales. *Fieldiana Geol. Mem.* 6:1–67.

ZANGERL, R., and M. E. WILLIAMS. 1975. New evidence on the nature of the jaw suspension in Paleozoic anacanthous sharks. *Palaeontology* 18(2):333–41.

ZHOU, X., W. J. SANDERS, and P. D. GINGERICH. 1992. Functional and behavioral implication of vertebral structure in *Pachyaena ossifraga* (Mammalia, Mesonychia). *Contrib. Mus. Paleontol. Univ. Michigan* 28:289–313.

ZHU, M., X. YU, and P. JANVIER. 1999. A primitive fossil fish sheds light on the origin of bony fishes. *Nature* 397:607–10.

ZIMMER, C. 1998. *At the Water's Edge.* Free Press.

———. 1999. Fossils give a glimpse of old mother lamprey. *Science* 286:1064–65.

———. 2000. In search of vertebrate origins: beyond brain and bone. *Science* 287:1576–79.

ZINSMEISTER, W. J., and A. E. OLEINIK. 1995. Discovery of a remarkably complete specimen of the giant cephalopod *Diplomoceras maximum* from the Late Cretaceous of Seymour Island, Antarctica. *Antarctic Jour. U.S.* 30(5):9–10.

INDEX

Page numbers in *italics* refer to illustrations.